Numerical Modeling and Experimental Studies of Two-Phase Flows

Numerical Modeling and Experimental Studies of Two-Phase Flows

Van-Tu Nguyen
Hemant J. Sagar

Basel • Beijing • Wuhan • Barcelona • Belgrade • Novi Sad • Cluj • Manchester

Editors

Van-Tu Nguyen
School of Mechanical
Engineering
Pusan National University
Busan
Republic of Korea

Hemant J. Sagar
Department of Hydro and
Renewable Energy (HRED)
Indian Institute of Technology
Roorkee
India

Editorial Office
MDPI AG
Grosspeteranlage 5
4052 Basel, Switzerland

This is a reprint of articles from the Special Issue published online in the open access journal *Fluids* (ISSN 2311-5521) (available at: https://www.mdpi.com/journal/fluids/special_issues/966T4PCYD5).

For citation purposes, cite each article independently as indicated on the article page online and as indicated below:

Lastname, A.A.; Lastname, B.B. Article Title. *Journal Name* **Year**, *Volume Number*, Page Range.

ISBN 978-3-7258-2136-5 (Hbk)
ISBN 978-3-7258-2135-8 (PDF)
doi.org/10.3390/books978-3-7258-2135-8

Cover image courtesy of Van-Tu Nguyen

© 2024 by the authors. Articles in this book are Open Access and distributed under the Creative Commons Attribution (CC BY) license. The book as a whole is distributed by MDPI under the terms and conditions of the Creative Commons Attribution-NonCommercial-NoDerivs (CC BY-NC-ND) license.

Contents

About the Editors vii

Preface ix

Van-Tu Nguyen and Hemant J. Sagar
Computational Fluid Dynamics Modeling and Experiments of Two-Phase Flows
Reprinted from: *Fluids* 2024, 9, 207, doi:10.3390/fluids9090207 1

Lingxi Han, Tianyuan Zhang, Di Yang, Rui Han and Shuai Li
Comparison of Vortex Cut and Vortex Ring Models for Toroidal Bubble Dynamics in Underwater Explosions
Reprinted from: *Fluids* 2023, 8, 131, doi:10.3390/fluids8040131 5

Yue Chen, Qichao Wang, Hongbing Xiong and Lijuan Qian
Vapor Bubble Deformation and Collapse near Free Surface
Reprinted from: *Fluids* 2023, 8, 187, doi:10.3390/fluids8070187 18

Jianyong Yin, Yongxue Zhang, Dehong Gong, Lei Tian and Xianrong Du
Dynamics of a Laser-Induced Cavitation Bubble near a Cone: An Experimental and Numerical Study
Reprinted from: *Fluids* 2023, 8, 220, doi:10.3390/fluids8080220 39

Dimitrios Kolokotronis, Srikrishna Sahu, Yannis Hardalupas, Alex M. K. P. Taylor and Akira Arioka
Bulk Cavitation in Model Gasoline Injectors and Their Correlation with the Instantaneous Liquid Flow Field
Reprinted from: *Fluids* 2023, 8, 214, doi:10.3390/fluids8070214 60

Taihei Onishi, Kaizheng Li, Hong Ji and Guoyi Peng
Shedding of Cavitation Clouds in an Orifice Nozzle
Reprinted from: *Fluids* 2024, 9, 156, doi:10.3390/fluids9070156 84

Yuxing Lin, Ebrahim Kadivar and Ould el Moctar
Experimental Study of the Cavitation Effects on Hydrodynamic Behavior of a Circular Cylinder at Different Cavitation Regimes
Reprinted from: *Fluids* 2023, 8, 162, doi:10.3390/fluids8060162 95

Mirza M. Shah
Prediction of Critical Heat Flux during Downflow in Fully Heated Vertical Channels
Reprinted from: *Fluids* 2024, 9, 79, doi:10.3390/fluids9030079 110

Arseniy Parfenov, Alexander Gelfgat, Amos Ullmann and Neima Brauner
Hartmann Flow of Two-Layered Fluids in Horizontal and Inclined Channels
Reprinted from: *Fluids* 2024, 9, 129, doi:10.3390/fluids9060129 126

Ediguer E. Franco, Sebastián Henao Santa, John J. Cabrera and Santiago Laín
Air Flow Monitoring in a Bubble Column Using Ultrasonic Spectrometry
Reprinted from: *Fluids* 2024, 9, 163, doi:10.3390/fluids9070163 155

Van-Tu Nguyen and Warn-Gyu Park
A Review of Preconditioning and Artificial Compressibility Dual-Time Navier–Stokes Solvers for Multiphase Flows
Reprinted from: *Fluids* 2023, 8, 100, doi:10.3390/fluids8030100 169

Zhiyong Wang, Bing Yan and Haoquan Wang
Application of Deep Learning in Predicting Particle Concentration of Gas–Solid
Two-Phase Flow
Reprinted from: *Fluids* **2024**, *9*, 59, doi:10.3390/fluids9030059 192

**Farshad Bolourchifard, Keivan Ardam, Farzad Dadras Javan, Behzad Najafi,
Paloma VegaPenichet Domecq, Fabio Rinaldi and Luigi Pietro Maria Colombo**
Pressure Drop Estimation of Two-Phase Adiabatic Flows in Smooth Tubes: Development of
Machine Learning-Based Pipelines
Reprinted from: *Fluids* **2024**, *9*, 181, doi:10.3390/fluids9080181 206

About the Editors

Van-Tu Nguyen

Dr. Van-Tu Nguyen is a Research Professor at the School of Mechanical Engineering, Pusan National University (PNU), South Korea. He earned his Master's and Ph.D. degrees in Mechanical Engineering from PNU between 2011 and 2017, following his Bachelor's degree in Mechanical Engineering from the University of Engineering and Technology, Hanoi National University, Vietnam.

Dr. Nguyen's expertise lies in computational fluid dynamics (CFD), with a focus on numerical methods, algorithms, and programming. His research interests encompass a wide range of topics within fluid dynamics, including multiphase flows, shock capturing, free surface flows, cavitation, bubble dynamics, and high-speed subsonic-supersonic flows. He has made significant contributions to the development of numerical coding programs tailored to these applications. Throughout his career, Dr. Nguyen has published over 80 international journal and conference papers. He has served as Principal Investigator (PI) or Co-PI on multiple research projects funded by the Korean and United States governments, such as the National Research Fund of Korea (NRF) and the Office of Naval Research Global (ONRG). His scholarly reputation is further established through his roles as a keynote speaker, Section Chairman, and Committee member at various international conferences. Additionally, he serves as an Associate Editor for *AIP Advances* and is a member of several editorial and reviewer boards for reputable scientific journals.

Dr. Nguyen's dedication to advancing the field of fluid dynamics and his extensive research contributions make him a respected figure in the scientific community.

Hemant J. Sagar

Hemant J. Sagar is a faculty member at the Department of Hydro and Renewable Energy in the Indian Institute of Technology (IIT) Roorkee. He studied and worked at the University of Duisburg-Essen since 2012 and Harbin Engineering University as a postdoctoral fellow. He is mainly engaged in teaching and research in the area of renewable energy and fluid mechanics. His areas of interest include the following: cavitation, bubble dynamics, and multiphase flow in both experimental and numerical aspects, but not limited to various interesting multiphase phenomena such as cavitation-silt erosion, laser-induced cavitation bubbles, hydrodynamics supercavitation, offshore wind turbines (floating and fixed), fluid–structure interaction, erosion-resistant coatings, acoustics, cavitation-based biomedical technology, and quantitative analysis of flow and hydrodynamics. His teaching assignments include renewable energy, hydromechanical equipment, cavitation erosion, and quantitative investigation of flows.

Preface

This reprint explores the most recent advancements in the field of two-phase flows, focusing extensively on both numerical modeling and experimental studies. The content presented here highlights the innovative approaches and methodologies that are shaping the future of research in this critical area of fluid dynamics. As two-phase flows are encountered in numerous industrial and natural processes, understanding their behavior is essential for improving efficiency, safety, and performance in various applications. The studies compiled in this volume represent the forefront of ongoing efforts to refine our ability to predict and control the complex dynamics of two-phase flows.

The editorial team has taken great care in curating a selection of high-quality studies that exemplify the latest research and practical applications in the field. These studies not only provide a deep dive into the current state of the art but also address emerging challenges and opportunities in the study of two-phase flows. By examining the interactions between different phases, investigating the effects of various parameters on flow behavior, and presenting new computational models, these contributions collectively offer a comprehensive overview of the progress being made in this domain. From fundamental research to applied studies, the articles presented here shed light on the multifaceted nature of two-phase flows and the ongoing innovations in this area.

In compiling this collection, we aim to offer a resource that will be invaluable to researchers, engineers, and practitioners working in the field of fluid dynamics. We hope that these studies will not only enhance the reader's understanding of two-phase flows but also inspire new ideas and directions for future research. By sharing these findings, we seek to foster collaboration across disciplines and promote the exchange of knowledge vital for this field's continued advancement. Ultimately, we believe that this volume will contribute to the ongoing efforts to improve the modeling, prediction, and management of two-phase flows in both academic and industrial settings.

We extend our gratitude to the contributors whose work is featured in this reprint. Their dedication and expertise are reflected in the high caliber of the studies included, and their efforts are helping to push the boundaries of what is possible in the study of two-phase flows. We are confident that their research will serve as a foundation for further advancements in the years to come.

Van-Tu Nguyen and Hemant J. Sagar
Editors

Editorial

Computational Fluid Dynamics Modeling and Experiments of Two-Phase Flows †

Van-Tu Nguyen [1],* and Hemant J. Sagar [2]

1 School of Mechanical Engineering, Pusan National University, Busan 46241, Republic of Korea
2 Department of Hydro and Renewable Energy, Indian Institute of Technology (IIT), Roorkee 247667, Uttarakhand, India; hemant.sagar@hre.iitr.ac.in
* Correspondence: vantunguyen@pusan.ac.kr
† This paper is part of the special issue, Numerical Modeling and Experimental Studies of Two-Phase Flows.

Two-phase flows are prevalent in natural phenomena, as well as a wide range of marine engineering and industrial applications. However, the study of two-phase systems has been limited due to the added complications of the two-phase interface, where mass transfer causes complex flow behaviors. These flows involve the interaction of two distinct phases within a system, leading to highly nonlinear dynamics. Some examples of these flows include free surface flows interacting with marine and offshore structures; cavitation; steam and water flow in power plants; oil and gas transportation in pipelines; boiling and condensation in heat exchangers; and other natural occurrences. Experimental studies of two-phase flows face significant difficulties—they are often expensive, time-consuming, and complex to set up, particularly when replicating extreme conditions or achieving detailed visualizations. Despite these challenges, experiments are essential for validating and confirming computational fluid dynamics (CFD) results. While CFD methods are crucial for simulating and analyzing two-phase flows due to their flexibility and cost-effectiveness, experimental data are essential for ensuring the accuracy and reliability of these numerical models. This Special Issue in Fluids, entitled "Numerical Modeling and Experimental Studies of Two-Phase Flows," focuses on recent advances in both numerical and experimental modeling of two-phase flows, providing deeper insights into the fundamental and physical aspects of these flows across several fields, including engineering and industry.

Bubble dynamics are a crucial phenomenon in fluid mechanics, impacting both natural processes and engineered systems. The study of bubble dynamics encompasses the formation, growth, oscillation, and collapse of bubbles within a fluid. These processes are influenced by factors such as pressure variations, surface tension, and the presence of interfaces between fluids [1]. The toroidal model and the ring shedding approach for toroidal bubble dynamics were numerically analyzed to address the discontinuous pressure field on the bubble interface, simulating behavior in terms of bubble geometry, internal gas pressure, and shock wave propagation [2]. Building on the framework of smoothed particle hydrodynamics (SPH) and incorporating the van der Waals (VDW) equation of state, the deformation and the collapse of a heated vapor bubble near a solid boundary were examined as the bubble ascended from the bottom surface [3]. Cavitation is a phenomenon that occurs when the pressure in a liquid falls below its vapor pressure, resulting in the formation of vapor-filled bubbles or cavities. These bubbles can rapidly expand and collapse, producing extreme local pressures and temperatures. Cavitation commonly arises in hydraulic systems and fluid machinery, including turbines, impellers, nozzles, and underwater propulsion systems. Although cavitation is often associated with equipment damage, such as material erosion and surface pitting, it can also be utilized in advantageous ways, such as in ultrasonic cleaning, water treatment, and medical applications such as lithotripsy [1]. The behavior of a cavitation bubble near a rigid conical surface is explored through a combination of numerical simulations and experimental observations, revealing

Citation: Nguyen, V.-T.; Sagar, H.J. Computational Fluid Dynamics Modeling and Experiments of Two-Phase Flows. *Fluids* **2024**, *9*, 207. https://doi.org/10.3390/fluids9090207

Received: 27 August 2024
Accepted: 30 August 2024
Published: 3 September 2024

Copyright: © 2024 by the authors. Licensee MDPI, Basel, Switzerland. This article is an open access article distributed under the terms and conditions of the Creative Commons Attribution (CC BY) license (https://creativecommons.org/licenses/by/4.0/).

key phenomena such as shock wave emissions, liquid jet formations, and localized high-pressure regions. In the experimental setup, a single cavitation bubble is produced using a pulsed laser, with its dynamics being recorded using a high-speed camera at 100,000 frames per second. For the numerical analysis, a compressible two-phase flow model that incorporates phase transitions and thermal effects is utilized, implemented in the OpenFOAM framework [4]. Utilizing high-resolution two-dimensional Particle Imaging Velocimetry, the internal flow dynamics of enlarged transparent models of multi-hole injectors were investigated. These measurements were aimed at comprehending the mechanisms behind bulk cavitation formation and their relationship to the injector's flow characteristics, subsequently providing insights into the flow field behaviors within the injector's internal structure [5]. The periodic behavior and formation mechanisms of cavitation clouds in a submerged water jet from an orifice nozzle were analyzed, where high-speed camera imaging and flow simulations reveal that pairs of ring-like clouds, consisting of a leading and a subsequent cloud, are periodically shed downstream [6]. The findings reveal that the leading cloud separates due to a shear vortex at the nozzle exit, while the subsequent cloud is released via a re-entrant jet following the collapse of a fully extended cavity. The impact of cavitation on the hydrodynamic properties of a circular cylinder in diverse cavitating flows was studied experimentally. The hydrodynamic forces acting on the cylinder were measured using a load cell, while a high-speed camera was employed to capture the cavitation dynamics occurring behind the cylinder. Subsequently, the cavitation behavior around the cylinder was analyzed for various cavitating regimes, including the inception of cavitation, partial cavitation, and cloud cavitation. [7]. During cavitation bubble collapse, shock waves and high-speed microjets, at speeds of up to thousands of meters per second, create high local energy concentrations, with temperatures exceeding tens of thousands of Kelvin and pressures reaching gigapascal levels. This extreme environment may lead to behaviors similar to those of supercritical fluids, including the release of non-condensable gasses, plasma formation, and chemical reactions. Both current computational fluid dynamics (CFD) models and experimental methods struggle to accurately predict these phenomena. Numerically and experimentally investigating the effect of non-condensable gas on the dynamics, pressure, and temperature could be challenging in the future. Understanding bubble dynamics is essential for numerous applications, from improving industrial processes to advancing biomedical technologies.

Flow within channels is a fundamental concept in fluid mechanics, describing the movement of fluids through confined pathways, such as pipes, ducts, and open channels. A comprehensive review of the literature on CHF in vertical downflow channels was conducted [8]. The behavior of boiling under downflow conditions in vertical channels, which is relevant to applications such as steam generators, power plants, and industrial cooling systems, has been thoroughly reviewed. A detailed comparison of existing correlations with experimental data and the prediction of critical heat flux (CHF) was presented, providing insights into flow stability and thermal performance. Additionally, the effect of a transverse magnetic field on a two-phase stratified flow in both horizontal and inclined channels was investigated [9]. The study in question explored how the magnetic field influences laminar stratified flows, particularly when the more dense, electrically conductive liquid occupies the lower layer, while the upper fluid acts as an electrical insulator. The introduction of a transverse magnetic field adds another dimension to flow control, particularly in systems with electrically conductive fluids, offering potential for enhanced flow stability. Meanwhile, the application of ultrasonic techniques for the real-time monitoring of bubbly flows presents promising opportunities for improving diagnostic capabilities in both industrial and research settings. In another investigation, the use of ultrasonic techniques to monitor bubbly flow and to determine bubble density in a water column was documented [10]. The quantity of bubbles was assessed by analyzing the performance of the positive displacement pump that was employed for air injection. The findings indicated that bubble density in the water column can be effectively monitored using the phase spectrum of the loss coefficient. Further research is warranted to refine these models and techniques, particularly in scaling them up for larger, more complex systems.

Two-phase flow research addresses significant real-world challenges such as understanding the breaking of waves and the overtopping of coastal structures, as well as analyzing the interactions of moving ships with extreme waves and green water on decks. Additionally, applications such as spray cooling, two-phase heat transfer, hydrodynamic cavitation, and dynamic bubble processes benefit from these advances, ultimately enhancing critical heat flux and the system's reliability. Despite the advantages of computational fluid dynamics (CFD) in studying two-phase flows, numerous challenges persist. Accurately modeling phase interactions, handling complex boundary conditions, ensuring numerical stability, and managing significant computational demands are some of the ongoing issues. Continuous advancements in computational methods, high-performance computing, and validation techniques are essential to fully leverage the potential of CFD in two-phase flow studies [11]. The integration of machine learning (ML) and deep learning (DL) into CFD represents a major step forward in optimizing fluid dynamics simulations. Traditional CFD models, while powerful, can be computationally expensive and time-consuming, especially when dealing with complex two-phase flows. As highlighted in [12], the industrial market is becoming increasingly competitive, pushing companies to adopt advanced technologies to gain a strategic advantage. One key resource is simulation, which plays an essential role in Industry 4.0, particularly in layout reconfiguration, to enable flexible product customization and to optimize manufacturing processes. In this context, computational fluid dynamics (CFD) simulations offer a substantial competitive advantage for smart factories by leveraging emerging technologies. In addition to the continued development of CFD methods and modeling techniques for two-phase flows, recent years have seen the rise of a transformative technology (ML and DL). ML and DL techniques can provide faster predictions, potentially reducing computational costs while maintaining accuracy. These techniques are reshaping various domains, and their impact on CFD is expected to be significant. A noteworthy application involves using DL methods to predict particle concentration in gas–solid two-phase flows [13]. This study compares the effectiveness of three approaches—Back-Propagation Neural Networks (BPNNs), Recurrent Neural Networks (RNNs), and Long Short-Term Memory (LSTM) networks—in analyzing gas–solid two-phase flow data. Seven key parameters, including temperature, humidity, upstream and downstream sensor signals, delay, pressure difference, and particle concentration, were utilized to construct the dataset. The comparison of different neural network architectures demonstrates the versatility of these methods in handling multi-faceted datasets, enabling more accurate predictions of complex flow behaviors such as particle concentration in gas–solid systems. This advancement opens up new possibilities for real-time monitoring and control in industrial applications.

Another study, which explores the use of machine learning for analyzing and estimating pressure drop in two-phase flow dynamics within smooth tubes, has been documented [14]. This research begins with experimental measurements of pressure drop for a water–air mixture across various flow conditions in horizontally oriented smooth tubes. ML techniques are then applied to predict pressure drop values using dimensionless parameters derived from the experimental data. Feature selection methods are employed to identify crucial features, which aids in better understanding the underlying physical mechanisms and improving the accuracy of the models. Furthermore, a genetic algorithm is employed to optimize the selection of the machine learning model and its tuning parameters. As a result, the optimized pipeline achieves a low mean absolute percentage error on both the validation and test datasets. These ongoing studies on combining CFD and ML or DL approaches will continue to provide crucial insights and solutions for both natural and engineered systems, particularly in maritime contexts where efficient and innovative designs are paramount.

Acknowledgments: The Guest Editors would like to thank all contributing authors and reviewers for their invaluable efforts in enhancing the quality of this Special Issue. We are particularly grateful to the editors of *Fluids* for their steadfast support and encouragement throughout the editing process. The contributions of the anonymous reviewers were essential, and their dedication is greatly appreciated.

Conflicts of Interest: The authors declare no conflicts of interest.

References

1. Nguyen, V.-T.; Sagar, H.J.; Moctar, O.e.; Park, W.-G. Understanding cavitation bubble collapse and rebound near a solid wall. *Int. J. Mech. Sci.* **2024**, *278*, 109473. [CrossRef]
2. Han, L.; Zhang, T.; Yang, D.; Han, R.; Li, S. Comparison of Vortex Cut and Vortex Ring Models for Toroidal Bubble Dynamics in Underwater Explosions. *Fluids* **2023**, *8*, 131. [CrossRef]
3. Chen, Y.; Wang, Q.; Xiong, H.; Qian, L. Vapor Bubble Deformation and Collapse near Free Surface. *Fluids* **2023**, *8*, 187. [CrossRef]
4. Yin, J.; Zhang, Y.; Gong, D.; Tian, L.; Du, X. Dynamics of a Laser-Induced Cavitation Bubble near a Cone: An Experimental and Numerical Study. *Fluids* **2023**, *8*, 220. [CrossRef]
5. Kolokotronis, D.; Sahu, S.; Hardalupas, Y.; Taylor, A.M.K.P.; Arioka, A. Bulk Cavitation in Model Gasoline Injectors and Their Correlation with the Instantaneous Liquid Flow Field. *Fluids* **2023**, *8*, 214. [CrossRef]
6. Onishi, T.; Li, K.; Ji, H.; Peng, G. Shedding of Cavitation Clouds in an Orifice Nozzle. *Fluids* **2024**, *9*, 156. [CrossRef]
7. Lin, Y.; Kadivar, E.; el Moctar, O. Experimental Study of the Cavitation Effects on Hydrodynamic Behavior of a Circular Cylinder at Different Cavitation Regimes. *Fluids* **2023**, *8*, 162. [CrossRef]
8. Shah, M.M. Prediction of Critical Heat Flux during Downflow in Fully Heated Vertical Channels. *Fluids* **2024**, *9*, 79. [CrossRef]
9. Parfenov, A.; Gelfgat, A.; Ullmann, A.; Brauner, N. Hartmann Flow of Two-Layered Fluids in Horizontal and Inclined Channels. *Fluids* **2024**, *9*, 129. [CrossRef]
10. Franco, E.E.; Henao Santa, S.; Cabrera, J.J.; Laín, S. Air Flow Monitoring in a Bubble Column Using Ultrasonic Spectrometry. *Fluids* **2024**, *9*, 163. [CrossRef]
11. Nguyen, V.-T.; Park, W.-G. A Review of Preconditioning and Artificial Compressibility Dual-Time Navier–Stokes Solvers for Multiphase Flows. *Fluids* **2023**, *8*, 100. [CrossRef]
12. Silvestri, L. CFD modeling in Industry 4.0: New perspectives for smart factories. *Procedia Comput. Sci.* **2021**, *180*, 381–387. [CrossRef]
13. Wang, Z.; Yan, B.; Wang, H. Application of Deep Learning in Predicting Particle Concentration of Gas–Solid Two-Phase Flow. *Fluids* **2024**, *9*, 59. [CrossRef]
14. Bolourchifard, F.; Ardam, K.; Dadras Javan, F.; Najafi, B.; Vega Penichet Domecq, P.; Rinaldi, F.; Colombo, L.P.M. Pressure Drop Estimation of Two-Phase Adiabatic Flows in Smooth Tubes: Development of Machine Learning-Based Pipelines. *Fluids* **2024**, *9*, 181. [CrossRef]

Disclaimer/Publisher's Note: The statements, opinions and data contained in all publications are solely those of the individual author(s) and contributor(s) and not of MDPI and/or the editor(s). MDPI and/or the editor(s) disclaim responsibility for any injury to people or property resulting from any ideas, methods, instructions or products referred to in the content.

Article

Comparison of Vortex Cut and Vortex Ring Models for Toroidal Bubble Dynamics in Underwater Explosions

Lingxi Han [1], Tianyuan Zhang [1], Di Yang [2], Rui Han [3,*] and Shuai Li [1]

[1] College of Shipbuilding Engineering, Harbin Engineering University, Harbin 150001, China; hanlingxi@hrbeu.edu.cn (L.H.); tianyuanzhang@hrbeu.edu.cn (T.Z.); lishuai@hrbeu.edu.cn (S.L.)
[2] The 1st Research Laboratory, Wuhan Second Ship Design and Research Institute, Wuhan 430200, China; 13351204392@163.com
[3] Heilongjiang Province Key Laboratory of Nuclear Power System & Equipment, Harbin Engineering University, Harbin 150001, China
* Correspondence: hanrui@hrbeu.edu.cn

Abstract: The jet impact from a collapsing bubble is an important mechanism of structural damage in underwater explosions and cavitation erosion. The Boundary Integral Method (BIM) is widely used to simulate nonspherical bubble dynamic behaviors due to its high accuracy and efficiency. However, conventional BIM cannot simulate toroidal bubble dynamics, as the flow field transforms from single-connected into double-connected. To overcome this problem, vortex cut and vortex ring models can be used to handle the discontinuous potential on the toroidal bubble surface. In this work, we compare these two models applied to toroidal bubble dynamics in a free field and near a rigid wall in terms of bubble profile, bubble gas pressure, and dynamic pressure induced by the bubble, etc. Our results show that the two models produce comparable outcomes with a sufficient number of nodes in each. In the axisymmetric case, the vortex cut model is more efficient than the vortex ring model. Moreover, we found that both models improve in self-consistency as the number of bubble surface elements (N) increases, with $N = 300$ representing an optimal value. Our findings provide insights into the numerical study of toroidal bubble dynamics, which can enhance the selection and application of numerical models in research and engineering applications.

Keywords: bubble dynamics; BIM; toroidal bubbles; numerical simulation

1. Introduction

Various types of bubbles in nature and industrial applications have been extensively studied. These include cavitation bubbles generated by snapping shrimp [1,2], vapor bubbles leading to cavitation erosion [3], bubbles triggered by underwater explosions [4–6], and airgun bubbles generated during resource exploration [7,8]. Previous works have proposed several analytical methods to understand the dynamics of spherical bubbles [9–13], such as the Rayleigh–Plesset equation [9,11] and the Keller equation [12]. Prosperetti and Lezzi [13,14] derived the one-parameter family equation of the first and second-order Mach number, respectively. The first-order equation can be simplified to the Keller or Herring form with different coefficients. Zhang et al. [15] established a unified theory for spherical bubble dynamics, which takes into account the effects of viscosity, boundaries, bubble interaction, gravity, bubble migration, fluid compressibility, and more. In reality, most bubbles become nonspherical and the high-speed water jet forms under gravity or induced by varieties of boundaries. Such jets carry up to 31% of the energy of the bubble system [16] and thus can cause much more concentrated damage in underwater explosions [16]. Thus, nonspherical bubbles are of great research value. Blake et al. [17] solved the boundary integral equation with linear elements to obtain the velocity of the bubble surface and simulate the bubble oscillating process. The same authors [18] used the method mentioned above to study the interaction between a cavitation bubble and free surfaces.

Later, the Boundary Integral Method (BIM) became very popular in the community of bubble dynamics. Chahine and his collaborators [19] carried out Three Dimensional (3D) simulations for bubble interactions with underwater structures and floating bodies. Khoo and his collaborators [20,21] further improved the 3D BIM model and performed extensive studies for underwater explosion bubbles. Li et al. [22] used 3D BIM to study the strong interaction between a pulsating bubble and a movable sphere. Zhang and his collaborators [23,24] also studied the nonlinear interaction between oscillating bubbles and various boundaries. Manmi et al. [25] used BIM to investigate the dynamics of a microbubble oscillating between two curved rigid plates in a planar acoustic field. Three-dimensional dynamics of a transient bubble oscillating inside a rigid corner were studied numerically by Dadvand et al. [26] using BIM. Wang et al. [27,28] proposed weakly compressible BIM and applied it to investigate nonspherical bubble dynamics of underwater explosions.

Recently, Li et al. [29] discovered that the weakly compressible BIM [30] and an all-Mach method (AMM) produced very similar results when Ma was below 0.3, and the incompressible BIM could be safely used with appropriate initial parameter settings. However, the traditional BIM cannot simulate bubble oscillations after jet impact since the flow domain becomes doubly connected. To simulate toroidal bubbles, Lundgren and Mansour et al. [31] divided the velocity potential into two parts: a continuous part from a smooth dipole distribution over the bubble surface and a discontinuous part related to a vortex ring. Best [32] proposed a cut created by the impact, whose geometry can change as a material surface. In this model, the domain can still be considered singly connected after impact. Later, Best [33] introduced a relocated cut that is re-mapped to a simple disc and follows each advancement of the flow. Additionally, Zhang et al. [34] introduced a vortex sheet to divide the jet from the surrounding flow field. Wang et al. [35] proposed a vortex ring model, which can also be applied to simulate three-dimensional problems. Later, Zhang et al. [36] developed a three-dimensional vortex ring model based on the two-dimensional one.

To date, numerous studies have investigated nonspherical bubble dynamics using numerical approaches [37,38]. Various numerical methods, including the Finite Volume Method (FVM) [39–42], the Finite Element Method (FEM) [4,43–45], the Smoothed Particle Hydrodynamics (SPH) [46–48], and the Boundary Integral Method (BIM), have been used to understand bubble dynamics [49–53]. Among these, the BIM is widely used for its high computational efficiency and accuracy. In this context, it has become crucial to comprehensively compare these different models, which would provide an indispensable foundation for toroidal bubble dynamics research. Therefore, here, we conduct a comprehensive study of the vortex ring and vortex cut models based on the original axisymmetric BIM code. We adopt the two mentioned numerical models to simulate bubbles generated from underwater explosions (hereafter, 'UNDEX bubbles') in the toroidal phase. We explore the characteristics and performances of each model in terms of CPU cost, convergence, and accuracy, as well as the toroidal bubble dynamics obtained from each model in terms of bubble profile, bubble gas pressure, and dynamic pressure induced by the bubble, etc.

The remainder of this study is organized as follows. In Section 2, we present an introduction to the BIM, the vortex cut model, and the vortex ring model. Validation and self-consistency analysis of the two models are conducted. In Section 3, we study toroidal bubbles in a free field or near a rigid wall and compare the results between the models. Finally, we present conclusions in Section 4.

2. Theories and Numerical Methods

2.1. Boundary Integral Method

The BIM has been widely used to research bubble dynamics as a high-efficiency numerical method [24,28,29,37,51,54,55], especially for nonspherical bubbles near boundaries. We assume the flow surrounding the bubble to be inviscid and irrotational. The Mach numbers of all cases presented herein are within 0.3, so we can safely use the incompressible BIM to study the nonspherical bubble dynamics with a proper setting of initial parameters [29].

The Peclet number (defined as $R_m^2/T_{osi}D$, where T_{osi} is the bubble period and D is the thermal diffusivity) in our work is $O(10^8)$. Thus, the bubble gas is considered to be adiabatic throughout the bubble life. We neglect heat and mass transfer across the bubble surface. As the maximum radius of a UNDEX bubble is $O(m)$, and the average velocity is $O(10 \text{ m/s})$, the Reynolds number is greater than $O(10^7)$, and thus the viscosity of the surrounding liquid is negligible. Here, we define a cylindrical coordinate system (r, θ, z), in which the positive direction of the z axis is upward. The origin is set at the point of detonation. According to the above assumptions, the flow is governed by

$$\nabla^2 \phi = 0. \tag{1}$$

where ϕ is the velocity potential.

According to the Green function, the boundary integral equation can be obtained to solve Equation (1):

$$\lambda(r,t)\phi(r,t) = \iint_S \left[\frac{\partial \phi(q,t)}{\partial n} G(r,q) - \phi(q,t) \frac{\partial}{\partial n} G(r,q) \right] dS, \tag{2}$$

where λ is the solid angle, q is the source point, r is the field point, S denotes all boundaries, and $\partial/\partial n$ is the normal outward derivative from the boundary. To study an explosion bubble in a free field and near an infinite rigid wall, the Green functions are written as

$$G(r,q) = \frac{1}{|r-q|}, \tag{3}$$

$$G(r,q) = \frac{1}{|r-q|} + \frac{1}{|r-q'|}, \tag{4}$$

where q' is the reflected image across the rigid wall of q.

The position of the bubble surface can be updated using the kinematic boundary condition:

$$\frac{dr}{dt} = \nabla \phi, \tag{5}$$

where r denotes the position of the bubble surface. Based on the Bernoulli equation, the dynamic boundary condition is given by

$$\frac{d\phi}{dt} = \frac{1}{2}|\nabla \phi|^2 + \frac{p_\infty - p}{\rho} - gz, \tag{6}$$

where p_∞ is the ambient pressure of the liquid at $z = 0$, p is the liquid pressure at the bubble surface, g is the acceleration of gravity, and ρ denotes the density of the liquid, which is 1024 kg/m^3 in this paper. Equation (6) can be used to update the velocity potential at the bubble surface.

For explosion bubbles, the surface tension is ignorable because the Weber number is much larger than 1. Thus, according to the adiabatic equation, the liquid pressure p at the bubble surface can be obtained by

$$p = p_b = p_c + p_0 \left(\frac{V_0}{V} \right)^\kappa, \tag{7}$$

where p_c refers to the vapor pressure, V is the bubble volume, and the subscript 0 denotes the initial state. κ is the adiabatic index, and $\kappa = 1.25$ for TNT charge.

The initial velocity of the bubble surface is 0. Following Klaseboer et al. [56], the initial bubble radius R_0 can be obtained by

$$\frac{1.39 \times 10^5}{p_\infty} \left(\frac{3W}{4\pi R_m^3} \right)^\kappa \left[1 - \left(\frac{R_0}{R_m} \right)^{-3(\kappa-1)} \right] = (\kappa - 1) \left[\left(\frac{R_0}{R_m} \right)^3 - 1 \right], \tag{8}$$

where W denotes the weight of the charge, and R_m is the maximum bubble radius. Following Cole [57], R_m can be obtained from empirical formulas.

In this paper, parameters are nondimensionalized for convenience. The superscript * denotes that the quantity is dimensionless. The ambient pressure of the liquid p_∞ (at the plane of $z = 0$), the maximum bubble radius R_m, and the density of the liquid ρ are considered the pressure, length, and density scales, respectively. The stand-off, buoyancy, and strength parameters are thus defined as:

$$\gamma = \frac{L}{R_m}, \tag{9}$$

$$\delta = \sqrt{\frac{\rho g R_m}{p_\infty}}, \tag{10}$$

$$\epsilon = \frac{p_0}{p_\infty}. \tag{11}$$

It is worth mentioning that the conventional BIM cannot simulate bubble oscillation after the jet impact because the flow domain becomes doubly connected. Two models based on the conventional BIM, the vortex cut model [32,34,58–60] and the vortex ring model [36,38,61], have been developed and widely used by researchers to study toroidal bubble dynamics. More details are given below.

2.2. Vortex Cut Model

In the vortex cut model, a vortex cut T is introduced, including an upper surface T_+ and a lower surface T_-, thus the modified boundary integral equation [32] becomes

$$\lambda(r,t)\phi(r,t) = \iint_S \left[\frac{\partial \phi(q,t)}{\partial n} G(r,q) - \phi(q,t) \frac{\partial}{\partial n} G(r,q) \right] dS$$
$$+ \iint_T \left[\frac{\partial \phi(r,t)}{\partial n_+} G(r,q) - \phi_+(r,t) \frac{\partial}{\partial n_+} G(r,q) \right] dS \tag{12}$$
$$+ \iint_T \left[\frac{\partial \phi(r,t)}{\partial n_-} G(r,q) - \phi_-(r,t) \frac{\partial}{\partial n_-} G(r,q) \right] dS,$$

where $\partial/\partial n_+$ and $\partial/\partial n_-$ denote the normal outward derivative from T_+ and T_-, respectively. Because $\partial/\partial n_+ G(r,q) = -\partial/\partial n_- G(r,q)$, Equation (12) can be written as

$$\lambda(r,t)\phi(r,t) = \iint_S \left[\frac{\partial \phi(q,t)}{\partial n} G(r,q) - \phi(q,t) \frac{\partial}{\partial n} G(r,q) \right] dS - \Delta\phi \iint_T \frac{\partial}{\partial n_+} G(r,q) dS, \tag{13}$$

where $\Delta\phi$ is the jump of potential ϕ across T. The position and shape of the vortex cut are arbitrary as long as $\Delta\phi$ keeps constant.

The model is shown in the right part of Figure 1. In our model, we set an upper limit (80% of the bubble's overall height) and a lower limit for the movement of the cut to keep the calculation stable. We set the cut at the lower limit initially. During the computation process, we reset the cut to the lower limit once it reaches the upper limit. Then the cut will continue moving along with the bubble surface in the limited region until the end of the calculation. Refer to Pearson et al. [58] for more details on cut shifting.

In the numerical model, we set T_+ and T_- as two cuts very close to each other for calculation, each of which connects with two adjacent elements at the bubble surface. The velocity potential and location of the bubble surface are updated by the kinematic and dynamic boundary conditions mentioned above.

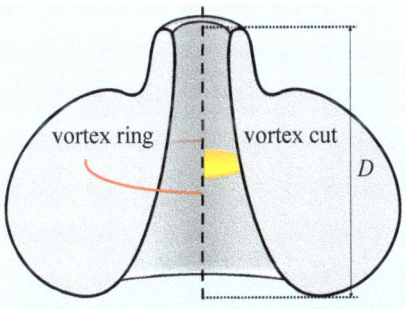

Figure 1. Vortex ring model (**left**) and vortex cut model (**right**).

2.3. Vortex Ring Model

In the vortex ring model, there exists a vortex ring in the bubble at (a, c), as shown in the left part of Figure 1. The velocity potential ϕ is divided into two parts, the induced potential by the vortex ring ϕ_{vor} and the remnant potential ϕ_{re}. Accordingly, the velocity in the flow is decomposed into the induced velocity by the vortex ring u_{vor} and the remnant velocity u_{re}, which can be calculated according to the Biot–Savart law and the boundary integral equation, respectively. The Biot–Savart law can be expressed as

$$u_{vor}(r,z) = \frac{\Gamma}{4\pi} \oint_C \frac{d\boldsymbol{l} \times \boldsymbol{r}}{r^3}, \tag{14}$$

where $d\boldsymbol{l}$ is the element of the vortex ring.

The induced velocity potential ϕ_{vor} can be obtained by using a semi-analytical method [62] with the known induced velocity u_{vor}:

$$\phi_{vor}(r,z) = \int_{(r,+\infty)}^{(r,z)} \boldsymbol{u}_{vor} \cdot \boldsymbol{e}_z dR_z = \frac{\Gamma}{4\pi} \oint \left(\frac{R_z}{|\boldsymbol{R}|} - 1\right) \frac{1}{R_r^2} \boldsymbol{e}_z \cdot (d\boldsymbol{l} \times \boldsymbol{R}), \tag{15}$$

$$\phi_{vor}(r,z) = \int_{(r,-\infty)}^{(r,z)} \boldsymbol{u}_{vor} \cdot \boldsymbol{e}_z dR_z = \frac{\Gamma}{4\pi} \oint \left(\frac{R_z}{|\boldsymbol{R}|} - 1\right) \frac{1}{R_r^2} \boldsymbol{e}_z \cdot (d\boldsymbol{l} \times \boldsymbol{R}), \tag{16}$$

where the integration path is the vortex ring and l is the infinitesimal value on it. $\boldsymbol{R}$ refers to the vector from the center of the infinitesimal to (r, z). Γ denotes the circulation of the vortex ring. It can be obtained from the jump in potential across the contact point during the jet impact process. $\boldsymbol{e}_z$ is the unit vector of the z axis. Equations (14) and (15) are suitable for the case that (r, z) is above or below the vortex ring, respectively.

The remnant potential ϕ_{re} can be updated via the dynamic boundary condition mentioned above. The remnant velocity u_{re} can be updated via the boundary integral equation. Then, we can sum the remnant velocity u_{re} and the induced velocity u_{vor}, yielding the total velocity u.

3. Results

3.1. Verification Analysis

We validate the BIM code first. We compare the spherical bubble dynamics obtained from the BIM simulation and analytical solution [15]. The initial conditions of the case are set as: $R_0^* = 0.15$, $\epsilon = 102.35$, and $\delta = 0$. Figure 2 shows the time evolution of the bubble radius in the first two bubble periods. No difference can be discerned between the two curves. The relative error of the maximum bubble radius is within 0.1%, which demonstrates the accuracy of our BIM model.

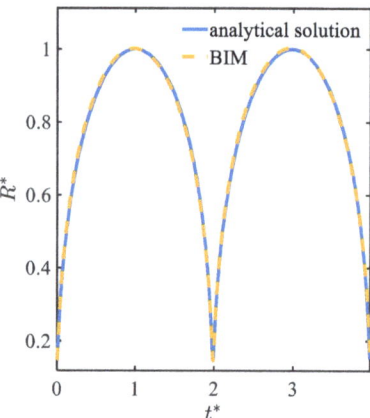

Figure 2. Comparison of a spherical bubble radius calculated from the analytical solution (blue solid line) and BIM (yellow dashed line).

Next, we carry out a self-consistency analysis for the bubble surface velocity u_r and u_z with different element numbers of the bubble surface N. The initial conditions are set as: $R_0^* = 0.15$, $\epsilon = 102.35$, and $\delta = 0.23$. We set $N = 100, 150, 200, 250, 300$, and 350 for the bubble surface and compare the velocity of each node before and after jet impact (Figure 3). At $N = 350$, the results of the two models are in excellent agreement with the velocities before the jet impact, except for the local position around the impact area. This suggests that each model has good self-consistency. Next, we study how the results change with N. We choose the node with the minimum u_r in each case and calculate the relative deviation of u_r before and after the impact moment. The velocity deviations of both u_r and u_z decrease as the number of bubble surface elements increases. The deviations reach the minimum values at $N = 300$ and remain constant after that. This demonstrates that the accuracy of both models improves as N increases, and achieves a convergence result at $N = 300$.

The element number of the vortex cut m can also affect the calculation. We set 5, 6, 7, 8, 9, and 10 elements for the cut and find that the computational process is unstable at too small an m. We obtain a stable and convergent result at $m = 8$ in the test of the code. Thus, we set eight elements for the vortex cut in this paper unless otherwise indicated. As for the vortex ring model, we set 50, 100, 200, and 400 elements for the ring to test the computational efficiency. The run time increases with more vortex ring elements while the results reach convergence at 200 elements, so we set 200 elements for the vortex ring to reduce the CPU cost.

3.2. Toroidal Bubble in a Free Field

In this section, we apply both models to study *UNDEX* bubble dynamics in a free field. The initial conditions are set as: $R_0^* = 0.15$, $\epsilon = 102.35$, and $\delta = 0.23$. Figure 4 demonstrates the time evolution of the bubble radius R^* obtained from the Rayleigh–Plesset equation (RPE) and BIM. The jet impacts the bubble surface after the minimum bubble volume moment. The two results are in good agreement. Gravity is ignored in the RPE, so there is a difference in the minimum bubble radius between the RPE and BIM simulation, which is 24%. However, the two numerical calculation results perfectly coincide with each other. This implies that both models can predict toroidal bubble oscillations well in terms of bubble volume.

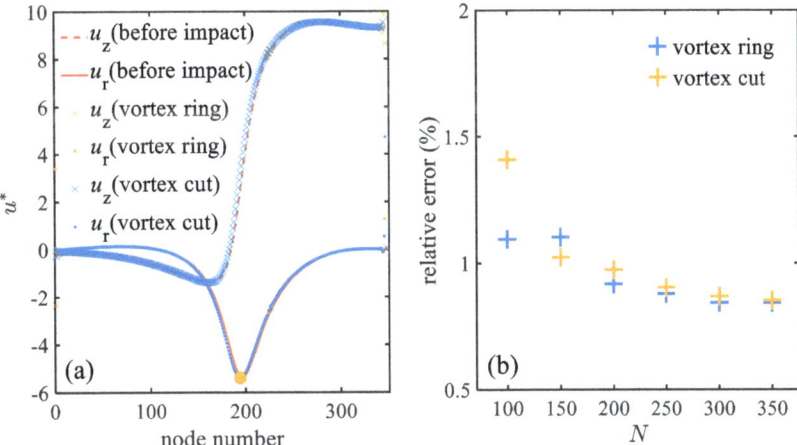

Figure 3. A self-consistency analysis for the bubble surface velocity. (**a**) The velocity at every node on the bubble surface before and after the jet impact moment obtained from the vortex ring and vortex cut models at $N = 350$. (**b**) The velocity deviations before and after the impact of a typical node (as marked in frame a) with a different number of bubble surface elements.

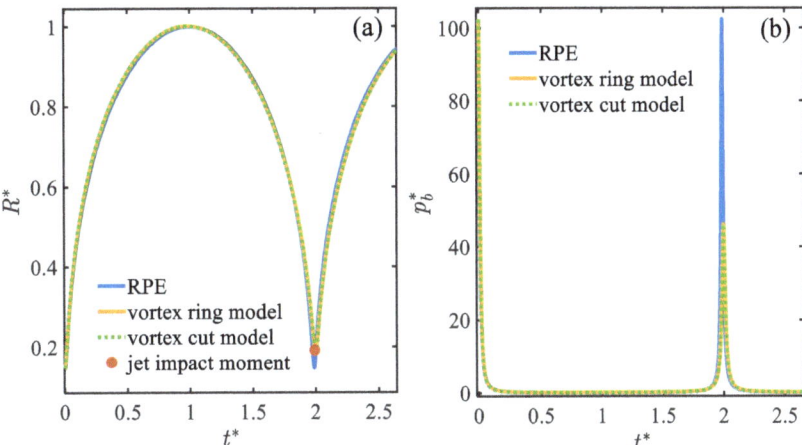

Figure 4. Comparison of (**a**) bubble radius and (**b**) bubble gas pressure calculated from RPE (blue solid lines), the vortex ring model (yellow solid lines), and the vortex cut model (green dashed lines).

Next, we compare bubble gas pressure p_b^* obtained analytically and numerically for a better illustration of the bubble dynamics (Figure 4b). There is an obvious difference at the pressure peak, corresponding to the difference at the minimum bubble radius (shown in Figure 4a). The difference in the bubble pressure peak is 55%. This suggests that gravity plays a vital role in large-scale UNDEX bubbles, and theories of spherical bubbles are not useful for describing the detailed behaviors of nonspherical bubbles and the induced pressure waves. As expected, the results from the two models are in excellent agreement. This suggests the high similarity between the two models in regard to toroidal bubble dynamics in a free field.

Figure 5 compares the bubble profiles at different moments ($t^* = 1.993$, 2.000, 2.016, 2.044, 2.130, and 2.309). The results of the two models are identical. After the jet impact moment, the bubble surface remains smooth and gradually collapses to the minimum volume with an annular jet inward. Thereafter, the bubble expands with a protrusion on the top of the surface, and the annular jet moves downward along the bubble surface (frames b–d). Meanwhile, the water jet becomes thinner as the bubble expands to the maximum volume (frames e and f). All of the details are predicted well by both models, including the annular jet and the protrusion.

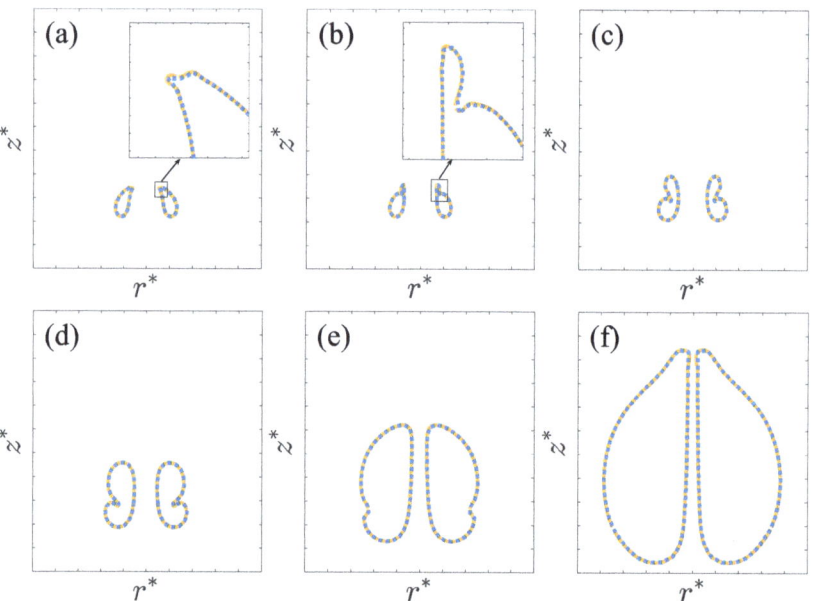

Figure 5. Toroidal bubble oscillations in a free field simulated by the vortex ring model (blue dashed lines) and the vortex cut model (yellow dashed lines) at $t^* = 1.993, 2.000, 2.016, 2.044, 2.130$, and 2.309. The scale ranges of the r axis and z axis are $(-1, 1)$ and $(-0.2, 2)$, respectively. (**a**) Bubble reaching the minimum volume at $t^* = 1.993$. (**b**) A protrusion appearing on the top of the bubble at $t^* = 2.000$. (**c**) Bubble rebounding with a larger protrusion at $t^* = 2.016$. (**d**) Bubble rebounding at $t^* = 2.044$. (**e**) Bubble becoming full at $t^* = 2.130$. (**f**) Bubble reaching the maximum volume at $t^* = 2.309$.

Next, we turn to the dynamic pressure in the flow field (Figure 6). In both models, two local pressure peaks are observed, marked A and B, which correspond to frames a and b in Figure 5, respectively. It is clear that the strong jet impact causes a growing protrusion at the top of the bubble. This leads to pressure peaks in the flow field. Similar phenomena and detailed analysis can be found in the study by Li et al. [61]. During the ascending and descending stages, the pressures calculated by the two approaches agree well with each other, with slight differences of 2.7% and 2.6% at local pressure peaks A and B, respectively. However, the pressure peaks are transient, so the comparison lacks universal significance. Next, we calculate the time integral of the dynamic pressure from the jet impact moment to $t^* = 2.2$. The difference in the pressure impulse is negligible at 0.6%. At field point $(2, 0)$, the results are virtually identical between the two models except for the two local pressure peaks. The differences are 1.7% and 2.5%, respectively. The difference in the pressure impulse calculated from the jet impact moment to $t^* = 2.2$ is 0.4%.

In sum, an overall match is obtained for toroidal bubble dynamics in a free field, except for some tiny differences in dynamic pressure in the flow field.

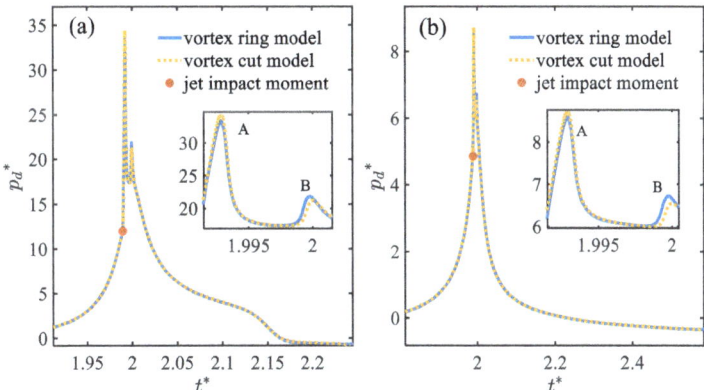

Figure 6. Comparison of the dynamic pressure at (**a**) point (0, 1.1) and (**b**) point (2, 0) obtained by the vortex ring model (blue solid line) and the vortex cut model (yellow dashed line).

3.3. Toroidal Bubble near a Rigid Wall

Next, we study the toroidal bubble dynamics near an infinite rigid wall using the two numerical models. The initial conditions are set as: $R_0^* = 0.19$, $\epsilon = 50.28$, $\delta = 0.11$ and $\gamma = 1.4$. Figure 7 compares the bubble profiles was obtained from the two models at $t^* = 2.298, 2.312, 2.359, 2.414, 2.616$, and 2.853. A growing protrusion forms at the top of the toroidal bubble in the initial stage (frames a–c) while the bubble moves towards the rigid wall. An annular water jet forms and develops inwards (frames c and d). After the bubble reaches the wall, its surface is flattened, and the contact surface gradually enlarges (frames e and f). The results of the two models are identical.

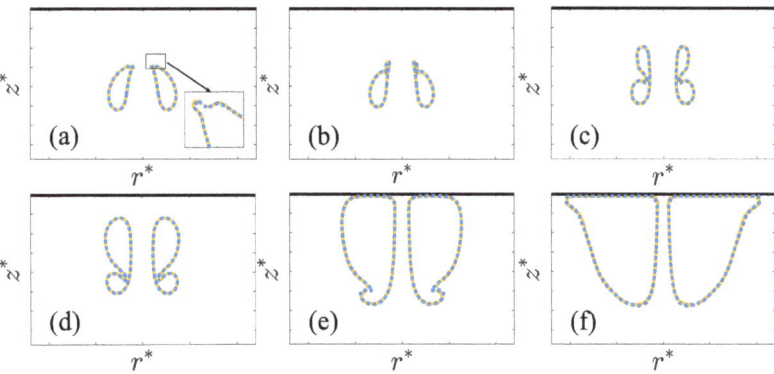

Figure 7. Toroidal bubble oscillations near a rigid wall simulated by the vortex ring model (yellow dashed lines) and the vortex cut model (blue dashed lines) captured at $t^* = 2.298, 2.312, 2.359, 2.414, 2.616$, and 2.853. The scale ranges of the r axis and z axis are $(-1.2, 1.2)$ and $(-0.15, 1.4)$, respectively. (**a**) Bubble reaching the minimum volume at $t^* = 2.298$. (**b**) A protrusion appearing on the top of the bubble at $t^* = 2.312$. (**c**) Bubble rebounding with a larger protrusion at $t^* = 2.359$. (**d**) An annular jet developing inwards at $t^* = 2.414$. (**e**) Bubble flattened by the wall at $t^* = 2.616$. (**f**) Bubble reaching the maximum volume at $t^* = 2.853$.

In the study of a UNDEX bubble near a solid boundary, the dynamic pressure induced at the boundary is of great significance and can be used to evaluate the damage ability. The evolution of dynamic pressure at the rigid wall p_d^* obtained by each model is shown in Figure 8. The moments of A and B correspond to frame a and b in Figure 7. After the water jet penetrates the upper surface of the bubble, p_d^* increases rapidly and reaches two local peaks (marked A and B). After the bubble becomes toroidal, the high-speed water jet violently impacts the rigid wall, which leads to the pressure peaks. The results of the two models coincide well except for some tiny differences in the pressure peaks (e.g., that at Peak A being 8.35% higher for the vortex ring model vs. the vortex cut model). The differences are 7.49% and 5.7% in group 2 and group 3, respectively. We integrate the dynamic pressures from the jet impact moment to $t^* = 2.65$ to obtain the pressure impulse. The results of the two models are both 3.79. This demonstrates the excellent agreement between the two models.

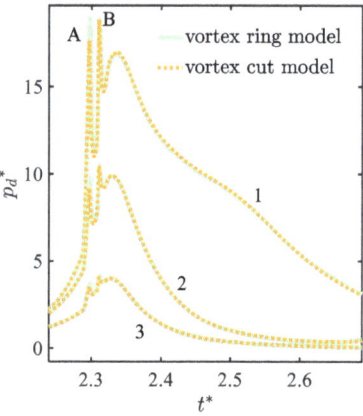

Figure 8. Comparison of dynamic pressure at the rigid wall calculated by the vortex ring model (green solid lines) and the vortex cut model (yellow dashed lines). The measuring points are set at (0, 1.4), (0.8, 1.4), and (2.5, 1.4), marked as groups 1, 2, and 3.

Finally, the computational efficiency of each model is assessed. We run the codes for different computation time steps on a personal laptop with 11th Gen Intel (R) Core (TM) i5-1135G7 CPU at 1.40 GHz). As shown in Figure 9, the vortex cut model has a significant advantage over the vortex ring model. The average difference in CPU cost for the four groups of time steps is 36%.

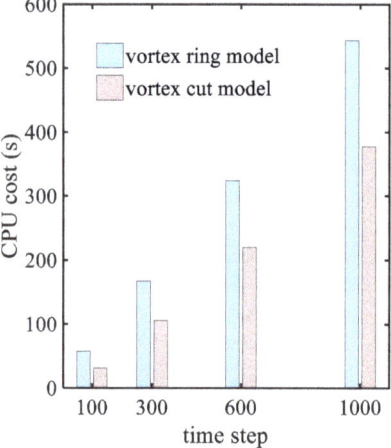

Figure 9. CPU cost of the vortex ring model (blue bars) and the vortex cut model (pink bars). The element number of the bubble surface, the vortex ring, and the vortex cut are 300, 50, and 8, respectively. The initial conditions are: $R_0^* = 0.11$, $\epsilon = 216.43$, $\delta = 0.44$ and $\gamma = 1.2$.

4. Conclusions

In this paper, we employ the BIM to study UNDEX bubble dynamics. After jet impact, the flow field becomes doubly connected; thus, the vortex ring and vortex cut models are proposed to study the toroidal bubble dynamics. We discuss the efficiency and accuracy of the two models and compare the bubble dynamics obtained from them in terms of bubble radius, bubble gas, and pressure impulse. The main findings are as follows:

(1) The self-consistency of each model improves as the number of bubble surface elements N increases, and both are optimal at $N = 300$. The computational process becomes more stable as the number of vortex ring and vortex cut elements increases.

(2) As for a bubble in a free field, the bubble radius and the bubble gas pressure obtained from RPE and the numerical models are compared. We find significant differences at the moment of minimum bubble volume. This suggests that the effect of gravity plays a vital role in large-scale UNDEX bubbles, and theories of spherical bubbles cannot be used to describe the detailed behaviors of nonspherical bubbles and the induced pressure waves. The two models coincide well at all time points. As for the dynamic pressure in the flow field, the two models have very close results with only slight differences at local pressure peaks.

(3) As for a bubble near an infinite rigid wall, the bubble oscillations are predicted well by the two numerical models. Regarding the dynamic pressure at the rigid wall, there are some differences at the pressure peaks of each model, which decrease with measuring-point distance from the center of the wall. However, the pressure peaks are transient, so the time integrals of pressure obtained from the two models are the same.

(4) The vortex ring model is more computationally demanding than the vortex cut model, with a 36% greater average CPU cost.

Author Contributions: Conceptualization, S.L.; methodology, S.L. and L.H.; software, S.L. and L.H.; validation, L.H.; formal analysis, L.H.; investigation, L.H. and S.L.; writing—original draft preparation, L.H.; writing—review and editing, T.Z., R.H. and S.L.; visualization, L.H. and S.L.; supervision, S.L.; funding acquisition, S.L., R.H. and D.Y. All authors have read and agreed to the published version of the manuscript.

Funding: This research was funded by the National Natural Science Foundation of China (12072087) and the Heilongjiang Provincial Natural Science Foundation of China (YQ2022E017).

Data Availability Statement: The data that support the findings of this study are available within the article.

Acknowledgments: The authors gratefully acknowledge A-Man Zhang for insightful discussions.

Conflicts of Interest: The authors declare no conflict of interest.

References

1. Versluis, M.; Schmitz, B.; von der Heydt, A.; Lohse, D. How snapping shrimp snap: Through cavitating bubbles. *Science* **2000**, *289*, 2114–2117. [CrossRef] [PubMed]
2. Lohse, D.; Schmitz, B.; Versluis, M. Snapping shrimp make flashing bubbles. *Nature* **2001**, *413*, 477–478. [CrossRef]
3. Lindau, J.W.; Boger, D.A.; Medvitz, R.B.; Kunz, R.F. Propeller cavitation breakdown analysis. *J. Fluids Eng.* **2005**, *127*, 995–1002. [CrossRef]
4. Liu, Y.L.; Zhang, A.M.; Tian, Z.L.; Wang, S.P. Investigation of free-field underwater explosion with Eulerian finite element method. *Ocean Eng.* **2018**, *166*, 182–190. [CrossRef]
5. Ghoshal, R.; Mitra, N. Underwater explosion induced shock loading of structures: Influence of water depth, salinity and temperature. *Ocean Eng.* **2016**, *126*, 22–28. [CrossRef]
6. Tian, Z.L.; Liu, Y.L.; Zhang, A.M.; Tao, L.B.; Chen, L. Jet development and impact load of underwater explosion bubble on solid wall. *Appl. Ocean Res.* **2020**, *95*, 102013. [CrossRef]
7. Li, S.; Prosperetti, A.; van der Meer, D. Dynamics of a toroidal bubble on a cylinder surface with an application to geophysical exploration. *Int. J. Multiph. Flow* **2020**, *129*, 103335. [CrossRef]
8. Li, S.; van der Meer, D.; Zhang, A.M.; Prosperetti, A.; Lohse, D. Modelling large scale airgun-bubble dynamics with highly non-spherical features. *Int. J. Multiph. Flow* **2020**, *122*, 103143. [CrossRef]
9. Plesset, M.S. The dynamics of cavitation bubbles. *J. Appl. Mech.* **1949**, *16*, 277–282. [CrossRef]

10. Brenner, M.P.; Hilgenfeldt, S.; Lohse, D. Single-bubble sonoluminescence. *Rev. Mod. Phys.* **2002**, *74*, 425. [CrossRef]
11. Rayleigh, L. On the pressure developed in a liquid during the collapse of a spherical cavity. *Lond. Edinb. Dublin Philos. Mag. J. Sci.* **1917**, *34*, 94–98. [CrossRef]
12. Keller, J.B.; Kolodner, I.I. Damping of underwater explosion bubble oscillations. *J. Appl. Phys.* **1956**, *27*, 1152–1161. [CrossRef]
13. Prosperetti, A.; Lezzi, A. Bubble dynamics in a compressible liquid. Part 1. First-order theory. *J. Fluid Mech.* **1986**, *168*, 457–478. [CrossRef]
14. Lezzi, A.; Prosperetti, A. Bubble dynamics in a compressible liquid. Part 2. Second-order theory. *J. Fluid Mech.* **1987**, *185*, 289–321. [CrossRef]
15. Zhang, A.M.; Li, S.M.; Cui, P.; Li, S.; Liu, Y.L. A unified theory for bubble dynamics. *Phys. Fluids* **2023**, *35*, 033323. [CrossRef]
16. Pearson, A.; Blake, J.R.; Otto, S.R. Jets in bubbles. *J. Eng. Math.* **2004**, *48*, 391–412. [CrossRef]
17. Blake, J.R.; Taib, B.B.; Doherty, G. Transient cavities near boundaries. Part 1. Rigid boundary. *J. Fluid Mech.* **1986**, *170*, 479–497. [CrossRef]
18. Blake, J.R.; Gibson, D.C. Cavitation bubbles near boundaries. *Annu. Rev. Fluid Mech.* **1987**, *19*, 99–123. [CrossRef]
19. Kalumuck, K.M.; Duraiswami, R.; Chahine, G.L. Bubble dynamics fluid-structure interaction simulation by coupling fluid BEM and structural FEM codes. *J. Fluids Struct.* **1995**, *9*, 861–883. [CrossRef]
20. Wang, C.; Khoo, B.C. An indirect boundary element method for three-dimensional explosion bubbles. *J. Comput. Phys.* **2004**, *194*, 451–480. [CrossRef]
21. Klaseboer, E.; Hung, K.C.; Wang, C.; Wang, C.W.; Khoo, B.C.; Boyce, P.; Debono, S.; Charlier, H. Experimental and numerical investigation of the dynamics of an underwater explosion bubble near a resilient/rigid structure. *J. Fluid Mech.* **2005**, *537*, 387–413. [CrossRef]
22. Li, S.; Zhang, A.M.; Han, R.; Ma, Q. 3D full coupling model for strong interaction between a pulsating bubble and a movable sphere. *J. Comput. Phys.* **2019**, *392*, 713–731. [CrossRef]
23. Zhang, A.M.; Wu, W.B.; Liu, Y.L.; Wang, Q.X. Nonlinear interaction between underwater explosion bubble and structure based on fully coupled model. *Phys. Fluids* **2017**, *29*, 082111. [CrossRef]
24. Han, R.; Zhang, A.M.; Tan, S.C.; Li, S. Interaction of cavitation bubbles with the interface of two immiscible fluids on multiple time scales. *J. Fluid Mech.* **2021**, *932*, A8. [CrossRef]
25. Manmi, K.M.A.; Aziz, I.A.; Arjunan, A.; Saeed, R.K.; Dadvand, A. Three-dimensional oscillation of an acoustic microbubble between two rigid curved plates. *J. Hydrodyn.* **2021**, *33*, 1019–1034. [CrossRef]
26. Dadvand, A.; Manmi, K.M.A.; Aziz, I.A. Three-dimensional bubble jetting inside a corner formed by rigid curved plates: Boundary integral analysis. *Int. J. Multiph. Flow* **2023**, *158*, 104308. [CrossRef]
27. Curtiss, G.A.; Leppinen, D.M.; Wang, Q.X.; Blake, J.R. Ultrasonic cavitation near a tissue layer. *J. Fluid Mech.* **2013**, *730*, 245–272. [CrossRef]
28. Wang, Q.X. Non-spherical bubble dynamics of underwater explosions in a compressible fluid. *Phys. Fluids* **2013**, *25*. [CrossRef]
29. Li, S.; Saade, Y.; van der Meer, D.; Lohse, D. Comparison of Boundary Integral and Volume-of-Fluid methods for compressible bubble dynamics. *Int. J. Multiph. Flow* **2021**, *145*, 103834. [CrossRef]
30. Wang, Q.X.; Blake, J.R. Non-spherical bubble dynamics in a compressible liquid. Part 1. Travelling acoustic wave. *J. Fluid Mech.* **2010**, *659*, 191–224. [CrossRef]
31. Lundgern, T.S.; Mansour, N.N. Vortex ring bubbles. *J. Mech.* **1991**, *224*, 177–196.
32. Best, J.P. The Dynamics of Underwater Explosions. Ph.D. Thesis, The University of Wollongong, Wollongong, Australia, 1991.
33. Best, J.P. *The Rebound of Toroidal Bubbles*; Bubble Dynamics and Interface Phenomena; Springer: Berlin/Heidelberg, Germany, 1994.
34. Zhang, S.; Duncan, J.H.; Chahine, G.L. The final stage of the collapse of a cavitation bubble near a rigid wall. *J. Fluid Mech.* **1993**, *257*, 147–181. [CrossRef]
35. Wang, Q.X.; Yeo, K.S.; Khoo, B.C.; Lam, K.Y. Nonlinear interaction between gas bubble and free surface. *Comput. Fluids* **1996**, *25*, 607–628. [CrossRef]
36. Zhang, Y.L.; Yeo, K.S.; Khoo, B.C.; Wang, C. 3D jet impact and toroidal bubbles. *J. Comput. Phys.* **2001**, *166*, 336–360. [CrossRef]
37. Wang, Q.X. Multi-oscillations of a bubble in a compressible liquid near a rigid boundary. *J. Fluid Mech.* **2014**, *745*, 509–536. [CrossRef]
38. Wang, Q.X.; Yeo, K.S.; Khoo, B.C.; Lam, K.Y. Vortex ring modelling of toroidal bubbles. *Theor. Comput. Fluid Dyn.* **2005**, *19*, 303–317. [CrossRef]
39. Lauterborn, W.; Lechner, C.; Koch, M.; Mettin, R. Bubble models and real bubbles: Rayleigh and energy-deposit cases in a Tait-compressible liquid. *IMA J. Appl. Math.* **2018**, *83*, 556–589. [CrossRef]
40. Li, T.; Wang, S.; Li, S.; Zhang, A.M. Numerical investigation of an underwater explosion bubble based on FVM and VOF. *Appl. Ocean Res.* **2018**, *74*, 49–58. [CrossRef]
41. Li, T.; Zhang, A.M.; Wang, S.; Li, S.; Liu, W. Bubble interactions and bursting behaviors near a free surface. *Phys. Fluids* **2019**, *31*, 042104. [CrossRef]
42. Zeng, Q.; Cai, J. Three-dimension simulation of bubble behavior under nonlinear oscillation. *Ann. Nucl. Energy* **2014**, *63*, 680–690. [CrossRef]

43. Tian, Z.L.; Liu, Y.L.; Zhang, A.M.; Tao, L.B. Energy dissipation of pulsating bubbles in compressible fluids using the Eulerian finite-element method. *Ocean Eng.* **2020**, *196*, 106714. [CrossRef]
44. Liu, N.N.; Zhang, A.M.; Liu, Y.L.; Li, T. Numerical analysis of the interaction of two underwater explosion bubbles using the compressible Eulerian finite-element method. *Phys. Fluids* **2020**, *32*, 046107.
45. He, M.; Liu, Y.L.; Zhang, S.; Zhang, A.M. Research on characteristics of deep-sea implosion based on Eulerian finite element method. *Ocean Eng.* **2022**, *244*, 110270. [CrossRef]
46. Sun, P.; Le Touzé, D.; Oger, G.; Zhang, A.M. An accurate SPH Volume Adaptive Scheme for modeling strongly-compressible multiphase flows. Part 2: Extension of the scheme to cylindrical coordinates and simulations of 3D axisymmetric problems with experimental validations. *J. Comput. Phys.* **2021**, *426*, 109936. [CrossRef]
47. Sun, P.N.; Pilloton, C.; Antuono, M.; Colagrossi, A. Inclusion of an acoustic damper term in weakly-compressible SPH models. *J. Comput. Phys.* **2023**, *483*, 112056. [CrossRef]
48. Wang, P.; Zhang, A.M.; Fang, X.; Khayyer, A.; Meng, Z. Axisymmetric Riemann–smoothed particle hydrodynamics modeling of high-pressure bubble dynamics with a simple shifting scheme. *Phys. Fluids* **2022**, *34*, 112122. [CrossRef]
49. Nguyen, V.T.; Phan, T.H.; Duy, T.N.; Kim, D.H.; Park, W.G. Modeling of the bubble collapse with water jets and pressure loads using a geometrical volume of fluid based simulation method. *Int. J. Multiph. Flow* **2022**, *152*, 104103. [CrossRef]
50. Nguyen, Q.T.; Nguyen, V.T.; Phan, T.H.; Duy, T.N.; Park, S.H.; Park, W.G. Numerical study of dynamics of cavitation bubble collapse near oscillating walls. *Phys. Fluids* **2023**, *35*, 013306. [CrossRef]
51. Ni, B.Y.; Zhang, A.M.; Wu, G.X. Numerical and experimental study of bubble impact on a solid wall. *J. Fluids Eng.* **2015**, *137*, 031206. [CrossRef]
52. Li, S.; Li, Y.B.; Zhang, A.M. Numerical analysis of the bubble jet impact on a rigid wall. *Appl. Ocean Res.* **2015**, *50*, 227–236. [CrossRef]
53. Zhang, Z.; Wang, C.; Zhang, A.M.; Silberschmidt, V.V.; Wang, L. SPH-BEM simulation of underwater explosion and bubble dynamics near rigid wall. *Sci. China Technol. Sci.* **2019**, *62*, 1082–1093. [CrossRef]
54. Wang, J.X.; Zong, Z.; Sun, L.; Li, Z.R.; Jiang, M.Z. Numerical study of spike characteristics due to the motions of a non-spherical rebounding bubble. *J. Hydrodyn.* **2016**, *28*, 52–65. [CrossRef]
55. Wang, Q.X.; Yang, Y.X.; Tan, D.S.; Su, J.; Tan, S.K. Non-spherical multi-oscillations of a bubble in a compressible liquid. *J. Hydrodyn.* **2014**, *26*, 848–855. [CrossRef]
56. Klaseboer, E.; Khoo, B.C.; Hung, K.C. Dynamics of an oscillating bubble near a floating structure. *J. Fluids Struct.* **2005**, *21*, 395–412. [CrossRef]
57. Cole, R.H. *Underwater Explosions*; Princeton University Press: Princeton, NJ, USA, 1948.
58. Pearson, A.; Cox, E.; Blake, J.R. Bubble interactions near a free surface. *Eng. Anal. Bound. Elem.* **2004**, *28*, 295–313. [CrossRef]
59. Zhang, S.; Duncan, J.H. On the nonspherical collapse and rebound of a cavitation bubble. *Phys. Fluids* **1994**, *6*, 2352–2362. [CrossRef]
60. Tong, R.P.; Schiffers, W.P.; Shaw, S.J.; Blake, J.R.; Emmony, D.C. The role of 'splashing' in the collapse of a laser-generated cavity near a rigid boundary. *J. Fluid Mech.* **1999**, *380*, 339–361. [CrossRef]
61. Li, S.; Han, R.; Zhang, A.M.; Wang, Q.X. Analysis of pressure field generated by a collapsing bubble. *Ocean Eng.* **2016**, *117*, 22–38. [CrossRef]
62. Zhang, A.M.; Liu, Y.L. Improved three-dimensional bubble dynamics model based on boundary element method. *J. Comput. Phys.* **2015**, *294*, 208–223. [CrossRef]

Disclaimer/Publisher's Note: The statements, opinions and data contained in all publications are solely those of the individual author(s) and contributor(s) and not of MDPI and/or the editor(s). MDPI and/or the editor(s) disclaim responsibility for any injury to people or property resulting from any ideas, methods, instructions or products referred to in the content.

Article

Vapor Bubble Deformation and Collapse near Free Surface

Yue Chen [1], Qichao Wang [1], Hongbing Xiong [1,*] and Lijuan Qian [2,*]

[1] State Key Laboratory of Fluid Power and Mechatronic Systems, Zhejiang University, Hangzhou 310027, China; cheny9924@163.com (Y.C.); 21824016@zju.edu.cn (Q.W.)
[2] College of Mechanical and Electrical Engineering, China Jiliang University, Hangzhou 310018, China
* Correspondence: hbxiong@zju.edu.cn (H.X.); qianlj@cjlu.edu.cn (L.Q.)

Abstract: Vapor bubbles are widely concerned in many industrial applications. The deformation and collapse of a vapor bubble near a free surface after being heated and raised from the bottom wall are investigated in this paper. On the basis of smoothed particle hydrodynamics (SPH) and the van der Waals (VDW) equation of state, a numerical model of fluid dynamics and phase change was developed. The effects of fluid dynamics were considered, and the phase change of evaporation and condensation between liquid and vapor were discussed. Quantitative and qualitative comparisons between our numerical model and the experimental results were made. After verification, the numerical simulation of bubbles with the effects of the shear viscosity η_s and the heating distance L were taken into account. The regularity of the effect of the local Reynolds number (Re) and the Ohnesorge number (Oh) on the deformation of vapor bubbles is summarized through a further analysis of several cases, which can be summarized into four major patterns as follows: umbrella, semi-crescent, spheroid, and jet. The results show that the Re number has a great influence on the bubble deformation of near-wall bubbles. For Re > 1.5×10^2 and Oh < 3×10^{-4}, the shape of the bubble is umbrella; for Re < 5×10^0 and Oh > 10^{-3}, the bubble is spheroidal; and for 5×10^0 < Re < 1.5×10^2, 3×10^{-4} < Oh < 10^{-3}, the bubble is semi-crescent. For liquid-surface bubbles, the Re number effect is small, and when Oh > 5×10^{-3}, the shape of the bubble is jet all the time; there is no obvious difference in the bubble deformation, but the jet state is more obvious as the Re decreases. Finally, the dynamic and energy mechanisms behind each mode are discussed. The bubble diameter, bubble symmetry coefficient, and rising velocity were analyzed during their whole processes of bubble growth and collapse.

Keywords: vapor bubble; deformation; SPH method; phase change; Re number

Citation: Chen, Y.; Wang, Q.; Xiong, H.; Qian, L. Vapor Bubble Deformation and Collapse near Free Surface. *Fluids* **2023**, *8*, 187. https://doi.org/10.3390/fluids8070187

Academic Editors: D. Andrew S. Rees, Hemant J. Sagar and Nguyen Van-Tu

Received: 13 March 2023
Revised: 17 June 2023
Accepted: 21 June 2023
Published: 22 June 2023

Copyright: © 2023 by the authors. Licensee MDPI, Basel, Switzerland. This article is an open access article distributed under the terms and conditions of the Creative Commons Attribution (CC BY) license (https://creativecommons.org/licenses/by/4.0/).

1. Introduction

A vapor bubble is a kind of gas that is generated by the instantaneous injection of high energy to liquid, such as laser, electricity, or other rapid heating methods. It has been applied in many industries; for example, the heat transfer of two-phase heat exchangers, surface corrosion caused by shock waves, injection without needles, and destruction of biological tissue using liquid jet superheat transfer [1–4]. In recent years, the study of vapor bubbles has attracted a lot of attention, but the dynamic process of vapor bubbles is nonlinear and highly complex [5,6]. It includes bubble oscillation and interface fluctuation during bubble growth, shock wave impact, and cavitation noise during bubble collapse. Understanding the deformation and collapse mechanisms of vapor bubbles is the key to successfully address these application-related challenges.

In order to study the dynamics of vapor bubbles, many researchers were involved in experimental research and numerical simulations. In experiments, the usual way to create vapor bubbles is to use pulsed lasers or electric sparks to shoot them instantaneously, and to observe the interaction of the bubbles with the solid surface using high-speed cameras [7,8]. Gonzalez et al. [9] studied the dynamic process of laser-induced bubbles in liquid gaps at different heights. Sun et al. [10,11] concluded that the thermal effect plays an important role in the growth and collapse of vapor bubbles in microchannels. Kangude et al. [12] studied the cavitation mechanism of vapor bubbles on hydrophobic surfaces using the thermal imaging method. The experimental results of Tang showed that the hydrophobic coating has a significant influence on the growth process of laser-induced bubbles [13]. According to the three-dimensional view of the experiment, Hement et al. [14] found that the near-wall bubble collapses very rapidly, and that the tangential flow would lead to the formation of the ring cavity. Considering the influence of gravity and viscosity, Sangeeth et al. [15] studied the jet velocity resulting from bubble collapse at a liquid surface; the dependence of the dimensionless jet velocity, expressed in terms of the Weber number, on the Bond number, is determined by the dimensionless cavity depth. The Weber number is used to measure the relationship between the surface tension and the inertia force, and the Bond number is used to measure the relationship between the gravity and the surface tension. By using a spark-generated device and a high-speed camera, Zhang et al. [16] conducted an experiment on the dynamic process of vapor bubbles under water; six classical water burial phenomena were induced, and their forming mechanism was analyzed. Ma et al. [17] investigated the growth of vapor bubbles under different levels of gravity by conducting an experiment and found that gravity has a significant influence on the growth of a single vapor bubble.

Besides experiments, many numerical methods have also been reported to analyze the bubble dynamics, including the Lattice Boltzmann method (LBM), the finite difference method (FDM), the finite volume method (FVM), the volume of fluid method (VOF), the smoothed particle hydrodynamics method (SPH), and so on. Liu et al. [18] deduced the relationship between the various characteristic parameters of bubbles, then explored that different liquid parameters would have a significant impact on the cavitation process based on the FDM. Phan et al. [19] used a compressible homogeneous mixture model to numerically investigate the dynamics of an underwater explosion bubble, and the dynamic bubble motion including the bubble expansion, contraction, collapse, jet, and rebound. As for the VOF methods, they contain the algebraic VOF and geometric VOF methods. Owing to the merits of mass conservation, the latter was widely applied in the bubble simulation in [20]. By using the VOF method, Tang [13] simulated the hydrophobic wall surface by controlling the thickness of the air film at the solid–liquid interface, studied the oscillation behavior of the laser-induced bubble, and summarized the dynamic mechanism and law. Nguyen et al. [21] used a geometrical VOF algorithm based on the piecewise–linear interface calculation (PLIC) to numerically investigate the dynamic behavior of bubble collapses, water jets, and pressure loads during the collapse of the bubble near walls and a free surface; the results showed a good agreement between the simulation and experiment of the bubble dynamics during the collapse process. Erin et al. [22] used the SPH and VOF methods to simulate the rising of bubbles, and compared with previous experimental results, they concluded that both the VOF and SPH methods may be used to capture physically realistic transient and steady-state multi-phase systems; the SPH method could better capture the centroid of the bubble, while the VOF method better captured the rising velocity of the bubble.

SPH is a meshless method. It is uniquely capable of representing the dynamic evolution of complicated geometries without additional algorithmic complication, such as those found in multi-phase flows [22]. When simulating fluid dynamics problems, the SPH method discrete the flow field into moving fluid micro clusters, which can be regarded as a combination of a series of molecules with the same properties [23]. Unlike the traditional grid algorithm, SPH has no grid connection between the particles, and follows the interaction between the particles, which is suitable for any large deformation problem. On the other hand, the SPH method uses the Lagrange method to describe the flow field, which can be used to study some multiphase flow problems with discrete phases and has obvious advantages over the traditional grid algorithm in the study of large deformation and dynamic boundary problems.

The diffusion interface method (DIM), based on the SPH method, is more commonly used when involved in the precise capture of the gas–liquid phase flow interface [24]. Sigalotti et al. [25] first used the DIM, which treats the gas–liquid interface as continuous, added the Korteweg tensor to characterize the capillary forces, and used the SPH algorithm [23] to solve, which was proven to be useful in cavitation hydrodynamics. Gallo et al. [26] verified the study on the nucleation of vapor bubbles in metastable liquid by using the DIM. Wang [24] applied the DIM to numerically simulate the rising of vapor bubbles in static water, and proved that it is feasible to calculate the dynamics of vapor bubbles by using the DIM. Moreover, the relation between the shape of the rising vapor bubble and the dimensionless parameters such as the Reynolds number was also introduced.

Based on the gas–liquid DIM, the Navier–Stokes–Korteweg (NSK) equation considering the gas–liquid interfacial tension is derived, the van der Waals (VDW) equation of state is introduced, and the SPH algorithm is used for the numerical solution. The effects of the shear viscosity η_s and the heating distance L on the growth and collapse processes of the vapor bubble are taken into account. The regularity of the effect of the Re number and the Oh number on the deformation of vapor bubbles is summarized through a further analysis of several cases, which can be summarized into four major patterns. Then, the formation mechanism is analyzed, and the growth and collapse of the bubbles are studied. According to our results, it is possible to precisely control the deformation of vapor bubbles by adjusting the two dimensionless parameters, the Re number and the Oh number. This has a certain engineering guiding significance for the current application, for example, the avoidance of surface corrosion caused by nuclear boiling, the dispersion of poisonous droplets caused by the disintegration of vapor bubbles, and the effect of bubble deformation on EHD-enhanced boiling heat transfer [27].

2. SPH Modeling

In our model, compressible vapor and liquid are considered to be two-phase fluids with a continuous density gradient. In the Lagrange formula, the liquid and gas phases uniformly follow the conservation equations of mass, momentum, and energy as follows:

$$\frac{d\rho}{dt} = -\rho \nabla \cdot \mathbf{v} \tag{1}$$

$$\rho \frac{d\mathbf{v}}{dt} = \nabla \cdot \mathbf{M} + F_E \tag{2}$$

$$\frac{dU}{dt} = \frac{1}{\rho} \mathbf{M} : \nabla \mathbf{v} + \frac{\kappa}{\rho} \nabla^2 T \tag{3}$$

where ρ is the density, $\mathbf{v}$ is the velocity vector, $\mathbf{M}$ is the stress tensor, F_E is the external force of gravity, T is the temperature, U is the internal energy, and κ is the thermal conductivity. The stress tensor $\mathbf{M}$ includes the pressure terms, the shear and bulk viscosity terms, as well as an additional Korteweg tensor $\mathbf{M}_c$ of the gas–liquid diffusion interface, as follows:

$$\mathbf{M} = -p\mathbf{I} + \eta_s (\nabla \mathbf{v} + \nabla \mathbf{v}^T) + (\eta_v - \frac{2}{\dim}\eta_s)(\nabla \cdot \mathbf{v})\mathbf{I} + \mathbf{M}_C \tag{4}$$

where p represents the pressure, dim represents the dimension of space, and η_s and η_v are the shear and volume dynamic viscosity, respectively. The Korteweg tensor $\mathbf{M}_c$ can be used to simulate the capillary force on the interface due to the density gradient, expressed as follows:

$$\mathbf{M}_C = K(\rho \nabla^2 \rho + \frac{1}{2}|\nabla \rho|^2)\mathbf{I} - K \nabla \rho \nabla \rho \tag{5}$$

where K is the gradient energy coefficient for a given material.

According to the description of the SPH model, the density can be better described in the following summation density Equation (6), instead of the continuity Equation (1).

The summation density conserves mass exactly and guarantees second-order accuracy [25], which benefits the simulation of the liquid–vapor interface.

$$\rho_a = \sum_b \frac{m_b}{\rho_b} \rho_b W_{ab} = \sum_b m_b W_{ab} \tag{6}$$

where m is the particle mass, the subscript b indicates the neighbor particles around this particle a, the subscript ab denotes the variable difference between particle a and b, the subscript b represents the adjacent particles around particle a, and W_{ab} is a kernel function, which explains the particle distance between particles a and b.

The momentum and energy calculations are discretized into long-range and short-range terms, because the same smoothing length for all the force terms was unable to handle the surface tension effects and cause the interfacial instability [28] as follows:

$$\frac{dv_a}{dt} = \sum_b m_b \left(\frac{\mathbf{M}_a}{\rho_a^2} + \frac{\mathbf{M}_b}{\rho_b^2}\right) \cdot \nabla W_{ab} + \sum_b m_b \left(\frac{\mathbf{M}_a^H}{\rho_a^2} + \frac{\mathbf{M}_b^H}{\rho_b^2}\right) \cdot \nabla W_{ab}^H + F_E \tag{7}$$

$$\frac{dU_a}{dt} = \frac{1}{2}\sum_a m_a \left(\frac{\mathbf{M}_a}{\rho_a^2} + \frac{\mathbf{M}_b}{\rho_b^2}\right) : v_{ba} \nabla W_{ab} + \frac{1}{2}\sum_b m_b \left(\frac{\mathbf{M}_a^H}{\rho_a^2} + \frac{\mathbf{M}_b^H}{\rho_b^2}\right) : v_{ba}^H \nabla W_{ab}^H + U_E \tag{8}$$

where the short-range repulsive term has the smooth length of h with no mark, and the double smooth length of $H = 2h$ is used for the long-range attractive term and marked with superscript H. $\mathbf{M}$ and $\mathbf{M^H}$ are equal to $\mathbf{M} - \bar{a}\rho^2 \mathbf{I} - \mathbf{M}_C$ and $\mathbf{M} + \bar{a}\rho^2 \mathbf{I}$, respectively.

In this paper, we use the hyperbolic kernel function proposed by Yang [29], which ensures that the distribution of particles is more uniform in both the two-dimensional and three-dimensional problems and does not lead to the unstable growth of stress, so better results can be obtained in the simulation.

In order to close the momentum and energy equations, the VDW equation is chosen to describe the pressure state equation, which can describe the gas–liquid coexistence system. The expression of the van der Waals equation of state is as follows:

$$p = \frac{\rho \bar{k}_b T}{1 - \bar{\beta}\rho} - \bar{\alpha}\rho^2 \tag{9}$$

where $\bar{k}_b$ is the Boltzmann constant, $\bar{\alpha}$ is the parameter of attraction, and $\bar{\beta}$ is related to the size of the particles. And the critical state is expressed by these three parameters:

$$T_c = \frac{8\bar{\alpha}}{27\bar{k}_b\bar{\beta}}, \quad \rho_c = \frac{1}{3\bar{\beta}}, \quad P_c = \frac{\bar{\alpha}}{27\bar{\beta}^2} \tag{10}$$

where $\bar{k}_b$, $\bar{\alpha}$, and $\bar{\beta}$ are set as 1, 2, and 0.5 for the VDW fluid, respectively. Here, the gas or liquid phase is distinguished by the critical density of the VDW fluid. According to the VDW isothermal curve [30], when $\rho \to 0$, the VDW equation transforms into the ideal gas law. Therefore, if the fluid density is less than the critical density, it is the gas phase. Otherwise, it is the liquid phase.

Using the SPH discretization Equations (6) to (8) and coupled with the VDW EOS Equation (9), the liquid and the heated vapor are simultaneously simulated. After that, the bubble position, velocity, size, and other properties are analyzed. More key parameters could be characterized by the following dimensionless numbers: Re represents the Reynolds number, which is used to measure the relationship between the inertia force and the viscous force, and Oh represents the Ohnesorge number, which is used to measure the relationship between the viscous force, the inertia force, and the surface tension. Their expressions are as follows:

$$\text{Re} = \frac{\rho \bar{v} D_{\max}}{\eta_s} \tag{11}$$

$$\text{Oh} = \frac{\eta_s}{\sqrt{\rho \sigma R_{\max}}} \tag{12}$$

where ρ represents the liquid density, $\bar{v}$ represents the average velocity of the bubble, D_{max} represents the maximum diameter, R_{max} represents the maximum radius as $R_{max} = D_{max}/2$, σ represents the surface tension coefficient, and η_s represents the dynamic viscosity coefficient.

The NSK equation is simulated in a non-dimensional scale; thus, all the data are presented without specified units in the following part. It is convenient to examine those bubble and liquid characteristics with the key non-dimensional parameters of the Re and Oh numbers. In the meantime, these non-dimensionalized parameters could be referred to dimensional ones using the given material properties. Here, for the water and water vapor bubble, the reference values in the length, temperature, time, and mass scales are 5.33×10^{-8} m, 546 K, 1.36×10^{-10} s, and 7.33×10^{-20} kg, respectively [30].

3. Validation

3.1. Comparison Verification

In this section, we use the SPH method to simulate the water vapor bubble interaction with the free surface in two-dimension and compare it with the experimental results [16]. Our simulation setting is shown in Figure 1.

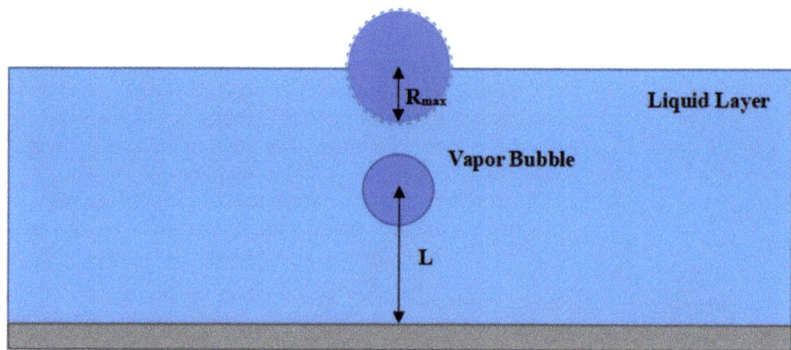

Figure 1. Vapor bubble heating geometry in our simulation.

Figure 2 shows the bubble deformation photos from Zhang's experiment with a bubble radius of about 15 mm. With the current computational efficiency and memory storage, it is still challenging to directly simulate the millimeter scale problem for our SPH simulation. Thus, we mimic the experimental setup in the following two critical parameters: the degree of super heat temperature ΔT, the relative heating distance γ_f.

Firstly, we estimated the power input into the bubble and, thereafter, the bubble's super heat temperature. The experiment was conducted using a capacitor discharge with U = 200 V before the discharge, and 150 V after the discharge, with a capacity of 6600 μF and a thermal efficiency of 2%. Thus, the heat input to the bubble was about 1.155 J [16]. Using the saturated water vapor density of 0.6 kg/m^3 and a heat capacity of 2.075 J/g·K, we obtained the super heat temperature for a 15 mm water vapor bubble of about 65,622 K, which is non-dimensional as $\Delta T = 12$ in our system.

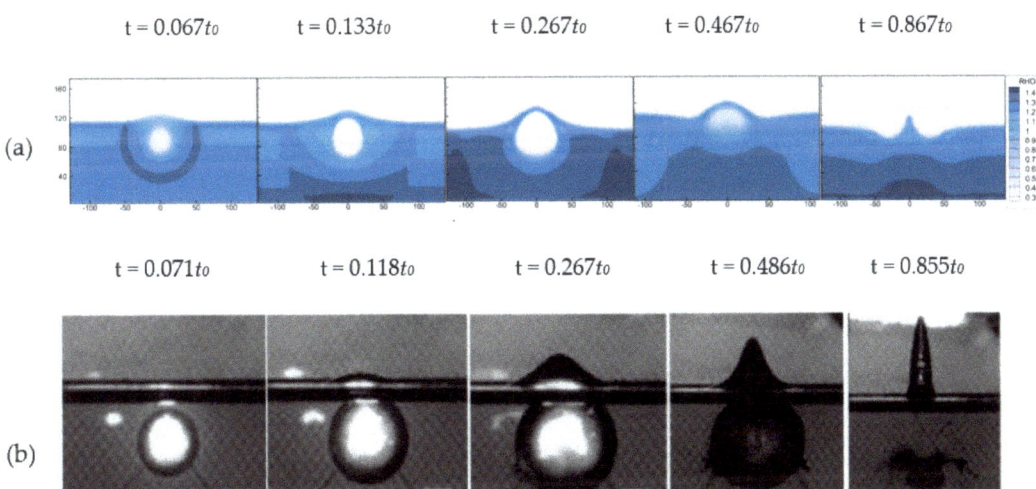

Figure 2. Comparison of the numerical results of SPH method and experiments. (**a**) Bubble deformation of $\gamma = 0.79$ in SPH simulation; (**b**) bubble deformation of $\gamma_f = 0.78$ in Zhang's experiment [16].

Then, we used the dimensionless distance between the burst point and the free liquid surface with the ratio to the bubble radius in Zhang's experiment as $\gamma_f = 0.78$ to arrange the heating position in our numerical simulation. We provide the data of two non-dimensional distances instead of one; $\varepsilon = H/R_{max}$ represents the dimensionless parameter of the liquid layer thickness, and $\lambda = L/R_{max}$ represents the dimensionless parameter of the bubble heating distance to the bottom of wall. Thus, the relative heating distance to the free surface could be obtained as $\gamma = \varepsilon - \lambda$, which has the same meaning as in Zhang's experiment of γ_f. Here, we used the height of the liquid layer $H = 160dx$ and the heating distance $L = 120dx$, where dx represents the initial distance between the SPH particles. After the simulation, we calculated $\varepsilon = H/R_{max} = 2.22$, $\lambda = L/R_{max} = 1.43$; therefore, $\gamma = \varepsilon - \lambda = 0.79$, which is approximately equal to the γ_f value of 0.78 in Zhang's experiment.

We compare our simulation bubble deformation with the experimental photos of [16] during the bubble growth and collapse near the free surface. The lifetime of the bubble during its first cycle period is set as t_0 either for the experiment or simulation. Figure 2 shows the typical bubble deformation from its generation to its collapse in our SPH simulation and Zhang's experiment under a similar heating degree and geometry. Figure 3 shows the dimensionless radius R/R_{max} during the first bubble cycle t_0, where R represents the local bubble radius, and R_{max} represents the maximum bubble radius. In the early period of the bubble growth, the bubble remains spherical; then, the bubble expands upwards along with the flow curve; finally, a clear upwelling column of water can be observed near the free surface. The results show that at the early stages before $0.3t_0$, our results are in good agreement with the experiments, either qualitatively or quantitatively. However, at the late stages after $0.3t_0$, there is a certain deviation between the numerical prediction and the experimental observation. For example, at $0.467t_0$, our simulation bubble has a more severe deformation, and a smaller jet height at $0.867t_0$, compared to the experiments. The reason is because the simulation accuracy is not that good for the SPH method; therefore, the computational error would be accumulated at the late stages. The advantages of SPH is the efficiency and flexibility in solving large deformation problems. In the future, we might strive to improve the accuracy of the SPH method. Currently, we believe that our SPH model is basically correct, and could be used to capture the bubble deformation during its deformation and collapse near the free surface.

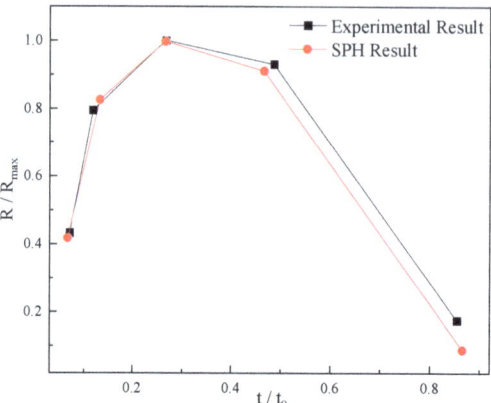

Figure 3. Comparison of SPH results with experiments for non-dimensional radius R/R_{max}.

3.2. Shock Wave

The second benchmark case is for the propagation of a shock wave. During the bubble's growth and collapse, the shock wave was found to be as important as the kinetic energy and thermal energy as shown in Figure 4. We tested the simulation of the discontinuity point in a typical one-dimensional shock wave problem using our SPH model.

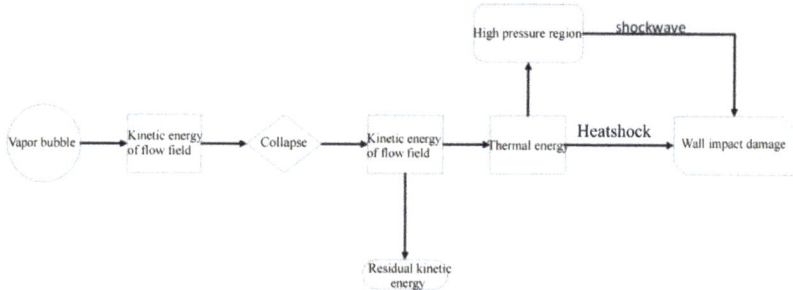

Figure 4. Energy transformation of vapor bubble.

Figure 5 shows the typical settings for a shock wave problem, and the discontinuity point is at $x = 2.5$. The initial state of the fluid on either side of the discontinuity point is $[\rho_L \ v_L \ p_L \ u_L] = [10 \ 0 \ 100 \ 25]$ and $[\rho_R \ v_R \ p_R \ u_R] = [1 \ 0 \ 1 \ 2.5]$. The specific heat capacity at constant volume is $c_v = 1$. The equation of state is as follows:

$$p = (\gamma - 1)\rho u = (\gamma - 1)c_v \rho T = 0.4 \cdot \rho T \tag{13}$$

where the subscripts L and R represent the left and right sides of the fluid. The total number of SPH particles is 1100, and the smooth length is $h = 1.8dx$.

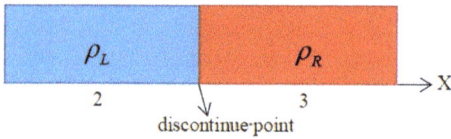

Figure 5. The settings of the shock wave problem.

The results for when the time $t = 0.4$ are shown in Figure 6. The red scatter is the result of the SPH, and the black line is the result of the exact Riemann solver. The SPH results agree well with the analytical solutions. This shows that our method is reliable in simulating the flow caused by the density and pressure difference.

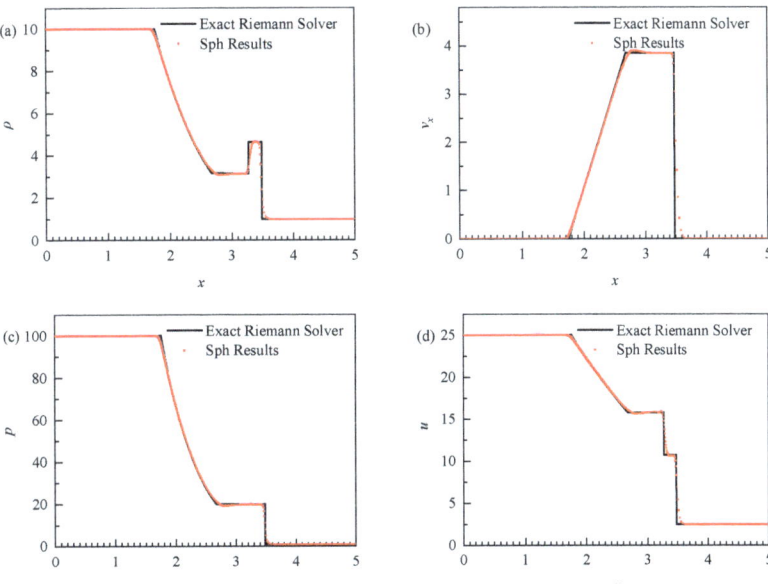

Figure 6. Comparison of the numerical results of the SPH method and the theoretical results of the exact Riemann solver. (**a**–**d**) show the distribution of density, velocity, pressure, and internal energy in the x direction, respectively.

4. Numerical Settings

The critical fluid density is introduced to distinguish between the liquid and vapor. The SPH liquid particle mass is $m = 0.6$, the stable density is $\rho = 1.2029$, the initial fluid temperature is $T_b = 1.01$, the heating height L above the solid wall is set to be $L = 15\sim120$, the shear viscosity is $\eta_s = 0.1\sim1.0$ as shown in Table 1, the volume viscosity is $\eta_v = 0.5\eta_s$, the 400×160 particles are arranged at the bottom area in the x and y directions, the heating radius is $r = 12dx$, and the excess heat is $\Delta T = 12$. The region is uniformly spherical after laser heating. The left and right boundaries are periodic boundaries, and the upper and lower wall are set as adiabatic solid boundaries, which can be referred to in our previously published study [30]. The vapor bubble heating geometry in our simulation is shown in Figure 1. The values of the three dimensionless parameters γ, ε, and λ, defined in Section 3.1, are shown in Table 2.

Table 1. The parameter settings of each case.

η_s	CASE ID	L/dx
1.0	C1	15
	C2	30
	C3	60
	C4	100
	C5	120
0.1	C6	15
	C7	30
	C8	60
	C9	100
	C10	120

Table 2. The dimensionless parameters of the bubble.

CASE ID	H/dx	L/dx	$\varepsilon = H/R_{max}$	$\lambda = L/R_{max}$	$\gamma = \varepsilon - \lambda$	N_{secb}
C1	160	15	4.31	0.32	3.99	0
C2	160	30	3.78	0.57	3.21	0
C3	160	60	3.55	1.06	2.48	0
C4	160	100	2.97	1.48	1.48	0
C5	160	120	2.22	1.43	0.79	0
C6	160	15	2.18	0.16	2.02	3
C7	160	30	2.36	0.35	2.01	3
C8	160	60	2.68	0.80	1.87	4
C9	160	100	2.78	1.39	1.39	0
C10	160	120	2.29	1.37	0.92	0

5. Results

5.1. Category of Bubble Deformation

First of all, the numerical simulations were performed for each case, and the effects of the shear viscosity η_s and the heating distance L on the growth and collapse processes of the vapor bubble are taken into account. The regularity of the effect of the Re number and the Oh number on the deformation of the vapor bubbles is obtained through a further analysis of several cases, which can be summarized into four major patterns. The following category between the Re and Oh numbers and the bubble deformation is drawn.

According to Figure 7, the liquid-surface vapor bubble takes on the jet shape, which is not greatly affected by the Re number. On the other hand, the near-wall vapor bubble varies in shape depending on the Re number and the Oh number. The shape can be divided into umbrella, semi-crescent, and spheroid. When $Re > 1.5 \times 10^2$ and $Oh < 3 \times 10^{-4}$, an umbrella shape is observed; when $Re < 5 \times 10^0$ and $Oh > 10^{-3}$, a spheroidal shape is present; and when $5 \times 10^0 < Re < 1.5 \times 10^2$, $3 \times 10^{-4} < Oh < 10^{-3}$, the bubble is categorized as semi-crescent.

Differences in the bubble deformation are notable between the liquid-surface bubbles (bubbles at high heating distance) and the near-wall bubbles (bubbles at medium and low heating distance). The high temperature in the vapor bubble of the near-wall bubble is absorbed by the solid wall and the thick liquid layer, leading to a less-pronounced deformation. Furthermore, the surface tension effect on the vapor bubble growth and collapse varies with the Re number and Oh number, indirectly influencing the limiting effect of the liquid, and ultimately leading to a differing bubble deformation.

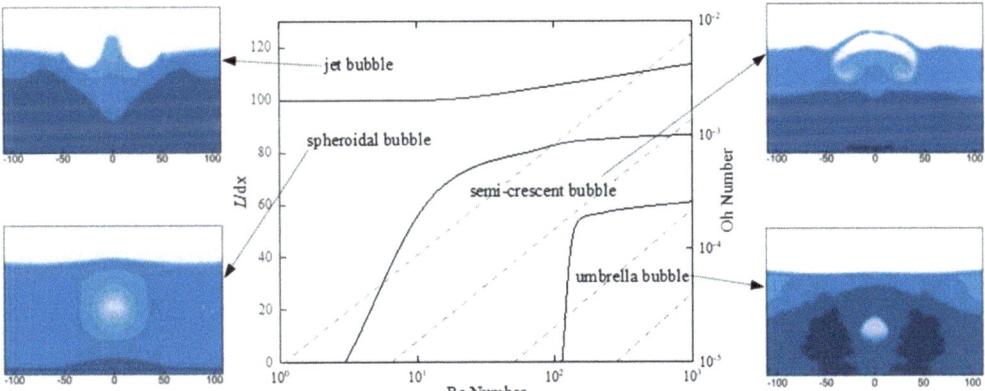

Figure 7. Category of bubble deformation of each mode.

5.2. Mechanism Discussion

5.2.1. Jet Bubble

An analysis of the liquid-surface bubbles is carried out. Figure 8 depicts the formation of the jet bubble in case C5. With Oh ≈ 2×10^{-3}, during the initial growth phase, the bubble retains a spherical shape, and the free surface curves at $t = 50$. However, due to the existence of hydraulic fluid and the top of the bubble being drawn into the liquid surface, the bubble eventually takes on an oblong shape with a sharp top. Subsequently, the bubble gradually collapses and fuses with the liquid surface. This fusion produces a noticeable up-welling column of water at the point of contact. While capillary waves diminish in magnitude as the Oh number increases, the boundary of the bubble is smooth, and the impulse at its bottom increases, leading to an escalation in the jet velocity; this phenomenon is referred to as the jet bubble. The upper level of the liquid experiences significant oscillations caused by the rapid evaporation process, which affects the amount of heat energy present.

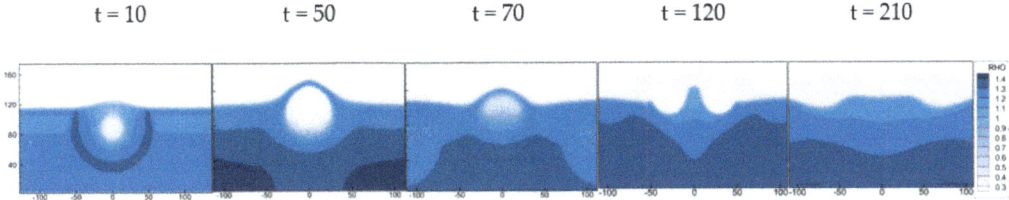

Figure 8. Deformation of jet bubble in case C5.

5.2.2. Spheroidal Bubble

The near-wall bubble is investigated for its sensitivity to the different Re and Oh numbers. The spheroidal bubble shape change diagram in case C3 (Figure 9) demonstrates that the bubble maintains a spherical shape during the growth and collapse processes while Re ≈ 1.5 and Oh ≈ 10^{-3}, with the Re number approaching 1 indicating that the bubble's viscosity force is nearly equal to its inertia force, and the bubble maintains a spherical shape in the growth and collapse processes. The bubble reaches the minimum shape when $t = 120$, while the fluid zone phase is completed with minimal fluctuation in the liquid level. Due to the distance between the vapor bubble's center and the top liquid layer, the energy exchange can only occur within the liquid layer. A slight oscillation occurs at the top of the liquid at the end of the bubble collapse.

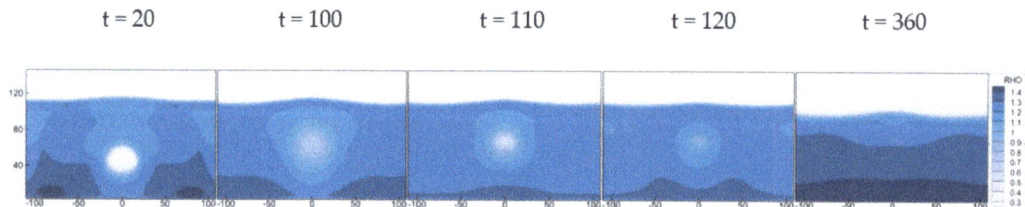

Figure 9. Deformation of spheroidal bubble in case C3.

5.2.3. Umbrella Bubble

For case C8, Figures 10 and 11 depict the density change and flow field vector diagrams of the umbrella bubble. The Re $\approx$ 150, and Oh $\approx 10^{-4}$. Initially, the bubble experiences extrusion pressure from both the left, right, and lower sides of the liquid, resulting in a longer flow field vector in the lower part of the bubble than in the upper part. As the Oh number reaches the minimum, due to the progressive damping of the capillary waves decreasing at $t = 50$, the bubble's edge becomes more irregular, causing it to transform from the ellipse to the umbrella shape. Throughout the process, buoyancy generates an upward force on the bubble, causing it to ascend. This results in the "umbrella handle" beneath the bubble to contract inwards, culminating in the formation of the fan bubble at $t = 70$.

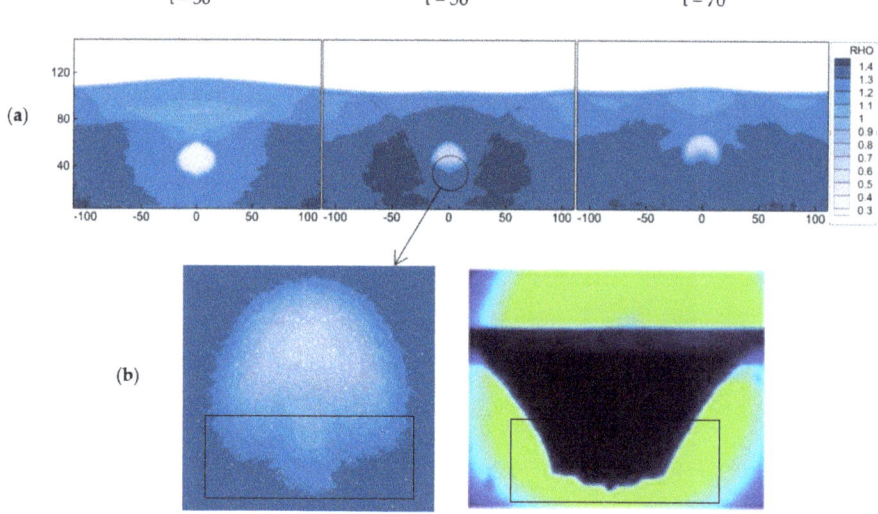

Figure 10. (a) Deformation of umbrella bubble in case C8; (b) capillary waves compared with Sangeeth's experiment [25].

Figure 11. The flow field vector diagram of umbrella bubble in case C8.

5.2.4. Semi-Crescent Bubble

Figure 12 shows the density change and flow field vector diagram of the semi-crescent bubble in case C8. The Re $\approx$ 80, and Oh $\approx 8 \times 10^{-4}$. We found that at $t = 230$, the gas in the bubble moved up to the liquid surface to form a liquid film, the liquid film did not break during the expansion, and a thin "liquid bridge" appeared on the top of the liquid surface. Subsequently, when $t = 270$, the surface tension of the "liquid bridge" prevented the bubble from breaking, and it started to shrink downwards due to the force of the "liquid bridge". The flow field vector diagram in Figure 12b shows that due to the high temperature gas in the vapor bubble interacting with the free liquid surface, the "liquid bridge" vaporized, and small vapor bubbles formed quickly. Eventually, the vapor bubble collapsed on the free surface.

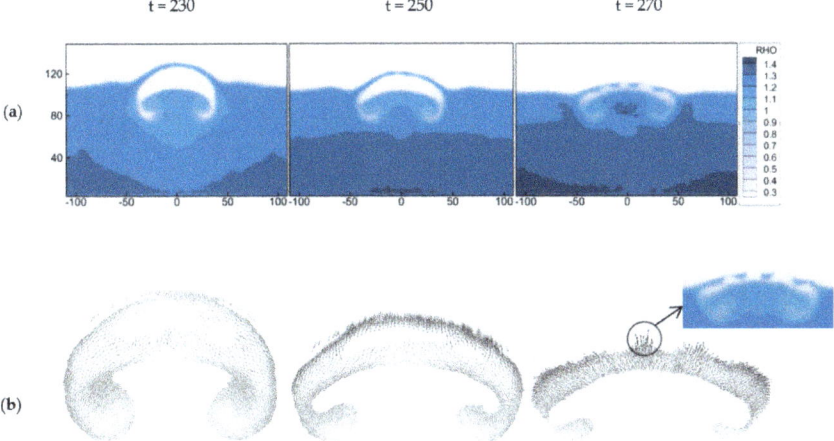

Figure 12. (a) Deformation of semi-crescent bubble; (b) the flow field vector diagram of semi-crescent bubble.

5.3. Bubble Deformation when Re < 10

The current analysis focuses on classical cases involving near-wall bubbles and liquid-surface bubbles when Re < 10, building on previous research. Figure 13 illustrates the variation in the bubble deformation and flow field density across cases C1~C5.

Figure 14 illustrates the variations in the longitudinal diameter of the near-wall bubbles and liquid-surface bubbles. The analysis indicates a marked difference in the deformation of the bubbles. For the near-wall bubbles (observed in cases C1~C3), the forces of viscosity and inertia are approximately equal, thereby enabling the bubbles to maintain their spherical shape. The bubbles undergo a growth stage, followed by a complete collapse throughout 3~4 cycles. For the liquid-surface bubbles (observed in cases C4~C5), the bubbles are subjected to an inertial force from below that causes them to make contact with the liquid surface and form the "liquid bridge"; then, by capillary wave damping, the impulse at the bottom of the bubbles increases, which eventually leads to the formation of the "jet". The two stages are discussed in detail below to allow for a better understanding of the bubbles' behavior.

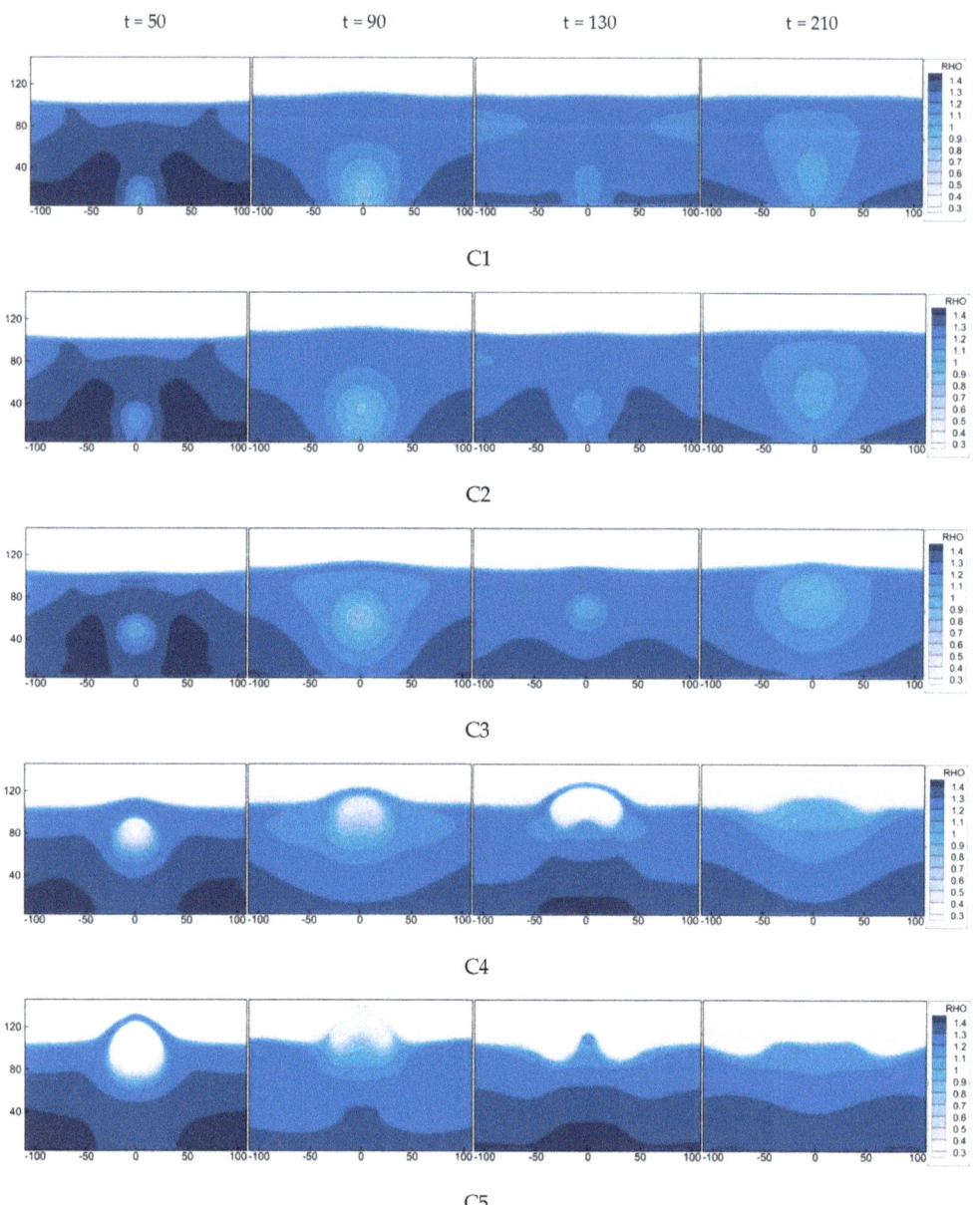

Figure 13. The density changes of near-wall bubbles (C1~C3) and liquid-surface bubbles (C4~C5).

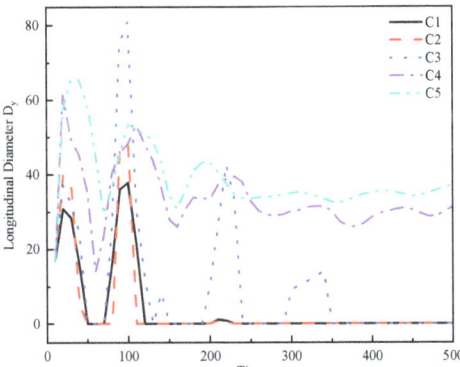

Figure 14. The longitudinal diameter changes of near-wall bubbles (C1~C3) and liquid-surface bubbles (C4~C5).

5.3.1. Deformation of Near-Wall Bubbles

The deformation of the near-wall bubbles for cases C1~C3 in Figure 13 shows that the vapor bubble increases significantly during the growing period. As it grows, the spherical shape appears to experience a little vertical compression, and it undergoes a period in which it repeatedly grows and collapses. At the beginning, though there are slight differences in the shape of the bubble, the radius of the bubble grows at the same speed. In the process of collapse, the collapse amplitude of the bubbles varies, and the maximum radius of the bubbles is also different, but the duration of the initial bubble is similar.

In Figure 14, for case C3, the bubble has four growth and collapse periods; for cases C1~C2, the bubbles have three growth and collapse cycles. The maximum diameter of bubbles in each period decreases with C1~C3, and the maximum diameter of the second bubbles is larger than that in the other periods.

The bubble symmetry coefficient β is introduced to measure the morphological changes of the bubble, where R_x is the transverse radius of the bubble, and R_y is the longitudinal radius of the bubble, as follows:

$$\beta = R_x/R_y \tag{14}$$

Figure 15 illustrates the variation in the symmetric coefficient β of the near-wall bubbles for cases C1~C3, and besides the various cycles, the change in β is similar. In the pro-phase of the bubble growth, $\beta > 1$, and the bubble is a longitudinal flat oval. In the middle process, as β decreases, the deformation of the bubbles undergoes a dramatic change in the collapse period, and β increases rapidly after the lowest point. In cases C1~C2, the bubble growth stage is attached to the wall surface; when the vapor bubble is shot out instantaneously, the heat exchange on the solid wall is faster, and when part of it is absorbed, the deformation of the bubble becomes more serious. In case C3, the symmetric coefficient β of the non-bonded vapor bubble has little fluctuation scope, and its shape remains spherical.

5.3.2. Deformation of Liquid-Surface Bubbles

In Figure 13, in cases C4~C5, the deformation of the liquid-surface bubble appears to extend upwards as soon as it comes into contact with the free liquid surface. When $t = 50$, the gas inside the bubble bends upwards, and the top portion is absorbed by the liquid surface, forming a distinct ellipsoidal tip. Then, the bubble keeps expanding and collapses into a spherical shape. While it is mixed with the liquid surface, a very distinct jet shape is generated at the bottom of the bubble when $t = 120$, as seen in case C5. Eventually, the gas inside the bubble drops quickly due to the gravity force, and then it is pushed back to the surface of the fluid, causing a small oscillation on the free liquid surface.

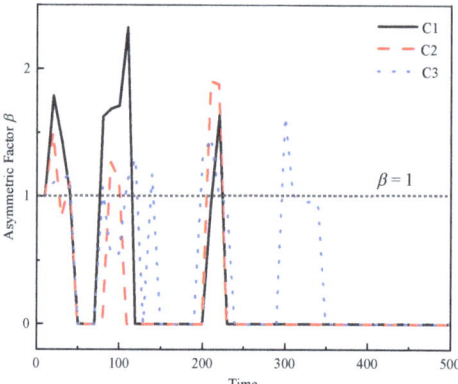

Figure 15. The changes of near-wall bubble symmetry coefficient.

Figure 16 shows a graph of the centroid of the Y-axis for the liquid-surface bubbles in cases C4~C5. The results show that after $t = 40$, when the vapor bubble moves nearer to the liquid surface, the higher the gradient of the curve will be. That is, the more rapidly the vapor bubble moves upwards, the more pronounced the jet shape generated by the impact with the free liquid surface. The reason is that as the bubble goes up, it becomes smaller and smaller as it moves closer to the surface. So much of its energy can be transformed into the jet's potential energy and the energy of the bubble's growth and collapse.

5.4. Bubble Deformation When Re ≥ 10

On the basis of former research, we analyzed several classical cases of near-wall bubbles and liquid-surface bubbles when Re $\geq$ 10. Figure 17 shows the change in the bubble deformation and the variation of the flow field density in cases C6~C10.

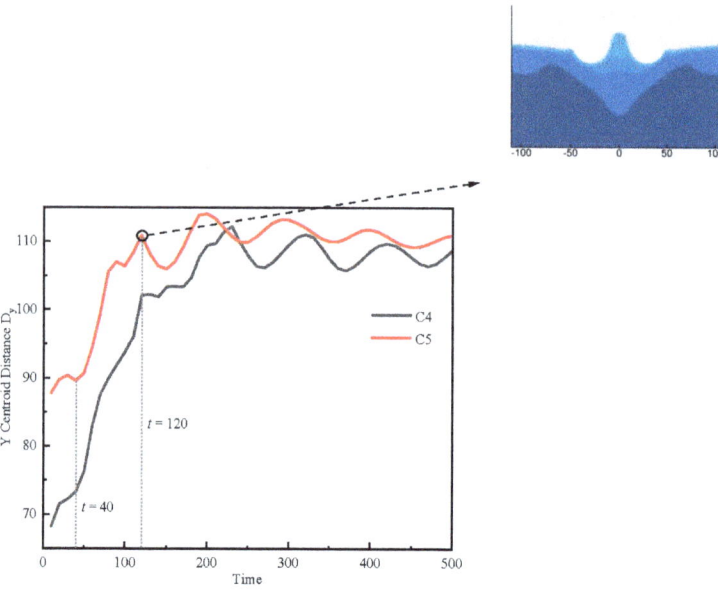

Figure 16. The changes of the liquid-surface bubble Y-axis centroid distance.

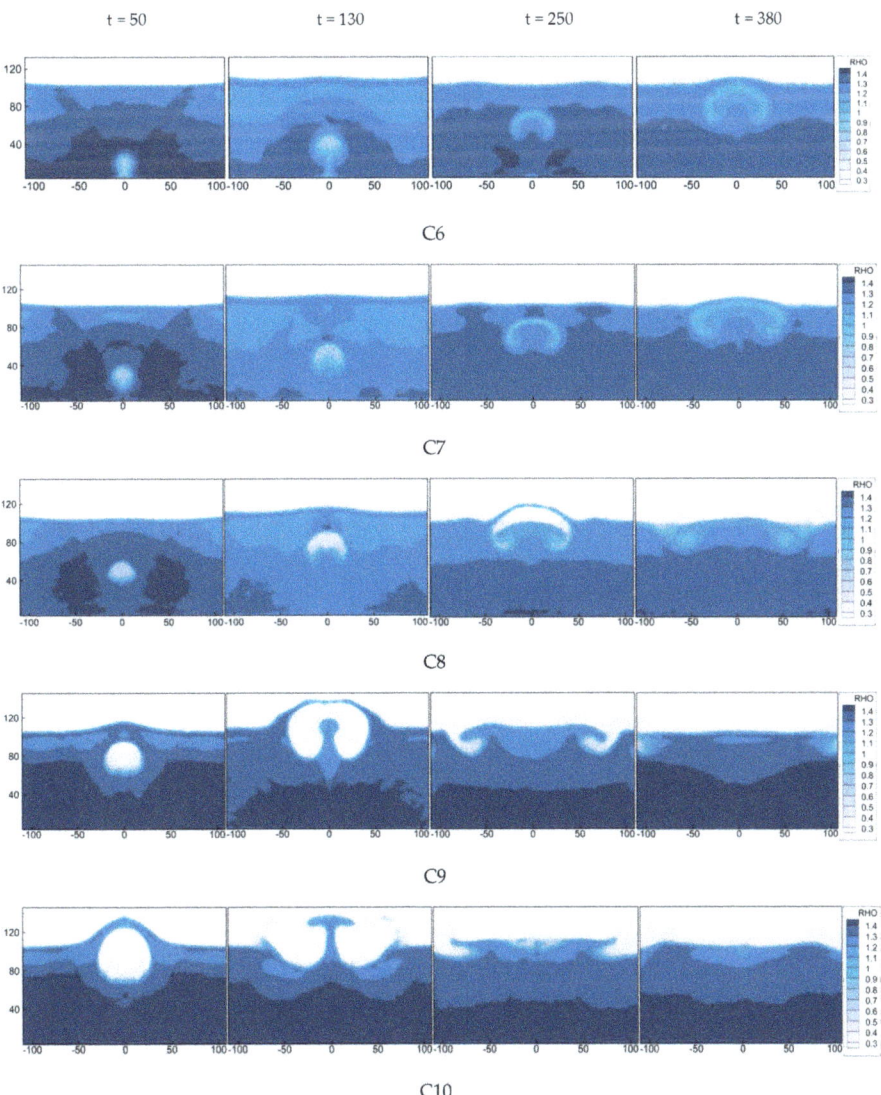

Figure 17. The density changes of near-wall bubbles (C6~C8) and liquid-surface bubbles (C9~C10).

The results show that when Re ≥ 10, the shape of the liquid-surface bubbles in cases C9~C10 is similar to that of cases C4~C5 when Re < 10, as seen in Figure 13. Figure 18 compares the changes in the liquid-surface bubble's longitudinal diameter at two conditions. The tendency of variation is quite similar, and there is no obvious difference among them. Therefore, the effect of the Re on the shape of the near-wall bubbles is not apparent. Furthermore, it is found that when Oh < 5×10^{-3}, at $t = 130$, the jet condition is more evident at case C5 in Figure 13 than at C10 in Figure 17, which is analogous to Sangeeth's experimental conclusion, which states that in the range Oh < 0.02, an increasing viscosity can increase the jet velocity through capillary wave damping [25].

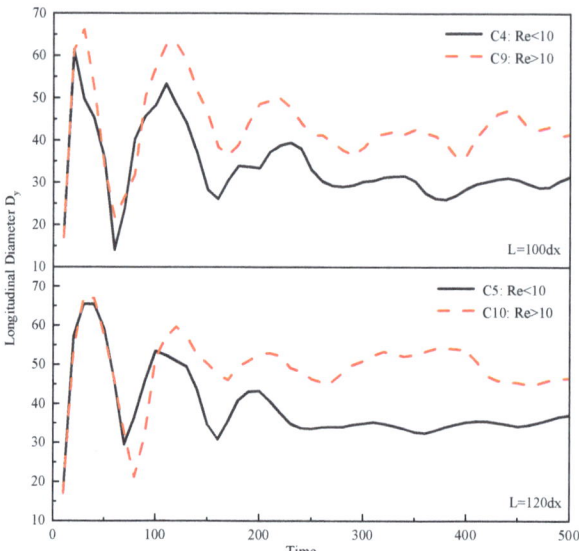

Figure 18. Comparison of the longitudinal diameter changes of liquid-surface bubbles with different Re numbers.

For cases C6~C8, we can differentiate the near-wall bubbles into several different stages based on their appearance, which is explained later.

Deformation of Near-Wall Bubbles

In cases C6~C8 of Figure 17, there is a process in the bubble deformation from the initial sphere to the umbrella bubble when $t = 50$ in case C8, and finally, a crack on the free liquid surface.

Figure 19 illustrates the variation of the rate of the near-wall bubbles in cases C6~C8. We find that there is initially a short but fast acceleration, then the bubble oscillates, and the increasing rate decreases as it moves upwards until it hits the free liquid surface. In addition, in cases C6~C8, when the vapor bubble is closer to the free surface, the speed of the bubble is faster, and the bubble oscillation is more intense. This is due to the fact that when the bubble approaches the free surface, there is a smaller amount of gravity potential energy, a larger amount of bubble kinetic energy is required, and a larger flow rate, which results in a more severe oscillation. In the later period, the rate is reduced. This is because the faster the bubble moves to the free surface, the faster it reaches the free surface. After reaching the free surface, the resistance is greater, which causes the increasing rate of the bubble to decrease gradually due to the pressure on the interface liquid and the surface tension of the liquid film. This also has a crucial effect on the subsequent analysis of the characteristics of the bubble deformation.

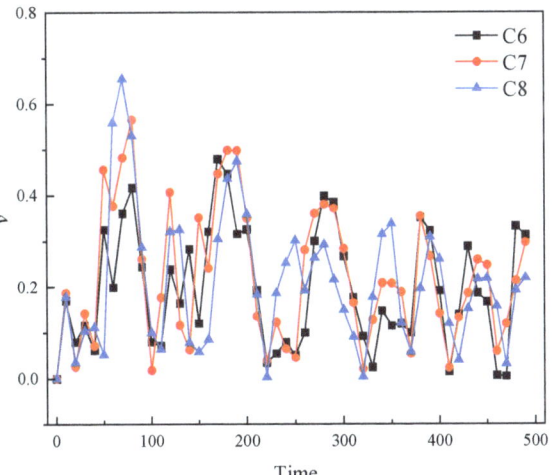

Figure 19. The changes of near-wall bubble rising velocity in cases C6~C8.

Based on the morphology characteristics of several stages, the graph of the symmetry coefficient β is drawn in Figure 20. Figure 21 shows the stage change chart of case C8, and it is used to classify the stage change of these three classical situations. From Figures 20 and 21, it can be seen that during the initial stage (a), the bubble is subject to extrusion pressure from both the left, right, and the lower sides of the liquid, the bubble changes from an oval to an umbrella shape, and it appears that $\beta > 1$ in the stage (b). Then, as the bubble goes up, the differential pressure in the interior and exterior of the gas–liquid interface at the bottom of the bubble becomes larger. Consequently, the lower part of the bubble quickly contracts inwards and presents a semi-crescent shape as shown in stage (c). Simultaneously, when the bubble approaches the free liquid surface, it pushes it up and forms a liquid film. While the fluid film expands without breaking, a thin "fluid bridge" is formed at the top of the bubble. In the final stage (d), because the surface tension of the "liquid bridge" prevents the bubble from breaking, it merges with the liquid surface.

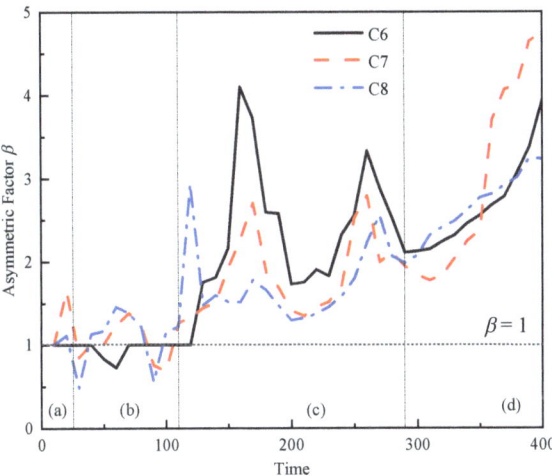

Figure 20. The changes of near-wall bubble symmetry coefficient in cases C6~C8.

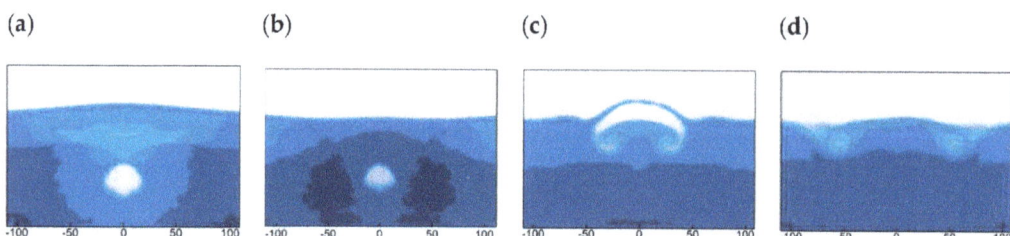

Figure 21. The stage change of classical case C8. (**a**) initial stage, (**b**) umbrella shape stage, (**c**) semi-crescent shape stage, (**d**) bubble collapse stage.

6. Conclusions

We used the SPH numerical simulation method to directly simulate the deformation and collapse of a vapor bubble near the free surface after being heated and raised from the bottom wall; the effects of the shear viscosity η_s and the heating distance L on the growth and collapse processes of the vapor bubble were taken into account. The regularity of the effect of the Re number and the Oh number on the deformation of the vapor bubbles was obtained through a further analysis of several cases, which can be summarized into four major patterns. The classification and mechanism were carried out according to the four major patterns of jet, umbrella, semi-crescent, and spheroid, and the category under each pattern was drawn. The main conclusions are as follows.

For liquid-surface bubbles, the Re number has little influence on them, as there is no significant difference in the specific deformation of the bubbles and the changes in the longitudinal diameter of the bubbles. When $Oh > 5 \times 10^{-3}$, all of them showed a jet shape, and the jet state is more obvious as the shear viscosity increases.

For near-wall bubbles, the Re number has a great influence on the bubble deformation; the shape can be categorized into umbrella, semi-crescent, and spheroid. For $Re > 1.5 \times 10^2$ and $Oh < 3 \times 10^{-4}$, the bubble appears to have an umbrella shape; for $Re < 5 \times 10^0$ and $Oh > 10^{-3}$, the bubble appears to be spheroidal; and for $5 \times 10^0 < Re < 1.5 \times 10^2$, $3 \times 10^{-4} < Oh < 10^{-3}$, the bubble appears to have a semi-crescent shape. Near-wall bubbles experience inhibited longitudinal growth and often collapse at the liquid surface without creating the jet shape. Additionally, the balance of the surface tension and inertia force, influenced by the Re and Oh numbers, contributes to the formation of different bubble shapes.

The spheroidal bubble (cases C1~C3) underwent 3~4 cycles of growth and collapse. As the bubble approached the free surface, its shape became less influenced by the solid wall. The maximum radius of the bubble decreased with each growth and collapse cycle, resulting in less fluctuation in the symmetrical coefficient. Ultimately, the bubble tended to become rounder in shape.

The jet bubble (cases C4~C5) experienced less gravitational potential energy loss as it approached the free surface, resulting in a faster upward movement of the bubble. This, in turn, led to a more prominent jet that occurred upon impact with the free liquid surface.

Author Contributions: Conceptualization, Y.C., H.X. and L.Q.; methodology, Q.W. and Y.C.; software, Q.W. and Y.C.; validation, Q.W. and Y.C.; formal analysis, Y.C., H.X. and L.Q.; investigation, Y.C.; resources, Y.C.; data curation, Y.C.; writing—original draft preparation, Y.C.; writing—review and editing, H.X. and L.Q.; visualization, Y.C.; supervision, H.X. and L.Q.; project administration, H.X. and L.Q.; funding acquisition, H.X. and L.Q. All authors have read and agreed to the published version of the manuscript.

Funding: This research was funded by the National Natural Science Foundation of China (No. 11972321, No. 91852102).

Institutional Review Board Statement: Not applicable.

Informed Consent Statement: Not applicable.

Data Availability Statement: The data presented in this study are available upon request from the corresponding authors. The data are not publicly available due to privacy reasons.

Conflicts of Interest: The authors declare no conflict of interest.

References

1. Krefting, D.; Mettin, R.; Lauterborn, W. High-speed observation of acoustic cavitation erosion in multibubble systems. *Ultrason. Sonochem.* **2004**, *11*, 119–123. [CrossRef] [PubMed]
2. Shao, J.; Xuan, M.; Dai, L.; Si, T.; Li, J.; He, Q. Near-Infrared-Activated Nanocalorifiers in Microcapsules: Vapor Bubble Generation for In Vivo Enhanced Cancer Therapy. *Angew. Chem. Int. Ed.* **2015**, *54*, 12782–12787. [CrossRef] [PubMed]
3. Oyarte Gálvez, L.; Fraters, A.; Offerhaus, H.L.; Versluis, M.; Hunter, I.W.; Rivas, D.F. Microfluidics control the ballistic energy of thermocavitation liquid jets for needle-free injections. *J. Appl. Phys.* **2020**, *127*, 104901. [CrossRef]
4. Chudnovskii, V.M.; Yusupov, V.I.; Dydykin, A.V.; Nevozhai, V.; Kisilev, A.Y.; Zhukov, S.A.; Bagratashvili, V.N. Laser-induced boiling of biological liquids in medical technologies. *Quantum Electron.* **2017**, *47*, 361–370. [CrossRef]
5. Rekhviashvili, S.S. Single-bubble sonoluminescence model. *Tech. Phys. Lett.* **2008**, *34*, 1072–1074. [CrossRef]
6. Prosperetti, A. Vapor Bubbles. *Annu. Rev. Fluid Mech.* **2017**, *49*, 221–248. [CrossRef]
7. Robinson, P.B.; Blake, J.R.; Kodama, T.; Shima, A.; Tomita, Y. Interaction of cavitation bubbles with a free surface. *J. Appl. Phys.* **2001**, *89*, 8225–8237. [CrossRef]
8. Zhang, Y.; Qiu, X.; Zhang, X.; Tang, N. Collapsing dynamics of a laser-induced cavitation bubble near the edge of a rigid wall. *Ultrason. Sonochemistry* **2020**, *67*, 105157. [CrossRef]
9. Gonzalez, S.R.; Klaseboer, E.; Khoo, B.C.; Ohl, C.-D. Cavitation bubble dynamics in a liquid gap of variable height. *J. Fluid Mech.* **2011**, *682*, 241–260. [CrossRef]
10. Sun, C.; Can, E.; Dijkink, R.; Lohse, D.; Prosperetti, A. Growth and collapse of a vapour bubble in a microtube: The role of thermal effects. *J. Fluid Mech.* **2009**, *632*, 5–16. [CrossRef]
11. Edel, Z.J.; Mukherjee, A. Experimental investigation of vapor bubble growth during flow boiling in a microchannel. *Int. J. Multiph. Flow* **2011**, *37*, 1257–1265. [CrossRef]
12. Kangude, P.; Srivastava, A. Understanding the growth mechanism of single vapor bubble on a hydrophobic surface: Experiments under nucleate pool boiling regime. *Int. J. Heat Mass Tran.* **2020**, *154*, 119775. [CrossRef]
13. Tang, J.K. Experimental and Simulation Research on the Oscillation Characteristics of Laser-Induced Cavitation Bubble Near a Hydrophobic Wall Surface. Master Thesis, Jiangsu University, Zhenjiang, China, 2020.
14. Sagar, H.J.; El Moctar, O. Numerical simulation of a laser-induced cavitation bubble near a solid boundary considering phase change. *Ship Technol. Res.* **2018**, *65*, 163–179. [CrossRef]
15. Krishnan, S.E.J.P. On the scaling of jetting from bubble collapse at a liquid surface. *J. Fluid Mech.* **2017**, *822*, 791–812. [CrossRef]
16. Zhang, A.M.; Wang, C.; Wang, S.P. Experimental study of interaction between bubble and free surface. *Acta Phys. Sinina-Chin. Ed.* **2012**, *61*, 300–312. [CrossRef]
17. Ma, Y.; Chung, J.N. A study of bubble dynamics in reduced gravity forced-convection boiling. *Int. J. Heat Mass Tran.* **2001**, *44*, 399–415. [CrossRef]
18. Liu, X.M.; He, J.; Lu, J.; Ni, X.-W. Effect of liquid viscosity on the behavior of laser-induced cavitation bubbles. *J. Optoelectron. Laser* **2006**, *157*, 985–988.
19. Phan, T.; Nguyen, V.; Park, W. Numerical study on strong nonlinear interactions between spark-generated underwater explosion bubbles and a free surface. *Int. J. Heat Mass Tran.* **2020**, *163*, 120506. [CrossRef]
20. Wang, R.Y.; Chen, P.P.; Ban, C.Y. Keeping Volume Fraction of Fluid in Reconstructing Moving-interfaces of VOF on Rectangular Meshes. *Chin. J. Comput. Phys.* **2008**, *122*, 431–436.
21. Nguyen, V.T.; Phan, T.H.; Duy, T.N.; Kim, D.H.; Park, W.G. Modeling of the bubble collapse with water jets and pressure loads using a geometrical volume of fluid based simulation method. *Int. J. Multiph. Flow* **2022**, *152*, 104103. [CrossRef]
22. Arai, E.; Villafranco, D.; Grace, S.; Ryan, E. Simulating bubble dynamics in a buoyant system. *Int. J. Numer. Methods Fluids* **2020**, *92*, 169–188. [CrossRef]
23. Liu, M.B.; Liu, G.R.; Li, S. Smoothed particle hydrodynamics: A meshfree method. *Comput. Mech.* **2004**, *33*, 491. [CrossRef]
24. Wang, Y. Numerical Simulation and Experimental Research on Motion and Growth of Single Bubble Based on Diffuse Interface Method. Ph.D. Thesis, South China University of Technology, Guangzhou, China, 2019.
25. Sigalotti, L.D.G.; Troconis, J.; Sira, E.; Peña-Polo, F.; Klapp, J. Diffuse-interface modeling of liquid-vapor coexistence in equilibrium drops using smoothed particle hydrodynamics. *Phys. Rev. E* **2014**, *90*, 013021. [CrossRef] [PubMed]
26. Gallo, M.; Magaletti, F.; Casciola, C.M. Thermally activated vapor bubble nucleation: The Landau-Lifshitz–Van der Waals approach. *Phys. Rev. Fluids* **2018**, *3*, 053604. [CrossRef]
27. Dong, W.; Yu, H.L.; Li, R.Y. Bubble Deformation in an electric field. *J. Eng. Thermophys.* **2006**, *27*, 265–267.
28. Nugent, S.; Posch, H.A. Liquid drops and surface tension with smoothed particle applied mechanics. *Phys. Rev. E* **2000**, *62*, 4968–4975. [CrossRef]

29. Yang, X.; Liu, M.; Peng, S. Smoothed particle hydrodynamics modeling of viscous liquid drop without tensile instability. *Comput. Fluids* **2014**, *92*, 199–208. [CrossRef]
30. Xiong, H.B.; Zhang, C.Y.; Yu, Z.S. Multiphase SPH modeling of water boiling on hydrophilic and hydrophobic surfaces. *Int. J. Heat Mass Tran.* **2019**, *130*, 680–692. [CrossRef]

Disclaimer/Publisher's Note: The statements, opinions and data contained in all publications are solely those of the individual author(s) and contributor(s) and not of MDPI and/or the editor(s). MDPI and/or the editor(s) disclaim responsibility for any injury to people or property resulting from any ideas, methods, instructions or products referred to in the content.

Article

Dynamics of a Laser-Induced Cavitation Bubble near a Cone: An Experimental and Numerical Study

Jianyong Yin [1,*], Yongxue Zhang [2,3], Dehong Gong [1], Lei Tian [2,3] and Xianrong Du [1]

[1] Electrical Engineering College, Guizhou University, Guiyang 550025, China; dhgong@gzu.edu.cn (D.G.); gs.xrdu22@gzu.edu.cn (X.D.)
[2] College of Mechanical and Transportation Engineering, China University of Petroleum-Beijing, Beijing 102249, China; zhyx@cup.edu.cn (Y.Z.); 2020310319@student.cup.edu.cn (L.T.)
[3] Beijing Key Laboratory of Process Fluid Filtration and Separation, China University of Petroleum-Beijing, Beijing 102249, China
* Correspondence: jyyin@gzu.edu.cn

Abstract: A bubble's motion is strongly influenced by the boundaries of tip structures, which correspond to the bubble's size. In the present study, the dynamic behaviors of a cavitation bubble near a conical tip structure are investigated experimentally and numerically. A series of experiments were carried out to analyze the bubble's shape at different relative cone distances quantitatively. Due to the crucial influence of the phase change on the cavitation bubble's dynamics over multiple cycles, a compressible two-phase model taking into account the phase change and heat transfer implemented in OpenFOAM was employed in this study. The simulation results regarding the bubble's radius and shape were validated with corresponding experimental photos, and a good agreement was achieved. The bubble's primary physical features (e.g., shock waves, liquid jets, high-pressure zones) were well reproduced, which helps us understand the underlying mechanisms. Meanwhile, the latent damage was quantified by the pressure load at the cone apex. The effects of the relative distance γ and cone angle θ on the maximum temperature, pressure peaks, and bubble position are discussed and summarized. The results show that the pressure peaks during the bubble's collapse increase with the decrease in γ. For a larger γ, the first minimum bubble radius increases while the maximum temperature decreases as θ increases; the pressure peak at the second final collapse is first less than that at the first final collapse and then much greater than that one. For a smaller γ, the pressure peaks at different θ values do not vary very much.

Keywords: cavitation bubble dynamics; cone; phase change and heat transfer; OpenFOAM; pressure peak

Citation: Yin, J.; Zhang, Y.; Gong, D.; Tian, L.; Du, X. Dynamics of a Laser-Induced Cavitation Bubble near a Cone: An Experimental and Numerical Study. *Fluids* 2023, 8, 220. https://doi.org/10.3390/fluids8080220

Academic Editors: Nguyen Van-Tu, Hemant J. Sagar and D. Andrew S. Rees

Received: 7 July 2023
Revised: 24 July 2023
Accepted: 27 July 2023
Published: 29 July 2023

Copyright: © 2023 by the authors. Licensee MDPI, Basel, Switzerland. This article is an open access article distributed under the terms and conditions of the Creative Commons Attribution (CC BY) license (https://creativecommons.org/licenses/by/4.0/).

1. Introduction

Cavitation, the process of a tiny bubble's formation, growth, and implosive transient collapse, is a critical physics problem that has drawn considerable attention since it is at the heart of many applications in industry, fluid machinery, biomass treatment, and other fields [1–5]. The transient energy released during the collapse of a cavitation bubble can induce potential damage to nearby surfaces [3,6]. Thus, studies of the cavitation collapsing process are beneficial for the further protection of structural surfaces from cavitating damage or a reduction in damage. Moreover, the presence of a structural surface changes the symmetry of the pressure field around the cavitation bubble, resulting in an unequal interface acceleration and thus inducing a high-speed microjet. In the literature, cavitation bubbles near different structures (e.g., a rigid flat surface [7,8], a rigid curved surface [9], a free surface [10], an ice surface [11], and an elastic surface [12]) have been intensively investigated, and many prominent features of cavitation bubbles have been revealed. However, these structures' surfaces are not smooth enough, and many tips are exposed after the cavitation damage appears. In addition, artificially designed functional boundaries with grooves and tips are employed for the optimization or the extension of a fluid's engineering applications [13]. The structural boundaries of the tip correspond to the bubble's size with a strong influence on the bubble's dynamics. In the present paper, the boundary of a fluid's tip is considered to be cone-shaped. Thus, the bubble's dynamic behaviors near a cone are investigated in depth experimentally and numerically.

Experiments of cavitation bubbles near boundaries of various shapes have been executed extensively to study their relevant dynamics [14–18]. Generally, a cavitation bubble near a planar rigid boundary forms a directed microjet flow towards the boundary, and its dynamical behaviors are closely related to the stand-off distance γ (defined as L/R_{max}, where L is the distance from the bubble's center to the rigid boundary and R_{max} is the maximum bubble radius). A recent study by Saini et al. [19] has revealed that the dynamics of a bubble in contact with a rigid wall hinge on the effective contact angle in the instant before its collapse. They found that when the contact angle is less than 90°, the conventional jet directed towards the wall can be observed, whereas if the contact angle is greater than 90°, an annular re-entrant jet parallel to the boundary occurs. Simultaneously, some research has been conducted to investigate bubble dynamic behaviors near non-flat rigid boundaries. Tomita et al. [9] investigated bubbles near a variety of curvatures of a solid wall and found that the velocity of the jet and the duration of the flow were closely related to the deviation of the curvature. Moreover, the jet velocity increased as the shape of the wall changed from concave to convex. Požar et al. [20] studied the interaction between a nanosecond laser-induced cavitation bubble near a concave surface and observed that the concave wall could refocus the shock wave and then induce secondary cavitation. Ebrahim et al. [21] reported the dynamical features of the cavitation bubble near a rigid cylindrical rod and found that a mushroom-shaped bubble could be generated at a smaller relative distance, which may cause a lower jet impact load.

Moreover, Brujan et al. [22] investigated a bubble near a corner (two perpendicular walls) and showed that the jet angle was proportional to the ratio of the distance from the walls to the bubble. Furthermore, Tagawa and Peters [23] studied the dynamics of a bubble near a corner with different angles. They obtained the analytical solution for the jet direction by using the method of images. Li et al. [24] found that the jet deviated from the horizontal direction within a much shorter range when a bubble near two connected walls with an obtuse angle. Zhang et al. [25] investigated the dynamics of a bubble near a triangular prism array and observed that the bubble's behaviors were strongly influenced by the relative positions of the bubble and the prisms, bubble size, and the distance from the bubble to the array.

On the other hand, numerical simulations are a well-known way to investigate bubble dynamics [26–31]. Based on the potential flow theory and boundary integral method (BIM), Wang et al. [32] simulated a bubble at a corner and found that the bubble migrated away from the near wall and the corner during its expansion and moved back toward them during its collapse. Moreover, a high-pressure region was formed during the late stages of its collapse. Trummler et al. [33] numerically studied the collapse of a bubble above rigid, notched walls and showed that the surface crevices had a significant influence on the collapse dynamics, jet formation, and wave dynamics. Shervani-Tabar and Rouhollahi [34] numerically investigated the effect of rigid concave walls on the motion of a bubble and found that the velocity of the liquid jet tends to increase with decreasing concavity. In addition, Li et al. [35] conducted numerical research on the collapsing behavior of a bubble near a solid conical wall and revealed the effect of the conical angles on the bubble's shape, collapse time, and liquid jet. They found that the collapse time increased as the cone angle increased.

In the above-mentioned literature, although the dynamic behaviors of a single cavitation bubble near various non-flat rigid boundaries (e.g., a corner, curved surface, cylindrical rod, crevice, and triangular prism array) are investigated experimentally, studies of a bubble near a tip (e.g., cone) are reported less often. Moreover, Li et al. [35] investigated a bubble near a cone through a numerical simulation, but they only showed the dynamics of the cavitation bubble in the first cycle due to a limitation of the model. The relevant parameters during the bubble's evolution (e.g., bubble position and shape, the temperature within the bubble, and the pressure load) have not yet been fully revealed. Thus, a profound analysis and investigation are imperative to reveal the underlying physical mechanisms.

In this study, experiments of a single laser-generated cavitation bubble near a cone are conducted and corresponding numerical simulations are performed. A pressure-based compressible model taking into account the mass and heat transfer developed in OpenFOAM is adopted in the current simulation. The numerical model has been validated in our previous works and employed for the simulation of a bubble near a solid wall [8,36]. The main aim of the current study is to examine the dynamics of the cavitation bubble in the nearby cone, which may be frequently encountered during the operation of hydraulic machinery. The following is a brief summary of the present paper. Section 2 describes the experimental methodology. In Section 3, a description of the governing flow equations and details about the numerical implementation are presented. The results from experiments and simulations under two different γ are shown in Section 4. Furthermore, the effects of γ and θ on the pressure load are further revealed in Section 5. Finally, the main conclusions are presented in Section 6.

2. Experimental Setup

In this study, a series of experiments were carried out to investigate the dynamical behaviors of a cavitation bubble near a cone. A schematic description of the experimental setup for the investigation of the interactions between the laser-generated cavitation bubble and the cone is shown in Figure 1. In the present experimental system, the laser beam was generated by a pulsed Nd:YAG laser generator (Penny-100-SH) with a laser wavelength of 532 nm and pulse duration of 5.4 ns. The laser beam (laser energy: 10 mJ, the outlet bean diameter: 3 mm) was focused using the LMH-10X-532 focusing lens (focal length: 15 mm) inside a transparent water tank (size: $100 \times 100 \times 100$ mm^3) filled with distilled water to induce a single cavitation bubble. The distance from the bubble to the water tank wall was far greater than the maximum radius of the cavitation bubble (about 1 mm in this paper). Thus, the effect of the tank wall on the bubble dynamics could be safely ignored. An LED light was employed for all lighting. The complete process of the interaction between the cavitation bubble and the cone was recorded by a high-speed camera (Phantom v1212) with a sampling rate of 100,000 frames/s, which is fast enough to show the primary characteristics [13]. The resolution of the recorded photos was 256×256 pixels, and the experimental data were processed in batches by MATLAB 2019b software.

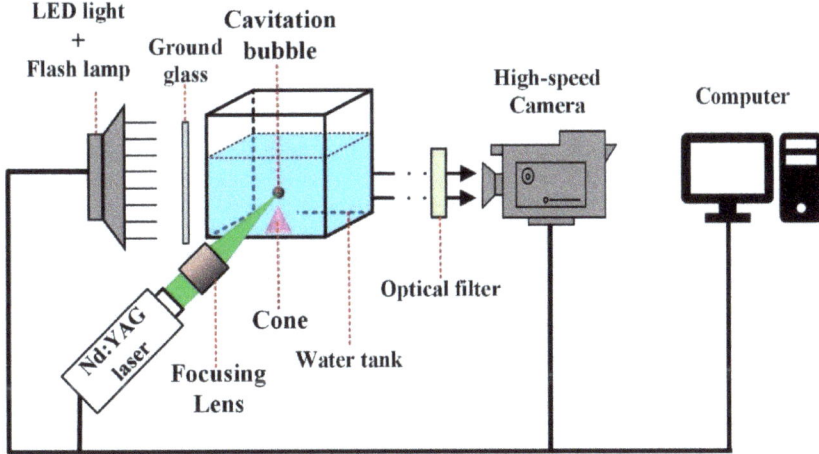

Figure 1. A schematic presentation of the experimental setup for the cavitation bubble–cone interactions.

3. Numerical Method

3.1. Governing Equations

A compressible two-phase flow model with consideration of heat and mass transfer is adopted for the present simulation, and a brief description of the compressible model is presented in the following.

A set of governing flow equations describing the dynamics of fluids can be solved within the entire computational domain, including the mass conservation equation, momentum conservation equation, and energy conservation equation [8], written as follows:

$$\frac{\partial(\alpha_i \rho_i)}{\partial t} + \nabla \cdot (\alpha_i \rho_i U) = \pm \dot{m} \quad (i = l, v) \tag{1}$$

$$\frac{\partial(\rho U)}{\partial t} + \nabla \cdot (\rho U U) = -\nabla p_{-rgh} - gH\nabla \rho + \sigma \kappa(\alpha_l)\nabla \alpha_l \\ + \mu \left[\nabla U + (\nabla U)^T - \frac{2}{3}(\nabla \cdot U)I\right] \tag{2}$$

$$\left[\frac{\partial(\rho T)}{\partial t} + \nabla \cdot (\rho T U)\right] + \left(\frac{\alpha_l}{C_{pl}} + \frac{\alpha_v}{C_{pv}}\right)\left[\frac{\partial(\rho K)}{\partial t} + \nabla \cdot (\rho K U)\right] \\ = \left(\frac{\alpha_l}{C_{pl}} + \frac{\alpha_v}{C_{pv}}\right)\left[\frac{\partial p_{-rgh}}{\partial t} + \nabla \cdot \left(\mu\left(\nabla U + (\nabla U)^T\right) \cdot U\right)\right] + \left(\frac{\alpha_l \lambda_l}{C_{pl}} + \frac{\alpha_v \lambda_g}{C_{pv}}\right)(\nabla^2 T) \tag{3}$$

where p_{-rgh}, U, T, K, g, H, and I are the total pressure, velocity, temperature, kinetic energy, gravity acceleration, height, and unit tensor, respectively; α, ρ, C_p, μ, and λ are the volume fraction, density, heat capacity, dynamic viscosity, and thermal conductivity for both water and vapor; σ is the surface tension coefficient; κ is the surface curvature; and $\dot{m} = \dot{m}^+ - \dot{m}^-$ is mass transfer source term, with $\dot{m}^+$ representing the condensation rate of the vapor and $\dot{m}^-$ representing the vaporization rate of water.

In addition, $\dot{m}$ is obtained by solving the phase change model. The Schnerr–Sauer phase change model considering the energy conservation equation [37] is employed and defined as follows:

$$\dot{m}^+ = \frac{3\rho_l \rho_v}{\rho}\alpha_l(1-\alpha_l)\frac{1}{R_b}\sqrt{\frac{2|p-p_v(T)|}{3\rho_l}} \tag{4}$$

$$\dot{m}^- = -\frac{3\rho_l \rho_v}{\rho}\alpha_l(1-\alpha_l+\alpha_{Nuc})\frac{1}{R_b}\sqrt{\frac{2|p-p_v(T)|}{3\rho_l}} \tag{5}$$

$$R_b = \left(\frac{3}{4\pi n}\frac{1-\alpha_l+\alpha_{Nuc}}{\alpha_l}\right)^{1/3} \tag{6}$$

$$\alpha_{Nuc} = \frac{n\pi(d_{Nuc})^3/6}{1+n\pi(d_{Nuc})^3/6} \tag{7}$$

where $p_v(T)$ is the saturation vapor pressure; R_b, α_{Nuc}, $n(2.1 \times 10^{11}/\text{m}^3)$, and $d_{Nuc}(2 \times 10^{-6}\text{ m})$ are the radius, volume fraction, number, and diameter of cavitation nucleation site, respectively.

Furthermore, α is solved by the volume of fluid (VOF) method widely used to capture the gas–liquid interface [38–40]. In the VOF method, the following transport equation, derived from the mass conservation equation for the water phase, is solved by the multidimensional universal limiter with the explicit solution solver [41,42].

$$\frac{\partial \alpha_l}{\partial t} + \nabla \cdot (\alpha_l U) + \nabla \cdot (U_r \alpha_l(1-\alpha_l)) \\ = \alpha_l(1-\alpha_l)\left(\frac{1}{\rho_v}\frac{d\rho_v}{dt} - \frac{1}{\rho_l}\frac{d\rho_l}{dt}\right) + \alpha_l \nabla \cdot U + \dot{m}\left(\frac{1}{\rho_l} - \alpha_l\left(\frac{1}{\rho_l} - \frac{1}{\rho_v}\right)\right) \tag{8}$$

where U_r is the relative velocity [38], and the third term on the left side is an artificial compression term [43].

Finally, the above set of governing equations is closed by adding the equations of state (EOS) and the corresponding compressibility (ψ) for the vapor and water. The EOS and compressibility for vapor are expressed as follows:

$$\rho_v = \frac{p}{R_v T} \qquad (9)$$

$$\psi_v = \frac{1}{R_v T} \qquad (10)$$

where R_v is the gas constant.

Accordingly, the EOS and compressibility for water [44] are defined as follows:

$$\rho_l = \frac{p + p_c}{K_c(T + T_c)} \qquad (11)$$

$$\psi_l = \frac{1}{K_c(T + T_c)} \qquad (12)$$

where p_c is the pressure constant with p_c = 1944.61 MPa; T_c is the temperature constant with T_c = 3867 K; and K_c is liquid constant with K_c = 472.27 J/(kg·K).

3.2. Numerical Setup

Several essential parameters in the present study are shown in Figure 2. Here, R is the radius of the cavitation bubble. L is the distance from the bubble center to the cone apex. h is the height of the cone and r is the radius of the conical bottom. θ is the cone angle. Furthermore, a brass cone (density: 8500 kg/m^3) with h = 14 mm and r = 7 mm ($\theta \approx 53°$) was used in the present experiment. To facilitate research, the relevant parameters are defined as follows:

$$\gamma = \frac{L}{R_{max}} \qquad (13)$$

$$\theta = 2\arctan\frac{r}{h} \qquad (14)$$

where R_{max} is the maximum bubble radius.

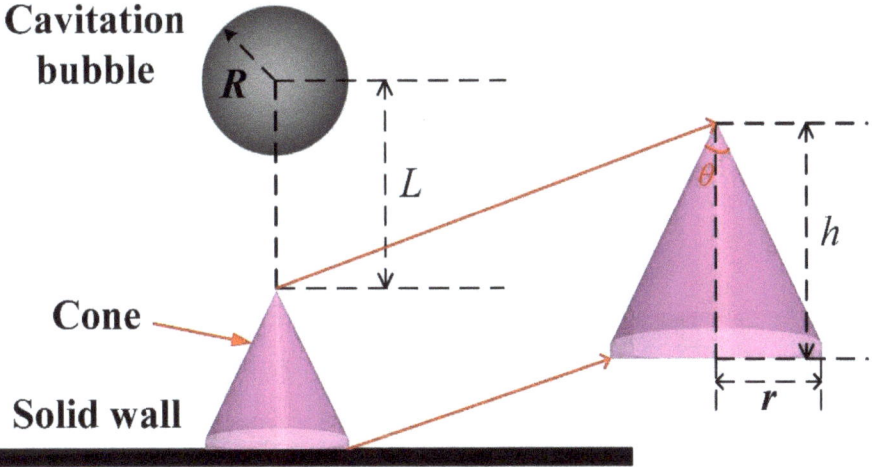

Figure 2. Definition of several essential parameters.

As shown in Figure 3a, an axisymmetric computational wedge domain of 50 mm × 60 mm is constructed to simulate the axisymmetric flow field problem, which is greater than

50 times the maximum bubble radius. The 5° wedge-shaped zone is adopted to save calculation time and achieve a reasonable computational cost. The structured mesh is implemented by ICEM CFD (ANSYS) with local refinement, as shown in Figure 3b. To guarantee consistent simulation results, a mesh sensitivity study was performed in our previous work [8]. Furthermore, a full mesh with 945,903 elements (the mesh size within the initial bubble is 3.3 μm) is subsequently applied in the current simulations. The no-slip boundary condition is employed for the right side and the bottom of the calculation zone. Because it is a stationary cone, the boundary condition for the cone is also set to be no-slip. In addition, the total pressure boundary condition is used for the top side of the region. The PIMPLE algorithm is applied to a couple of pressure and velocity. The first-order implicit Euler and the second-order Gaussian TVD schemes are employed to perform the time and spatial discretization, respectively. More detailed information for the numerical implementation (such as convergence criterion, Courant number, and time step selection) can be seen in our previous works [8,36]. Table 1 shows the values of physical properties used in this paper.

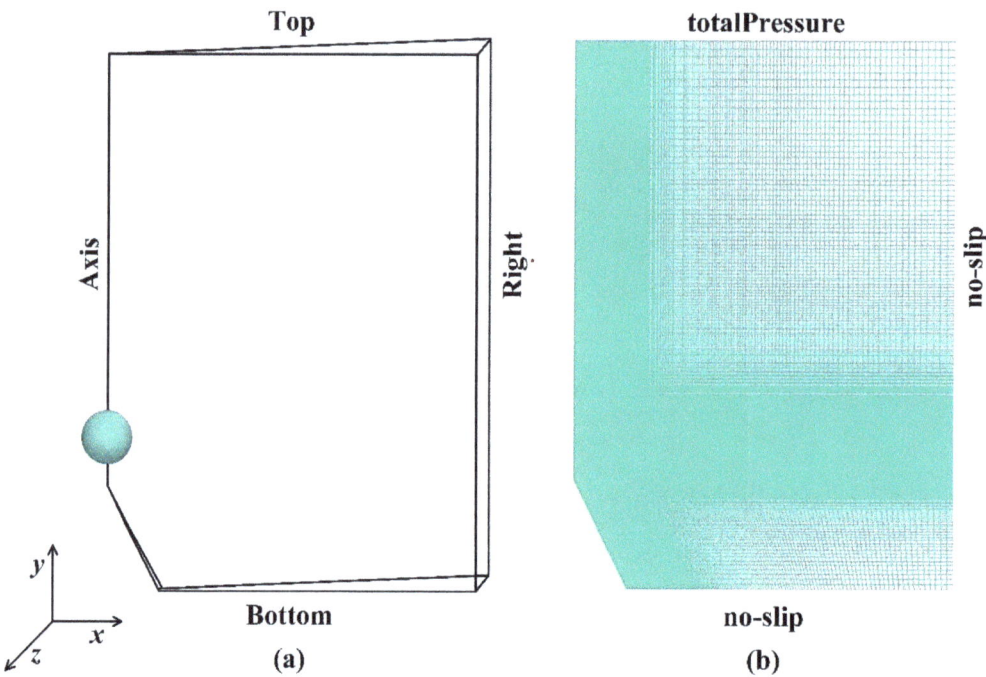

Figure 3. (a) The axisymmetric computational domain; (b) the applied boundary conditions and mesh distribution.

Table 1. The initial values of liquid- and vapor-phase properties.

	ρ (kg/m^3)	μ (Pa·s)	C_p (J/(kg·K))	λ (W/(m·K))	σ (N/m)
Vapor	0.0171	9.75×10^{-6}	1862.6	0.02	0.07
Water	998.16	9.982×10^{-4}	4180	0.677	

4. Experimental and Computational Results

4.1. Bubble Dynamics of Multiple Collapses with $\gamma = 1.3$

The bubble dynamics of multiple collapses near a cone with $\gamma = 1.3$ are investigated in our experiment and simulation. The distance between the bubble center and the cone apex

L is 1.3 mm in the experiment. The maximum experimental radius $R_{max,exp}$ is 0.99 mm ($\gamma = L/R_{max,exp} = 1.3$). For the numerical simulation, the initial parameters inside the bubble, i.e., the radius, pressure, and temperature, are set to be 40 MPa, 0.092 mm, and 593.15 K, respectively. For better comparison, the bubble radius (R) and physical time (t) are normalized by using the reference radius and time, i.e., $R^* = R/R_{max}$ and $t^* = t/t_{osc}$, where t_{osc} is the oscillation time from the initial moment to the first minimum radius.

Figure 4 presents the quantitative comparison of the bubble radius between the measured experimental data (black dot) and the simulation results (solid red line). As shown, the simulation results are in accordance with the measured data, including the first, second, and even third cycles. Meanwhile, the comparison of the bubble multi-period shape evolution is shown in Figure 5. Expressly, (a1)~(a14) denotes the experimental phenomenon, and (b1)~(b14) are the numerical prediction results. During the first bubble cycle ((a1)~(a6)), the cone apex has little influence on the bubble dynamics since the distance from the cone apex to the bubble center is larger than the maximum bubble radius. Furthermore, the bubble also does not touch the cone apex during the second and third collapse, as shown in frames (a7)~(a14). Such a phenomenon is entirely different from the case of the single bubble near the solid wall with $\gamma = 1.3$ [7], in which the bubble is in contact with the wall during the second collapse.

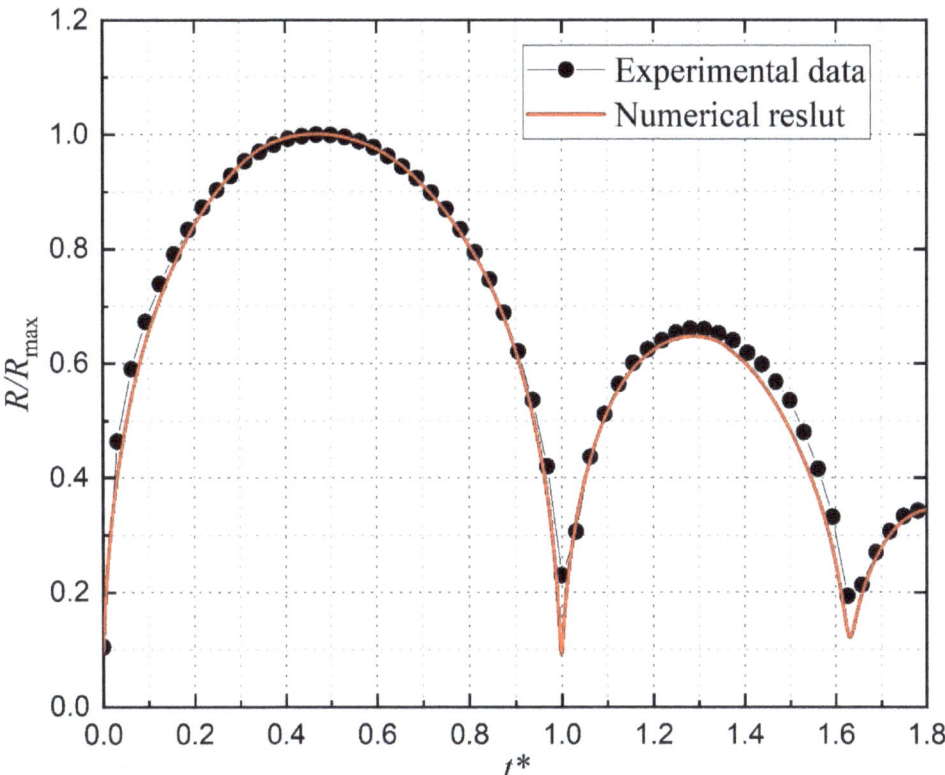

Figure 4. Comparison of the bubble equivalent radius between simulation (solid red line) and experiment (block dot) with $\gamma = 1.3$.

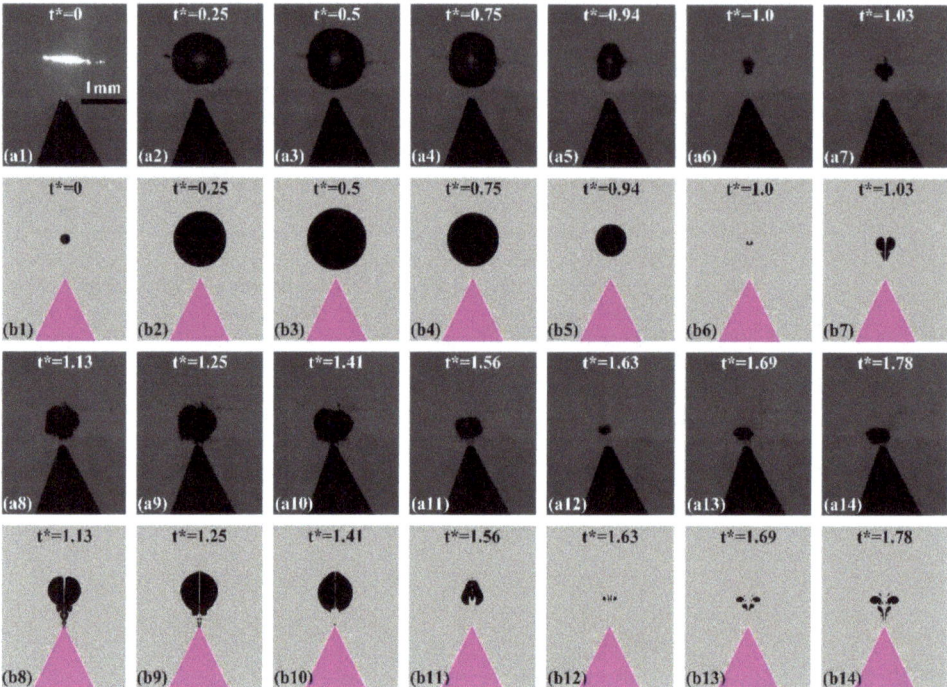

Figure 5. The evolution of bubble shape with $\gamma = 1.3$. ((**a1**)~(**a14**)) for the experiments; ((**b1**)~(**b14**)) for the numerical results.

In Figure 6, the pressure and velocity fields are shown for some typical moments to discuss the bubble dynamic behavior in detail. During the collapse, a locally high-pressure zone appears around the bubble, as shown in Figure 6a. Because the cone obstructs the flow, the pressure at the upper part of the bubble is slightly greater than that at the lower part of the bubble, resulting in a faster collapse of the upper surface of the bubble. Figure 6b shows that the liquid jet, driven by high pressure, enters the bubble from the upper surface. When the jet penetrates the bottom of the bubble, a strong impact such as the water hammer is formed, and a shock wave is emitted and propagated outwards. At the initial stage of the bubble re-expansion, a noticeable protrusion is generated on the lower part of the bubble due to the effect of the liquid micro-jet, as seen in Figure 6c. The protrusion continues to move towards the tip of the cone and impact it while the bubble further rebounds, as shown in Figure 6d. Figure 6e shows that the re-expansion bubble breaks into two parts of different volumes due to the velocity difference, in which the smaller one is close to the cone. Figure 6f,g show that the larger rebound bubble still collapses, while the smaller one eventually dissolves in the liquid. Finally, the remaining bubble collapses to its minimum radius (Figure 6h) and rebounds again (Figure 6i).

Figure 7 shows the time histories of the numerical results for the bubble radius, bubble centroid position, maximum temperature inside the bubble, and pressure at the apex of the cone. As seen, the maximum values of the temperature within the bubble and pressure at the apex of the cone always coincide with that for which the minimum volume is attained. More details on the source of the pressure peaks can be found in our previous work [8]. The sharp change in the bubble's centroid position occurs at the bubble's collapse stages.

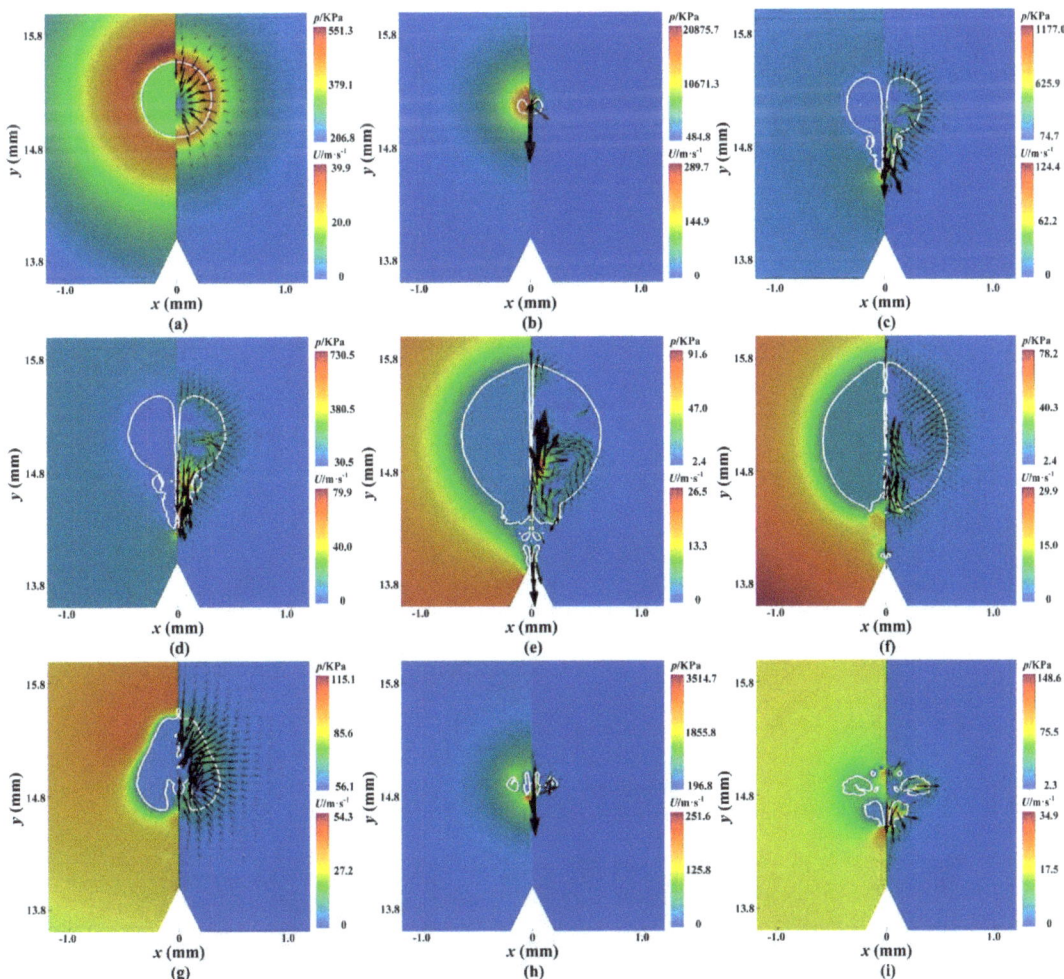

Figure 6. Evolution of pressure (left half) and velocity (right half) fields with bubble shape (white line) at $\gamma = 1.3$. (**a**) $t^* = 0.974$, (**b**) $t^* = 0.999$, (**c**) $t^* = 1.033$, (**d**) $t^* = 1.067$, (**e**) $t^* = 1.248$, (**f**) $t^* = 1.405$, (**g**) $t^* = 1.562$, (**h**) $t^* = 1.625$, (**i**) $t^* = 1.689$.

4.2. Bubble Dynamics of Multiple Collapses with $\gamma = 0.4$

When γ is reduced to 0.4, the bubble dynamics and their intensities are more influenced by the conical structure. Figure 8 shows a comparison of the bubble shape frame by frame between the numerical predictions and the experimental phenomena. As observed in the experiment (frame (a1)), the bubble is initialized closer to the apex of the cone. Thus, the upper part of the cone is gradually swallowed by the expanding bubble (see frames (a2)~(a4)). Due to the substantial restriction of the cone, the bubble interface becomes unstable and relatively rough during the rebound stage (see frames a7~a11). Eventually, the bubble almost disappears into the liquid, as seen in frames (a13)~(a14). Overall, the numerical predictions are in good agreement with the observed experimental results.

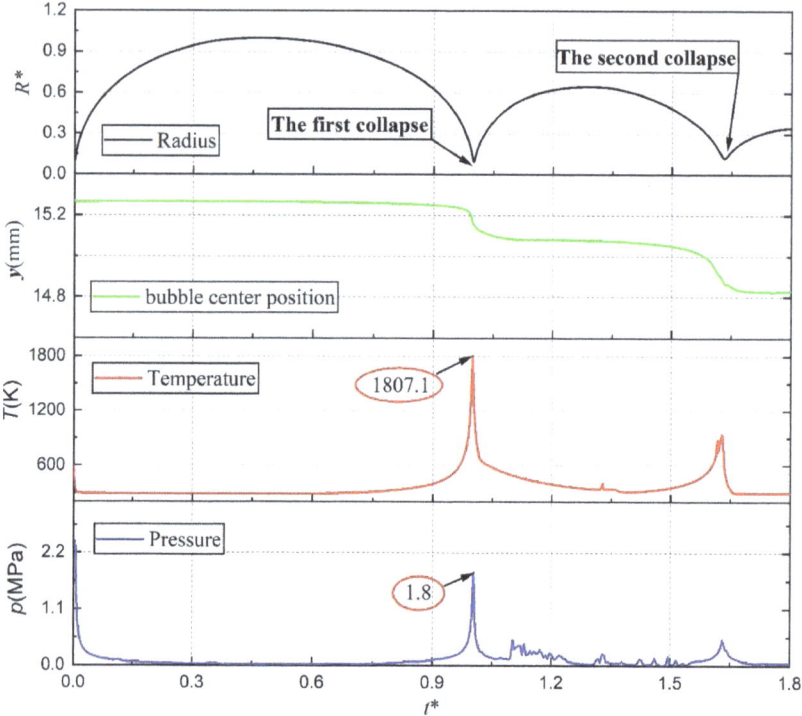

Figure 7. Time histories of the bubble radius (solid black line), bubble centroid position (solid green line), the maximum temperature within the bubble (solid red line), and pressure at the apex of the cone (solid blue line) with $\gamma = 1.3$.

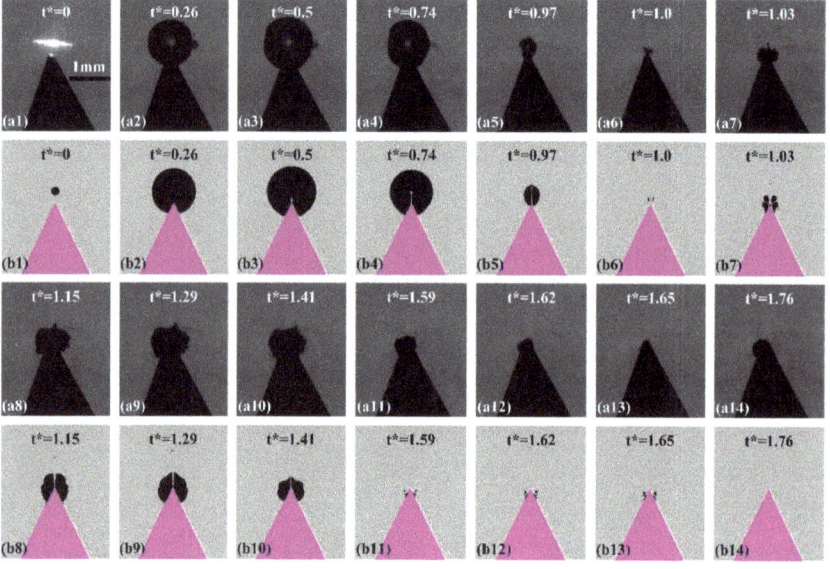

Figure 8. The evolution of bubble shape with $\gamma = 0.4$. ((**a1**)~(**a14**)) for the experiments; ((**b1**)~(**b14**)) for the simulations.

The pressure (left half) and velocity (right half) fields during the bubble collapse and rebound stages are presented in Figure 9. During the first collapse, high pressure is locally formed above the bubble, and the downward liquid jet is driven into the bubble, as shown in Figure 9a. In Figure 9b, the bubble shrinks to its minimum volume, and the high pressure is generated near the cone apex due to the liquid jet continuing to impact the cone. During its rebound, the flow along the conical wall is formed due to the cone apex redirecting the liquid jet, causing the expanding bubble along the conical wall, as shown in Figure 9c. Figure 9d shows the maximum rebound bubble; the bubble bottom is in direct contact with the conical wall. In addition, the internal pressure is lower than that outside the bubble. Thus, the rebound bubble collapses again. Figure 9e shows a high-pressure zone occurring above the bubble, which is similar to the first collapse (see Figure 9a). Figure 9f shows the bubble contracts to the minimum size again.

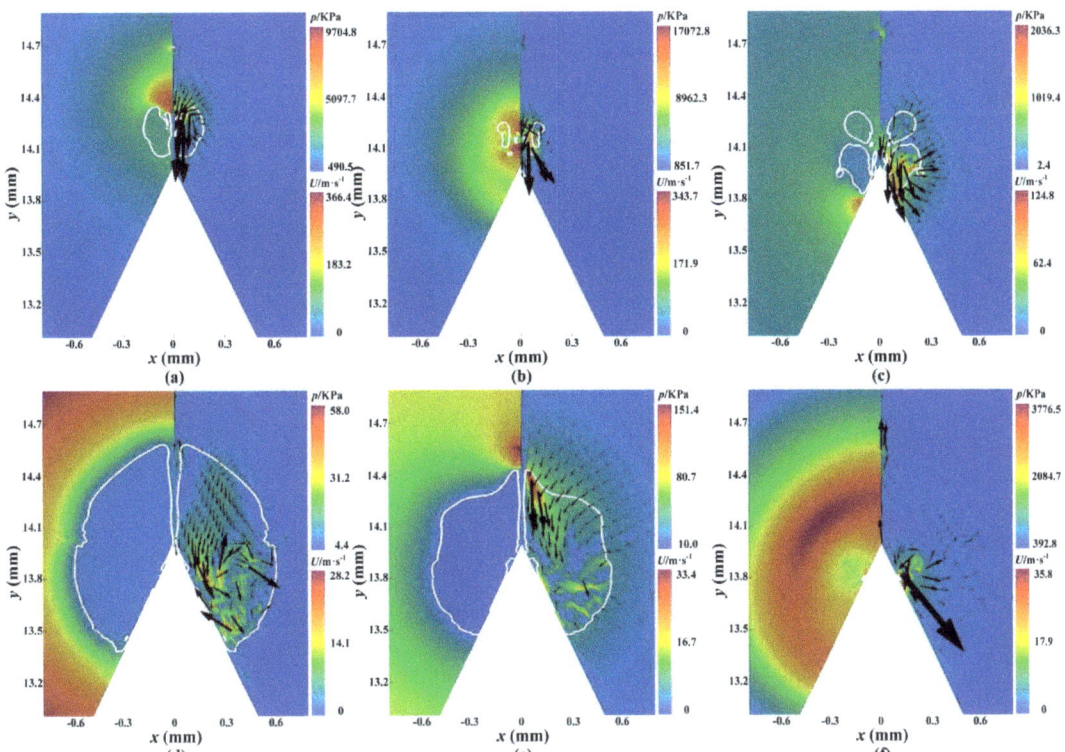

Figure 9. Evolution of pressure (left half) and velocity (right half) fields with bubble shape (white line) at $\gamma = 0.4$. (a) $t^* = 0.993$, (b) $t^* = 1.0$, (c) $t^* = 1.016$, (d) $t^* = 1.261$, (e) $t^* = 1.412$, (f) $t^* = 1.551$.

Figure 10 shows the shock wave distributions expressed with a density gradient using the numerical Schlieren results [45]. (a) It presents the incident shock wave and (b) indicates the shock wave of the first bubble collapse and (c) the second bubble collapse. The outward propagation of shock waves will promote the generation of low-pressure regions, leading to abrupt changes in the liquid density [46]. As a result, some disturbances could be seen in the experiments [47,48]. If the pressure value is lower than the vapor saturation pressure value, a secondary cavitation disturbance appears [49].

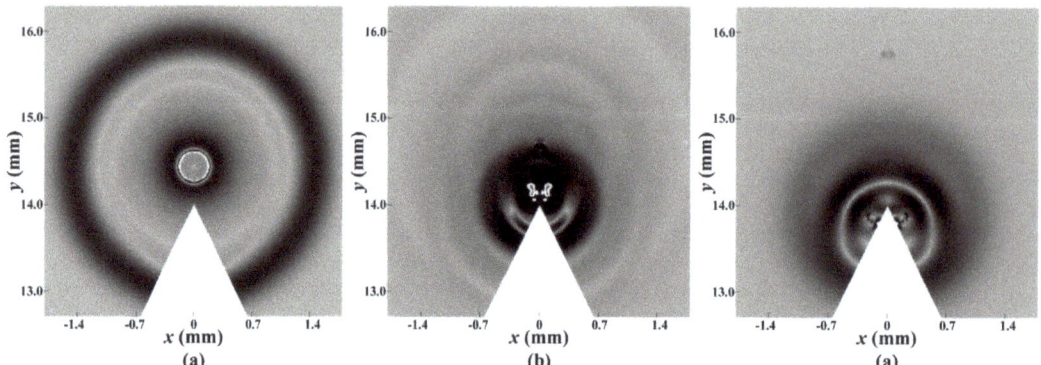

Figure 10. Shock wave distribution with density gradient for (**a**) $t^* = 0.005$, (**b**) $t^* = 1.0$, (**c**) $t^* = 1.55$ with $\gamma = 0.4$.

Figure 11 shows the time histories of the bubble radius, bubble centroid position, maximum temperature, and pressure load at the apex of the cone with $\gamma = 0.4$. As shown, the maximum temperature and pressure are obtained when the bubble first shrinks to the minimum volume. Significantly, the maximum pressure peak at the cone apex reaches 20.3 MPa, which is much greater than that shown in Figure 7. This is because the maximum pressure peak for $\gamma = 1.3$ mainly comes from the first collapse shock wave, while that for $\gamma = 0.4$ is attributed to the superposition of the collapse shock wave and the liquid jet.

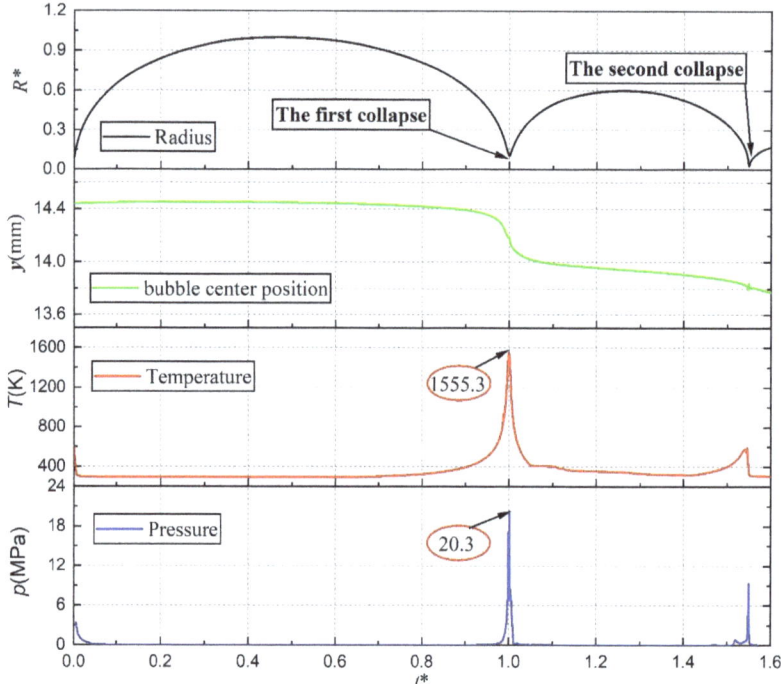

Figure 11. Time histories of the bubble radius (solid black line), bubble centroid position (solid green line), the maximum temperature within the bubble (solid red line), and pressure at the apex of the cone (solid blue line) with $\gamma = 0.4$.

5. Further Discussions

5.1. The Effect of the Distance from the Bubble to the Cone Apex

As discussed above, the distance from the bubble to the cone, i.e., γ, strongly affects the bubble dynamics. Thus, to further investigate and analyze the tendencies of bubble dynamics, more numerical simulation cases with different γ values were carried out. Figure 12a shows the evolution of the bubble center position over time under different γ values. The red dotted line indicates the position of the cone apex. As seen, the dramatic change in the bubble center position always occurs in the final collapse stages, including the first and second collapses. When $\gamma > 0.8$, the position of the bubble in the first and second collapses is higher than that of the cone apex, meaning a water layer still exists between the bubble and the cone apex (see Figure 5(a6,a12)) and prevents the liquid micro-jet from directly impacting the cone. When $0.5 \leq \gamma \leq 0.8$, it can be seen that the bubble center is above the cone apex in the first collapse but below the cone apex in the second collapse, indicating that the water layer gradually disappears during the second collapse and the bubble is in direct contact with the surface of the cone. When $0 < \gamma < 0.5$, the bubble is very close to or even lower than the apex of the cone (see Figure 8(a6,a13)).

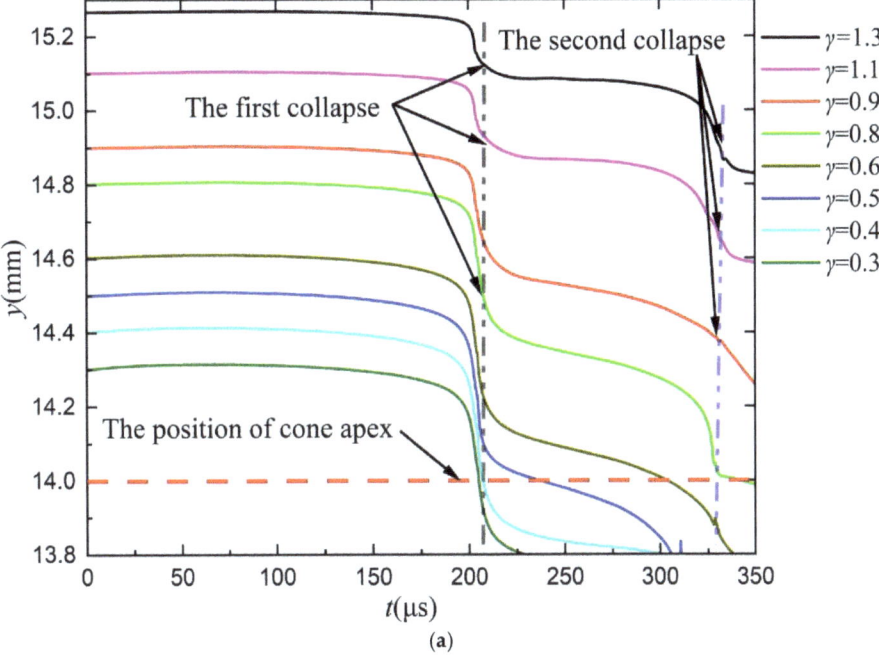

Figure 12. Cont.

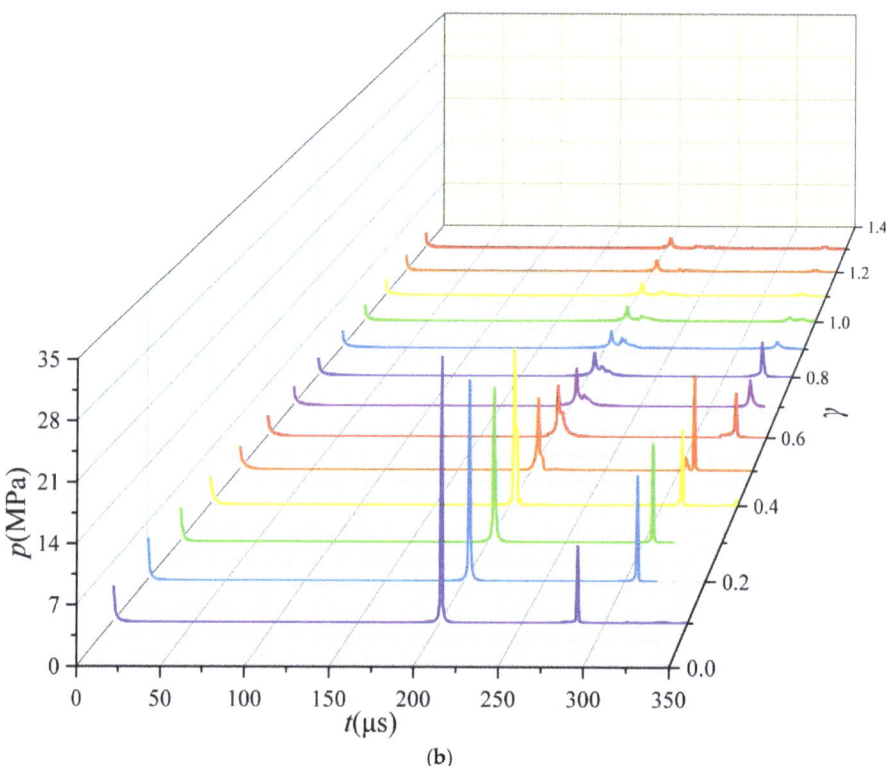

Figure 12. Time history of the bubble centroid migration (**a**) and pressure on the cone apex (**b**) with various γ values.

In addition, the evolution of the impact pressure on the cone apex over time under various γ values is shown in Figure 12b. As seen, the maximum pressure is achieved in the final stages of the bubble's collapse. Furthermore, the bubble's second collapse occurs earlier as γ decreases. In particular, the pressure peaks from the final collapse phase can be divided into three stages. In stage I ($\gamma > 0.8$), the bubble remains away from the cone during its collapse, having a weak impact on the liquid jet. Thus, the pressure peaks mainly come from the collapse shock waves, as shown in Figure 13c,f, and their values are almost constant as γ decreases. Moreover, the pressure peak of the first bubble collapse is slightly larger than that of the second collapse. In stage II ($0.5 \leq \gamma \leq 0.8$), the pressure peaks are larger than those in stage I due to the small distance between the bubble and cone (see Figure 13b,c), and their values increase as γ decreases. Remarkably, the bubble puts pressure on the cone during its second collapse, as shown in Figure 13e, causing a violent impact. It can be seen that the pressure peak generated by the second collapse is comparable to that from the first. In stage III ($0 < \gamma < 0.5$), the bubble clings to the cone apex during the first collapse, as shown in Figure 13a, leading to a more severe impact. Therefore, the pressure peak is much greater than those in stages I and II, and its value sharply increases with the decrease in γ. Furthermore, the pressure peak of the second bubble collapse cannot be ignored (see Figure 13d) and significantly different from the bubble's behavior near the solid wall (for which no apparent rebound phenomenon is observed with a range of $0 < \gamma < 0.5$ [8]).

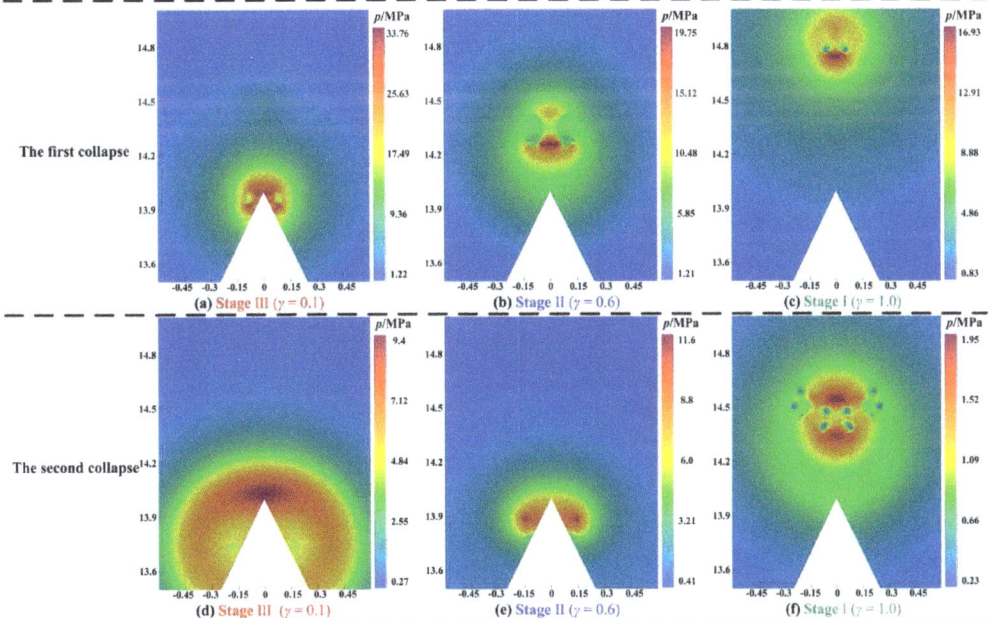

Figure 13. Pressure fields at the end of the bubble collapse for each stage.

5.2. The Effect of the Cone Angles (θ)

Figure 14 shows the schematic diagram of different θ values, which all have the same height (i.e., h = 14 mm). As shown in Figure 12b, the pressure peak is closely related to γ. In this section, a larger γ ($\gamma = 1.3$), indicating that the initial bubble stays away from the cone, is chosen to investigate the bubble dynamic features under different θ values. The influence of θ on the bubble center position is shown in Figure 15. Here, $\theta = 180°$ indicates the cone is a solid plane. As shown in the figure, the bubble gradually approaches the cone apex with the increase in θ, implying the effect of the cone is strengthened by the degree. To be specific, the bubble is still moving away from the cone in the first collapse but quickly approaches or even clings to the cone in the second collapse.

Figure 14. Schematic diagram of different θ.

The pressure peaks on the cone apex from the first and second collapses under different θ values are shown in Figure 16. The pressure peak of the second bubble collapse (magenta dots) increases rapidly as θ increases, while the pressure peak from the first (olive green triangles) increases slowly. Particularly, when $\theta < 90°$, the pressure peak of the second bubble collapse is slightly smaller than that of the first. Only a small part of the initial energy goes into the second collapse phase. Thus, the intensity of the shock wave generated by the second collapse is much smaller than that of the first collapse (see Figure 17a,c), causing a weak impact on the conical surface. When $\theta \geq 90°$, the bubble is close to or even attaches to the surface of the cone in the second collapse, resulting in a stronger impact on the cone due to the superposition of the liquid microjet and the collapse shock wave as shown in Figure 17b,d.

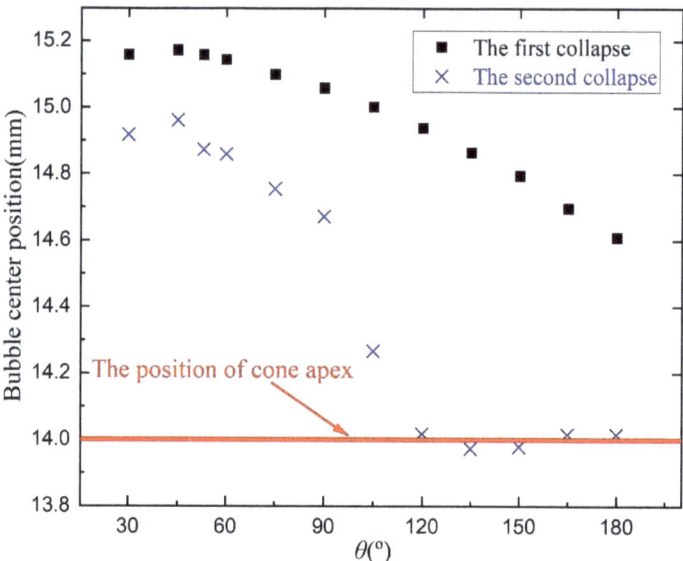

Figure 15. Bubble center position in the final collapse stages at $\gamma = 1.3$ with various θ values.

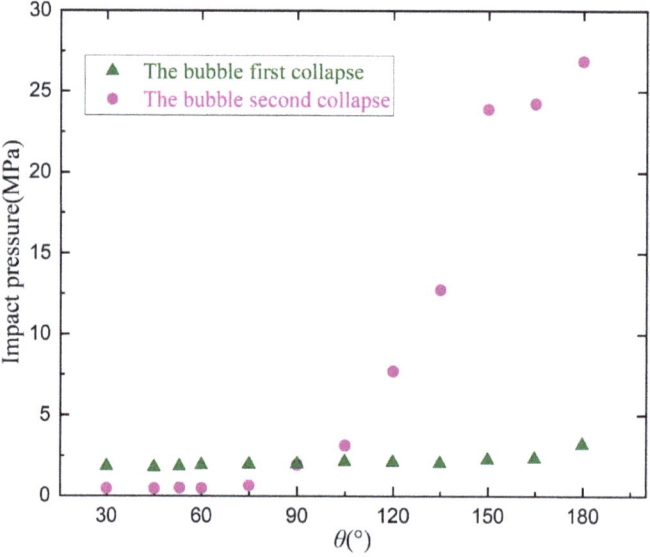

Figure 16. Pressure peaks on the cone apex in the final collapse stages at $\gamma = 1.3$ with various θ values.

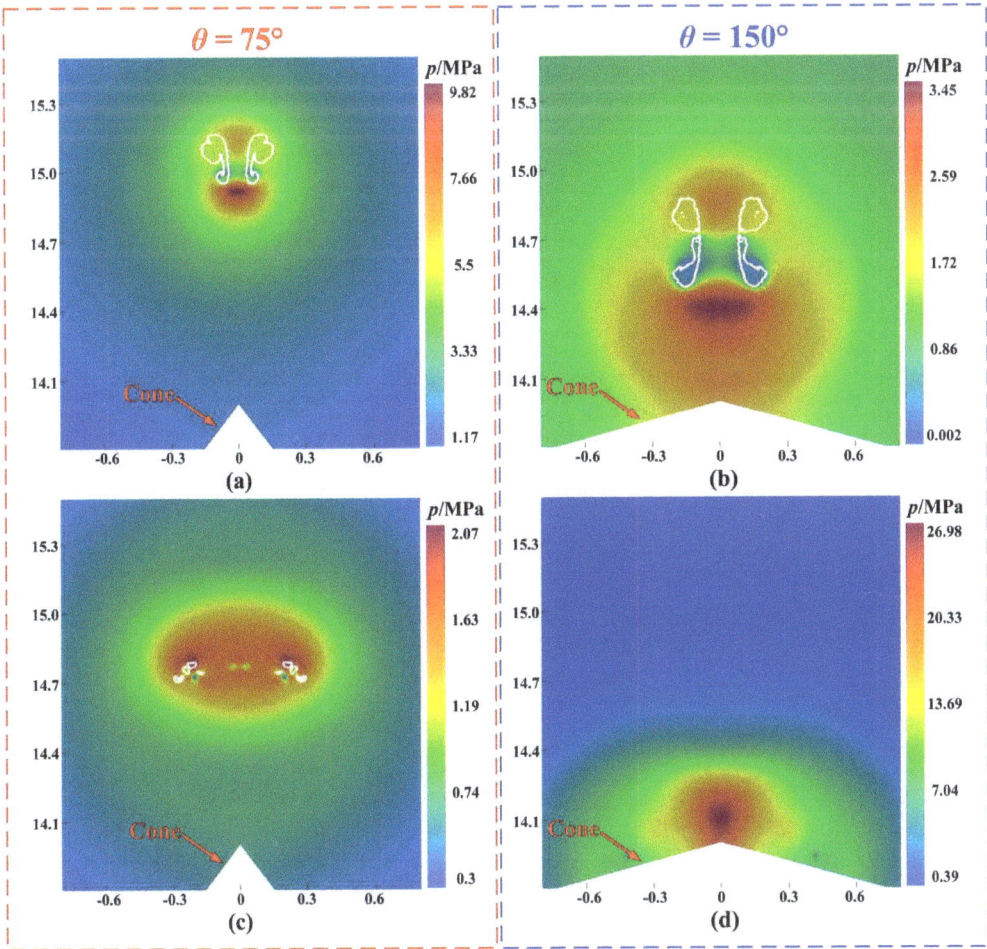

Figure 17. The typical pressure fields in the final bubble collapse stage under different θ values. (a,c) are the first collapse and second collapse for $\theta = 75°$, respectively; (b,d) are the first collapse and second collapse for $\theta = 150°$, respectively.

In addition, the effects of θ on the maximum temperature inside the bubble, the first minimum radius, and the dissipated energy during the first bubble collapse are shown in Figure 18. The dissipated energy in the bubble's first cycle decreases with the increase in θ, indicating the retardation effect of the conical surface is strengthened. Thus, a larger minimum radius at the first collapse can be expected, as shown by the blue/white dots in Figure 18. In general, the temperature peak value within the bubble is approximately proportional to the minimum collapse radius. For a single bubble collapsing in a free field, the temperature can reach an extremely high value (about 22,000 K) since it can collapse to a much smaller radius ($R_{min}/R_{max} \approx 0.004$) [36]. In this paper, the maximum temperature decreases as θ increases due to a larger minimum radius.

For a smaller γ ($\gamma = 0.1$), Figure 19 illustrates the variation in essential parameters (the bubble center position, pressure peaks, and dissipated energy) under various θ values. As shown in Figure 19a, the bubble center position in the first two final collapse stages touches the cone apex due to the smaller distance between the bubble and the cone. In addition, more than 80% of the initial bubble energy is lost in the first cycle, as shown in

Figure 19c. In other words, only a tiny portion of the bubble energy enters into the rebound. Thus, a larger pressure peak for the first collapse can be seen, as shown in Figure 19b. Simultaneously, the pressure peaks at different θ values do not vary very much, implying that the influence of θ is not significant for a smaller γ.

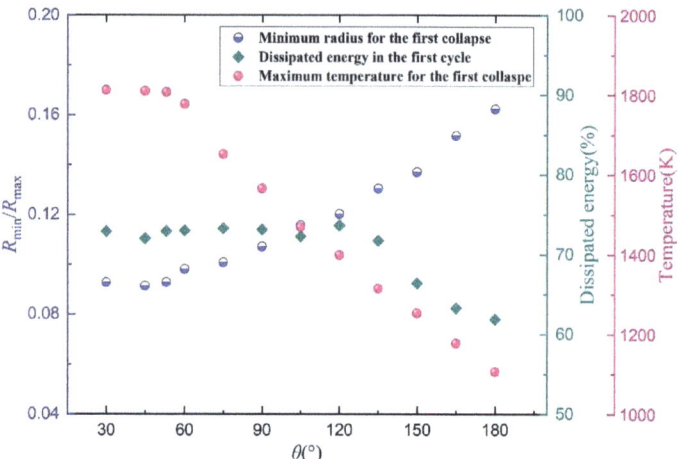

Figure 18. Maximum temperature inside the bubble (pink dots), the first minimum radius (white/blue dots), and the dissipated energy in the bubble's first cycle (dark cyan diamonds) at $\gamma = 1.3$ under various θ values.

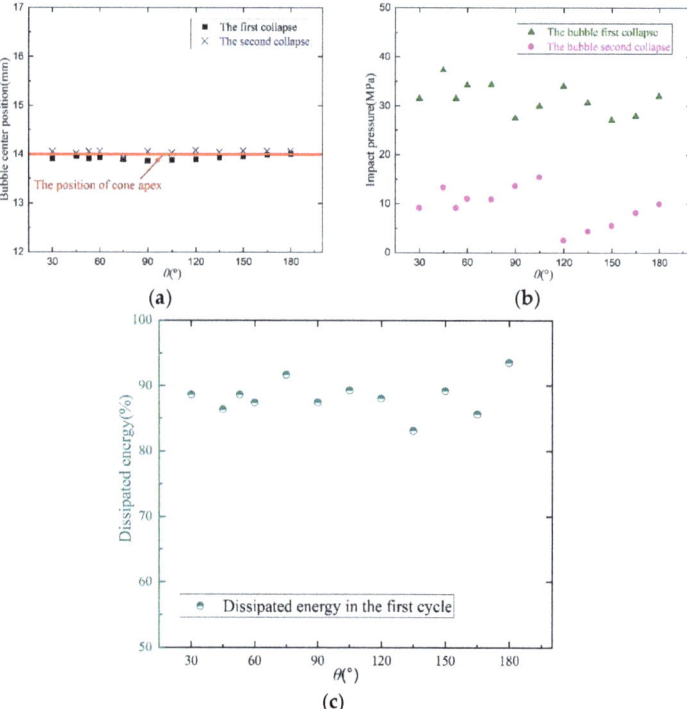

Figure 19. Bubble center position (**a**), pressure peaks (**b**), and dissipated energy (**c**) at $\gamma = 0.1$ under various θ values.

6. Conclusions

The dynamics of a cavitation bubble near a rigid conical boundary are studied through a numerical simulation and experiments. A pulsed laser generator is used to generate a single cavitation bubble, and a high-speed camera with a sampling rate of 100,000 frames/s is employed to capture the motion behavior of the bubble. In addition, a compressible flow solver, with consideration of the phase change and thermodynamic effect, is adopted to perform the corresponding numerical simulations. The numerical results for the bubble radius and bubble shape are in good agreement with the related experimental data. The bubble's primary physical features during the first two cycles are reproduced well based on our numerical model. Moreover, the physical implications in terms of the bubble shape, bubble radius, shock wave, pressure load, internal temperature, and dissipated energy during collapse are analyzed in detail to support the observed experimental phenomena. Generally, the peak values of the pressure and temperature occur almost at the moment of collapse once the bubble reaches its minimum volume. Finally, the effects of the relative distance γ and cone angle θ on the maximum temperature, bubble position, and pressure peaks are discussed. The main conclusions can be drawn as follows:

1. The pressure peaks from the first and second collapse increase with the decrease in γ. Moreover, the rate of increase of the pressure peak from the first collapse is much greater than that from the second.
2. For a larger γ, as θ increases, the first minimum bubble radius increases while the maximum temperature decreases. Additionally, the pressure peak of the second collapse is slightly smaller than that of the first one due to most of the bubble's energy being lost in the first cycle when $\theta < 90°$. The pressure peak at the second final collapse is much larger than that at the first because the bubble clings to the cone tip during the bubble's second collapse when $\theta \geq 90°$.
3. For a smaller γ, more energy is lost at the first collapse and the bubble always clings to the conical surface during the collapse. As a result, the pressure peak in the first final collapse is much greater than that in the second. The pressure peaks at different θ do not vary very much, and the influence of θ on the bubble's behavior is negligible.

Author Contributions: Conceptualization, J.Y. and Y.Z.; methodology, J.Y.; software, D.G.; validation, J.Y., L.T. and X.D.; formal analysis, J.Y.; investigation, J.Y.; resources, Y.Z.; data curation, X.D.; writing—original draft preparation, J.Y.; writing—review and editing, Y.Z.; visualization, D.G.; supervision, L.T.; project administration, J.Y. and Y.Z.; funding acquisition, J.Y. and Y.Z. All authors have read and agreed to the published version of the manuscript.

Funding: The work was supported by the National Natural Science Foundation of China (Grant No. 52179094), the Guizhou Provincial Basic Research Program (Grant No. ZK [2023] General 069), the Guizhou Provincial Key Technology R&D Program (Grant No. 2022[18]), and Scientific Research Foundation for the Talent Introduction of Guizhou University (2022[14]).

Data Availability Statement: The data supporting the findings of this study are available from the author upon reasonable request.

Acknowledgments: The authors would like to acknowledge the support of the National Natural Science Foundation of China, the Guizhou Provincial Basic Research Progra, the Guizhou Provincial Key Technology R&D Program, and Scientific Research Foundation for the Talent Introduction of Guizhou University.

Conflicts of Interest: The authors declare no conflict of interest.

References

1. Bhat, A.P.; Gogate, P.R. Cavitation-based pre-treatment of wastewater and waste sludge for improvement in the performance of biological processes: A review. *J. Environ. Chem. Eng.* **2021**, *9*, 104743. [CrossRef]
2. Chudnovskii, V.M.; Yusupov, V.I.; Dydykin, A.V.; Nevozhai, V.I.; Kisilev, A.Y.; Zhukov, S.A.; Bagratashvili, V.N. Laser-induced boiling of biological liquids in medical technologies. *Quantum Electron.* **2017**, *47*, 361. [CrossRef]
3. Brennen, C.E. Cavitation in medicine. *Interface Focus* **2015**, *5*, 20150022. [CrossRef] [PubMed]

4. Prado, C.A.; Antunes, F.A.; Rocha, T.M.; Sánchez-Muoz, S.; Barbosa, F.G.; Terán-Hilares, R.; Cruz-Santos, M.M.; Arruda, G.L.; da Silva, S.S.; Santos, J.C. A review on recent developments in hydrodynamic cavitation and advanced oxidative processes for pretreatment of lignocellulosic materials. *Bioresour. Technol.* **2022**, *345*, 126458. [CrossRef] [PubMed]
5. Wang, S.; Guo, Z.P.; Zhang, X.P.; Zhang, A.; Kang, J.W. On the mechanism of dendritic fragmentation by ultrasound induced cavitation. *Ultrason. Sonochemistry* **2019**, *51*, 160–165. [CrossRef]
6. Xu, W.L.; Wang, Q.F.; Wei, W.R.; Luo, J. Effects of air bubble quantity on the reduction of cavitation erosion. *Wear* **2021**, *482*, 203937. [CrossRef]
7. Philipp, A.; Lauterborn, W. Cavitation erosion by single laser-produced bubbles. *J. Fluid Mech.* **1998**, *361*, 75–116. [CrossRef]
8. Yin, J.; Zhang, Y.; Zhu, J.; Lv, L.; Tian, L. An experimental and numerical study on the dynamical behaviors of the rebound cavitation bubble near the solid wall. *Int. J. Heat Mass Transf.* **2021**, *177*, 121525. [CrossRef]
9. Tomita, Y.; Robinson, P.B.; Tong, R.P.; Black, J.R. Growth and collapse of cavitation bubbles near a curved rigid boundary. *J. Fluid Mech.* **2002**, *466*, 259–283. [CrossRef]
10. Li, T.; Zhang, A.M.; Wang, S.P.; Liu, W.T. Bubble interactions and bursting behaviors near a free surface. *Phys. Fluids* **2019**, *31*, 042104.
11. Cui, P.; Zhang, A.M.; Wang, S.P.; Khoo, B.C. Ice breaking by a collapsing bubble. *J. Fluid Mech.* **2018**, *841*, 287–309. [CrossRef]
12. Brujan, E.A.; Nahen, K.; Schmidt, P.; Vogel, A. Dynamics of laser-induced cavitation bubbles near elastic boundaries: Influence of the elastic modulus. *J. Fluid Mech.* **2001**, *433*, 283–314. [CrossRef]
13. Zhang, Y.; Qiu, X.; Zhang, X.Q.; Tang, N.N.; Zhang, Y.N. Collapsing dynamics of a laser-induced cavitation bubble near the edge of a rigid wall. *Ultrason. Sonochemistry* **2020**, *67*, 105157. [CrossRef] [PubMed]
14. Chudnovskii, V.M.; Guzev, M.A.; Yusupov, V.I.; Fursenko, R.V.; Okajima, J. Study of methods for controlling direction and velocity of liquid jets formed during subcooled boiling. *Int. J. Heat Mass Transf.* **2021**, *173*, 121250. [CrossRef]
15. Ren, Z.; Li, B.; Xu, P.; Wakata, Y.; Liu, J.; Sun, C.; Zuo, Z. Cavitation bubble dynamics in a funnel-shaped tube. *Phys. Fluids* **2022**, *34*, 093313. [CrossRef]
16. Sun, Y.; Yao, Z.; Wen, H.; Zhong, Q.; Wang, F. Cavitation bubble collapse in a vicinity of a rigid wall with a gas entrapping hole. *Phys. Fluids* **2022**, *34*, 073314. [CrossRef]
17. Wang, X.; Wu, G.; Zheng, X.; Du, X.; Zhang, Y. Theoretical investigation and experimental support for the cavitation bubble dynamics near a spherical particle based on Weiss theorem and Kelvin impulse. *Ultrason. Sonochemistry* **2022**, *89*, 106130. [CrossRef]
18. Zeng, Q.; Gonzalez-Avila, S.R.; Ohl, C.-D. Splitting and jetting of cavitation bubbles in thin gaps. *J. Fluid Mech.* **2020**, *896*, A28. [CrossRef]
19. Saini, M.; Tanne, E.; Arrigoni, M.; Zaleski, S.; Fuster, D. On the dynamics of a collapsing bubble in contact with a rigid wall. *J. Fluid Mech.* **2022**, *948*, A45. [CrossRef]
20. Požar, T.; Agrež, V. Laser-induced cavitation bubbles and shock waves in water near a concave surface. *Ultrason. Sonochemistry* **2021**, *73*, 105456. [CrossRef]
21. Kadivar, E.; Phan, T.H.; Park, W.G.; Moctar, O. Dynamics of a single cavitation bubble near a cylindrical rod. *Phys. Fluids* **2021**, *33*, 113315. [CrossRef]
22. Brujan, E.A.; Noda, T.; Ishigami, A.; Ogasawara, T.; Takahira, H. Dynamics of laser-induced cavitation bubbles near two perpendicular rigid walls. *J. Fluid Mech.* **2018**, *841*, 28–49. [CrossRef]
23. Tagawa, Y.; Peters, I.R. Bubble collapse and jet formation in corner geometries. *Phys. Rev. Fluids* **2018**, *3*, 081601. [CrossRef]
24. Li, S.M.; Cui, P.; Zhang, S.; Liu, W.; Peng, Y. Experimental and numerical study on the bubble dynamics near two-connected walls with an obtuse angle. *China Ocean. Eng.* **2020**, *34*, 828–839. [CrossRef]
25. Zhang, Y.; Li, S.; Zhang, Y.; Zhang, Y. Dynamics of the bubble near a triangular prism array. In Proceedings of the 10th International Symposium on Cavitation CAV, Baltimore, MA, USA, 14–16 May 2018.
26. He, M.; Wang, S.P.; Ren, S.F.; Zhang, S. Numerical study of effects of stand-off distance and gravity on large scale bubbles near a breach. *Appl. Ocean. Res.* **2021**, *117*, 102946. [CrossRef]
27. Morad, A.M.; Selima, E.S.; Abu-Nab, A.K. Thermophysical bubble dynamics in N-dimensional Al_2O_3/H_2O nanofluid between two-phase turbulent flow. *Case Stud. Therm. Eng.* **2021**, *28*, 101527. [CrossRef]
28. Morad, A.M.; Selima, E.S.; Abu-Nab, A.K. Bubbles interactions in fluidized granular medium for the van der Waals hydrodynamic regime. *Eur. Phys. J. Plus* **2021**, *136*, 306. [CrossRef]
29. Park, S.-H.; Phan, T.-H.; Park, W.-G. Numerical investigation of laser-induced cavitation bubble dynamics near a rigid surface based on three-dimensional fully compressible model. *Int. J. Heat Mass Transf.* **2022**, *191*, 122853. [CrossRef]
30. Zeng, Q.; An, H.; Ohl, C.-D. Wall shear stress from jetting cavitation bubbles: Influence of the stand-off distance and liquid viscosity. *J. Fluid Mech.* **2022**, *932*, A14. [CrossRef]
31. Zhang, J. Effect of stand-off distance on "counterjet" and high impact pressure by a numerical study of laser-induced cavitation bubble near a wall. *Int. J. Multiph. Flow* **2021**, *142*, 103706. [CrossRef]
32. Wang, Q.; Mahmud, M.; Cui, J.; Smith, W.R. Numerical investigation of bubble dynamics at a corner. *Phys. Fluids* **2020**, *32*, 053306.
33. Trummler, T.; Bryngelson, S.H.; Schmidmayer, K.; Schmidt, S.J.; Colonious, T.; Adams, N.A. Near-surface dynamics of a gas bubble collapsing above a crevice. *J. Fluid Mech.* **2020**, *899*, A16. [CrossRef]

34. Shervani-Tabar, M.; Rouhollahi, R. Numerical study on the effect of the concave rigid boundaries on the cavitation intensity. *Sci. Iran.* **2017**, *24*, 1958–1965. [CrossRef]
35. Li, B.B.; Jia, W.; Zhang, H.C.; Lu, J. Investigation on the collapse behavior of a cavitation bubble near a conical rigid boundary. *Shock. Waves* **2014**, *24*, 317–324. [CrossRef]
36. Yin, J.; Zhang, Y.; Zhu, J.; Zhang, Y.; Li, S. On the thermodynamic behaviors and interactions between bubble pairs: A numerical approach. *Ultrason. Sonochemistry* **2021**, *70*, 105297. [CrossRef] [PubMed]
37. Zhu, J.; Chen, Y.; Zhao, D.; Zhang, X. Extension of the Schnerr–Sauer model for cryogenic cavitation. *Eur. J. Mech.-B/Fluids* **2015**, *52*, 1–10. [CrossRef]
38. Koch, M.; Lechner, C.; Reuter, F.; Köhler, K.; Mettin, R.; Lauterborn, W. Numerical modeling of laser generated cavitation bubbles with the finite volume and volume of fluid method, using OpenFOAM. *Comput. Fluids* **2016**, *126*, 71–90. [CrossRef]
39. Reese, H.; Schädel, R.; Reuter, F.; Ohl, C.D. Microscopic pumping of viscous liquids with single cavitation bubbles. *J. Fluid Mech.* **2022**, *944*, A17. [CrossRef]
40. Yin, J.Y.; Zhang, Y.; Zhu, J.; Lv, L.; Li, S. Numerical investigation of the interactions between a laser-generated bubble and a particle near a solid wall. *J. Hydrodyn.* **2021**, *33*, 311–322. [CrossRef]
41. Boris, J.P.; Book, D.L. Flux-corrected transport. I. SHASTA, a fluid transport algorithm that works. *J. Comput. Phys.* **1973**, *11*, 38–69. [CrossRef]
42. Greenshields, C.J. *OpenFOAM User Guide, Version 5.0*; OpenFOAM Foundation Ltd.: London, UK, 2015; Volume 3, p. 47.
43. Weller, H.G. *A New Approach to VOF-Based Interface Capturing Methods for Incompressible and Compressible Flow*; Report TR/HGW; OpenCFD Ltd.: Reading, UK, 2008; Volume 4, p. 35.
44. Shin, B.; Iwata, Y.; Ikohagi, T. Numerical simulation of unsteady cavitating flows using a homogenous equilibrium model. *Comput. Mech.* **2003**, *30*, 388–395. [CrossRef]
45. Settles, G.S. *Schlieren and Shadowgraph Techniques: Visualizing Phenomena in Transparent Media*; Springer Science & Business Media: Berlin/Heidelberg, Germany, 2001.
46. Phan, T.-H.; Kadivar, E.; Nguyen, V.-T.; el Moctar, O.; Park, W.-G. Thermodynamic effects on single cavitation bubble dynamics under various ambient temperature conditions. *Phys. Fluids* **2022**, *34*, 023318. [CrossRef]
47. Luo, J.; Xu, W.; Khoo, B.C. Stratification effect of air bubble on the shock wave from the collapse of cavitation bubble. *J. Fluid Mech.* **2021**, *919*, A16. [CrossRef]
48. Supponen, O.; Obreschkow, D.; Farhat, M. Rebounds of deformed cavitation bubbles. *Phys. Rev. Fluids* **2018**, *3*, 103604. [CrossRef]
49. Zhang, M.; Chang, Q.; Ma, X.; Wang, G.; Huang, B. Physical investigation of the counterjet dynamics during the bubble rebound. *Ultrason. Sonochem.* **2019**, *58*, 104706. [CrossRef] [PubMed]

Disclaimer/Publisher's Note: The statements, opinions and data contained in all publications are solely those of the individual author(s) and contributor(s) and not of MDPI and/or the editor(s). MDPI and/or the editor(s) disclaim responsibility for any injury to people or property resulting from any ideas, methods, instructions or products referred to in the content.

Article

Bulk Cavitation in Model Gasoline Injectors and Their Correlation with the Instantaneous Liquid Flow Field

Dimitrios Kolokotronis [1], Srikrishna Sahu [2,*], Yannis Hardalupas [3], Alex M. K. P. Taylor [3] and Akira Arioka [4]

[1] Department of Mechanical Engineering, Aristotle University of Thessaloniki, 54636 Thessaloniki, Greece; dkolokotronis@meng.auth.gr
[2] Department of Mechanical Engineering, IIT Madras, Chennai 600036, India
[3] Department of Mechanical Engineering, Imperial College London, London SW7 2AZ, UK; y.hardalupas@imperial.ac.uk (Y.H.); a.m.taylor@imperial.ac.uk (A.M.K.P.T.)
[4] Keihin Corporation, Tochigi R&D Center, Tochigi 329-1233, Japan; akira-arioka@keihin-corp.co.jp
* Correspondence: ssahu@iitm.ac.in

Abstract: It is well established that spray characteristics from automotive injectors depend on, among other factors, whether cavitation arises in the injector nozzle. Bulk cavitation, which refers to the cavitation development distant from walls and thus far from the streamline curvature associated with salient points on a wall, has not been thoroughly investigated experimentally in injector nozzles. Consequently, it is not clear what is causing this phenomenon. The research objective of this study was to visualize cavitation in three different injector models (designated as Type A, Type B, and Type C) and quantify the liquid flow field in relation to the bulk cavitation phenomenon. In all models, bulk cavitation was present. We expected this bulk cavitation to be associated with a swirling flow with its axis parallel to that of the nozzle. However, liquid velocity measurements obtained through particle image velocimetry (PIV) demonstrated the absence of a swirling flow structure in the mean flow field just upstream of the nozzle exit, at a plane normal to the hypothetical axis of the injector. Consequently, we applied proper orthogonal decomposition (POD) to analyze the instantaneous liquid velocity data records in order to capture the dominant coherent structures potentially related to cavitation. It was found that the most energetic mode of the liquid flow field corresponded to the expected instantaneous swirling flow structure when bulk cavitation was present in the flow.

Keywords: cavitation; injector; PIV; POD; visualization

1. Introduction

Cavitation is an important factor for spray formation within diesel or gasoline injectors, and it appears to affect the properties of the resultant spray [1,2]. In addition, differences in the spray cone angle and tip penetration have been reported in [3–6] depending on the type of cavitation observed (with edge flow separation cavitation occurring close to the walls or bulk cavitation, which occurs within the flow far from the walls). While the potential role of edge flow separation cavitation on spray formation in nozzles has been thoroughly investigated, the formation and the effects of bulk cavitation are issues that need more attention in terms of acquiring quantitative flow field data. The phenomenon of bulk cavitation (or string cavitation) has been reported in diesel model injectors in references [7–9], as well as in gasoline injectors in references [10–12]. Recent studies on the visualization of string cavitation have attempted to explain the interaction of vortices between adjacent nozzles when bulk cavitation is present in diesel multi-hole injectors [13]. Additionally, these studies have explored the temporal evolution of this type of cavitation and the cycle-by-cycle variation in its shape [14]. However, neither study has provided quantitative flow field data for the bulk cavitation area. Only in [15] were some flow field data derived through a simulation based on the work of [16] provided. These simulations showed the presence of a vortical flow around the core of bulk cavitation.

Citation: Kolokotronis, D.; Sahu, S.; Hardalupas, Y.; Taylor, A.M.K.P.; Arioka, A. Bulk Cavitation in Model Gasoline Injectors and Their Correlation with the Instantaneous Liquid Flow Field. *Fluids* 2023, *8*, 214. https://doi.org/10.3390/fluids8070214

Academic Editors: D. Andrew S. Rees and Nguyen Van-Tu

Received: 30 May 2023
Revised: 1 July 2023
Accepted: 3 July 2023
Published: 22 July 2023

Copyright: © 2023 by the authors. Licensee MDPI, Basel, Switzerland. This article is an open access article distributed under the terms and conditions of the Creative Commons Attribution (CC BY) license (https://creativecommons.org/licenses/by/4.0/).

Based on computational pressure field results, it has also been suggested that the initiation of this type of cavitation is a result of gas-phase components that remain after previous injection events, with vortices acting as gas-phase carriers. These studies have provided valuable insight into this phenomenon. However, to our knowledge, computational fluid mechanic tools for calculating the flow field in relation to bulk cavitation, as presented in reference [16] and other computational studies ([17–19]), are not (yet) capable of predicting bulk cavitation. Whether it is vortex-induced cavitation or the elongated bubble clouds of the remaining gas phase, this phenomenon influences the surrounding flow field and, more specifically, leads to the redistribution of vorticity, as mentioned in reference [20]. In addition, it has been reported in reference [21] that vortex properties determine the dynamics (growth and collapse) and shape of bulk cavitation. This, along with the fact that bulk cavitation can affect spray properties, motivated our targeted flow field measurements to correlate with this phenomenon.

Reports on the flow field inside injectors, which could be responsible for the formation of bulk cavitation, are limited. An early reference providing quantitative PIV measurements inside diesel fuel injectors was reported in reference [22], followed by the laser Doppler velocimetry measurements (LDV) presented in reference [23]. PIV measurements on the internal flow field of fuel injectors are also presented in references [12,24–26]. In the last reference, the authors observed the presence of bulk cavitation in the same injector geometries as those examined in this paper. In a recent publication [4], 2D PIV was applied to a full-size diesel injector in the area just upstream of the nozzle exit, where bulk cavitation was initiated. However, although some vortical structures seemed to be present, detailed high-resolution experimental flow field data that could quantify the presence of vortices in that area were not provided. Finally, the study presented in references [27–29] experimentally demonstrated in model nozzles that downstream of the flow separation cavitation occurring at the nozzle entry, instantaneous vortices are initiated in the shear layer of the liquid flow. These vortices can lead to a low enough local pressure to cause bulk cavitation, which even extends into the liquid container attached downstream from the nozzle exit. However, the authors did not quantify the local velocity field associated with these structures. Therefore, the literature indicates that the local flow field may be able to induce bulk cavitation.

As a continuation of the work reported in reference [12], bulk cavitation was visualized in three gasoline multi-hole injectors. A two-dimensional micron resolution particle imaging velocimetry was employed to measure the internal flow field of 10:1 super-scale transparent models of multi-hole injectors in the vicinity of a region just upstream from the entrance to the holes of the injector plates, under conditions shortly after the onset of cavitation. This plane of measurement was parallel to the injector plates (which is normal to the notional axis of symmetry in the injector). In cases where bulk cavitation was present, we applied proper orthogonal decomposition (POD) to the measured instantaneous velocity data in order to capture the dominant coherent structures potentially related to cavitation. Our objective was not only to visualize cavitation but also to correlate the flow field upstream of the nozzles with the occurrence of bulk cavitation within the nozzles. In addition, the probability of bulk cavitation was calculated by applying POD to shadowgraph images of different nozzles of the same type of injector. Details of the experimental techniques and methods used are provided in the next section. Subsequently, the results are presented, and the paper concludes with a summary of the main findings.

2. Experimental Methods and Analysis

2.1. Injector Models, Experimental Setup, and Measurement Conditions

The schematics of the gasoline injector models are shown in Figure 1 for Type A and B models and in Figure 2 for the Type C model. Details of the significant differences in geometry between Types A and C are briefly described below; these details can be found in [12]. The main parts of the injector model are shown, and the planes of measurement are indicated with the grey area, with one occupied by a refractive index matching liquid.

These models, which are scaled up by a factor of 10, represent the parts of the prototype gasoline injectors, which are adjacent to the nozzles. The model was finished to the required optical surface quality (Kuwana Engineering Plastic Co., Ltd., Kuwana, Japan).

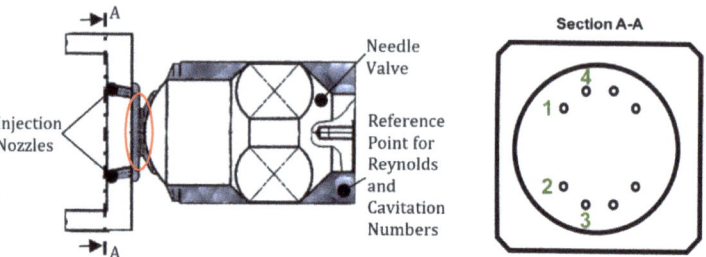

Figure 1. (**left**) Schematic of Type A and Type B injector model. Grey area indicates fluid flow. Flow is from right to left. (**right**) Section A-A shows the nozzle arrangement. Numbers in green identify nozzle numbers 1–4, some of which are referred to in the 'Results and Discussion' section.

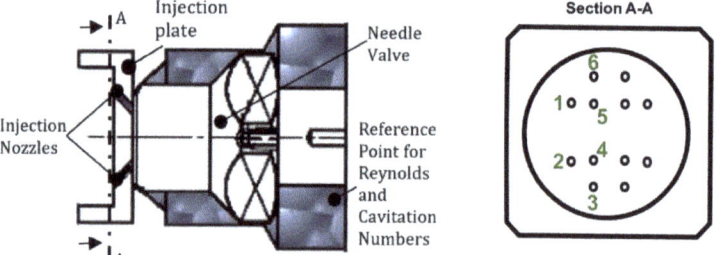

Figure 2. (**left**) Schematic of Type C injector model. Grey area indicates fluid flow. Flow is from right to left. (**right**) Section A-A shows the nozzle arrangement. Numbers in green identify nozzle numbers 1–6, some of which are referred to in the 'Results and Discussion' section.

It should be noted that Types A and B injectors have 8 nozzles, and Type C has 12. The locations of the nozzles are indicated in the respective injection plate in Figures 1 and 2. The differences in the geometry between Types A and B were minor, while for the case of Type C, the cylindrical sections were larger in diameter and, in combination with the needle valve geometry, led to more abrupt changes in the flow direction compared to the case of Type A model. In addition, in the Type C injector, the size of the holes is slightly smaller, and also, the distance between neighboring holes is smaller. The main difference between Type A and Type B is that, for the latter, the "neck" (see the encircled part in Figure 1) is longer. This was expected to induce differences in the hairpin type flow as the liquid enters the sections just upstream of the nozzles, which were validated with PIV, compared with the measurements shown in [25]. In terms of the needle valve, although the geometry is similar between Type A and Type B, the cylinder with the spherical valve end is shorter and has a larger diameter in the former. This allows shorter flow paths to the sections that include the above-mentioned part of the needle valve. The needle valve lift for all cases (Type A, B, and C) was set to 0.8 mm in the model, corresponding to the maximum needle valve lift of the prototype.

In all the model injectors, the flow is from right to left (refer to the grey area in Figures 1 and 2). The flow is contained in the annular passage between the needle valve and the needle valve seat until the exit nozzles. The hydraulic circuit used, which contained a refractive index matching fluid (31.65% v/v of 1,2,3,4–Tetrahydronaphthalene and 68.35% v/v of turpentine oil), is illustrated in Figure 3. All the tubing was made of stainless steel in order to withstand the corrosive mixture of the working fluid, which was connected to the model with a reinforced plastic tube. There were two pumps that circulated the working fluid. Pump 1 was of low capacity (1.5 m^3/h), and Pump 2 had a maximum flow rate of 8 m^3/h. Pump 1 was used for low Reynolds Number conditions and Pump 2 for high Reynolds Numbers, which were near the conditions of the initiation of cavitation. It should be noted that the temperature of the liquid was continuously monitored by a thermocouple. A cooler controlled the temperature of the liquid at 25 °C with an accuracy of 0.1 °C. This was performed because the fluid's refractive index was affected by temperature changes, and it only showed the desired refractive index (1.49, same as that of the acrylic plastic material of the model) at a temperature of 25 °C. Though not shown here and as explained in [12], downstream of the exit from the nozzles, the flow proceeded to a large liquid-filled plenum, from which it is redirected back to the liquid 'sump' tank: part of the hydraulic circuit. The outflow pattern was certainly affected by this setup, but it is not examined in this work, as only measurements from the internal flow field were acquired. Upstream of the injector body, the conditions of these measurements were set so that the internal flow could bear geometrical and dynamic similarity with the prototype injector.

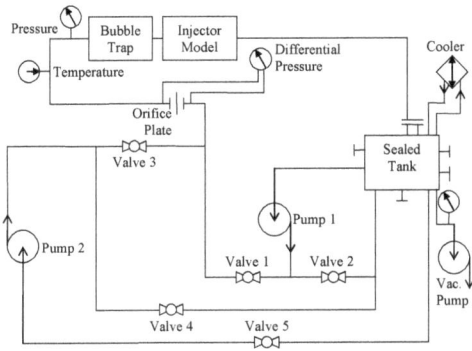

Figure 3. Hydraulic circuit of the refractive index matching rig (Aleiferis et al. [26]).

A bubble trap was used to capture gas bubbles, which were physically trapped in the model and could be present in the working fluid, either during the filling of the hydraulic circuit or when dissolved in the liquid and after cavitation was initiated for some flow conditions. This was a settling chamber with the inlet at the top and the outlet at the bottom so that any bubbles could flow to the top part of the chamber. A schematic of the bubble trap is given in Figure 4. It should also be noted that the flow downstream of the nozzle exit was liquid, and there was no chance to entrain air from outside the nozzle. Therefore, the reported observations are only due to local cavitation in the flow. A vacuum pump acted on the free surface of the liquid downstream of the model in the sealed tank so that the downstream static pressure of the liquid was controlled in order for the cavitation number to be matched between the real and large-scale models above the pressure limits (0.2–0.4 atm depending on the pump in operation). This was allowed by the Net Positive Suction Head (NPSH) of each pump. If we applied a lower pressure than the NPSH of the pump, we could induce cavitation upstream of the injector model, which was not desirable. More details about the operation of the hydraulic circuit can be found in [25,26]. To investigate the temporal development of cavitation in all three different geometries, we used a high-speed camera, "KODAK HS 4500", which had a frame rate of 4500 fps at a maximum resolution (256 pixels × 256 pixels), although velocities were not measured while acquiring high temporal resolution pictures because the available high resolution (2048 pixels × 2048 pixels) PIV system had a maximum acquisition frequency of 10 Hz.

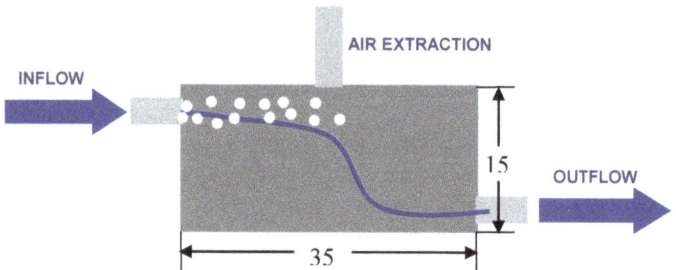

Figure 4. Schematic of the bubble trap (Aleiferis et al. [26]).

The optical configuration for the PIV measurements is shown in Figure 5. The system consisted of two double pulsed Nd: Yag lasers (New Wave Gemini PIV), a 12-bit CCD PIV camera (Kodak Megaplus ES 4.0), with an array of 2048 × 2048, and an image acquisition system (LaVision FlowMaster 2S, excluding the PIV software which was developed in–house) based on a dual processor computer (2 × Intel Pentium IV 2 GHz processors) with a programmable timing unit; this synchronized the lasers and the camera to obtain the PIV images. The fluorescent light emitted by the fluorescent 'seeding' particles used was transmitted through an optical window of the model injector and via a 45° mirror to the long-distance microscope (Davro Optical Systems DOS Model 77) camera system. Measurements were conducted just upstream of the nozzle at the plane intersected by the laser sheet, as shown in Figure 5 for the Type C model and for the Type B model. It was observed that the refractive index matching fluid slowly deteriorated on the surface quality of the injector models during the measurements. The flow fields of the Type B injector model were compared to the flow field after changing the needle valve with the Type A needle valve in [12], and it was found that they were similar.

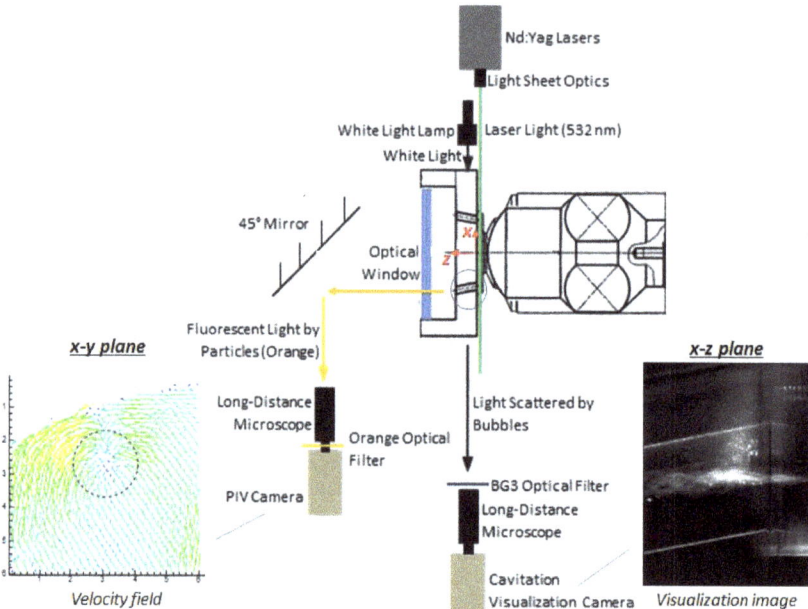

Figure 5. Optical arrangement for measurements of the flow field in the transverse (x-y) plane normalized to the notional axis of the injector upstream of a nozzle, and flow visualization within the nozzle (x-z plane). Sample results obtained from both techniques are demonstrated.

In order to obtain information for cavitation simultaneously with the flow field (after acquiring the high-speed cavitation images), a second camera (Cavitation visualization camera) with a long-distance microscope was necessary. This camera had a 12-bit conversion with a spatial resolution of 1378 pixels × 1040 pixels, and the long-distance microscope was similar to the one used for PIV measurements. Since a laser was used for the PIV flow velocity measurements, the cavitation images were saturated by the high-power scattered light and by bubbles crossing the laser sheet. For that reason, BG3 optical filters were used (which transmitted only blue light) with a white light lamp in order to provide wavelength separation between the cavitation visualization images and the PIV images, which used the fluorescent light emitted by the particles, and excited by the green light (532 nm) of the Nd: Yag laser. Although this arrangement worked satisfactorily, the lamp, which also emitted light at the fluorescent wavelength, induced some noise into the PIV images. This noise was minimized using neutral density filters optically and through the image processing software, where a median and the maximum filter routines were incorporated to de-noise the images [30]. The use of fluorescent particles was necessary (nominal diameter 1–20 μm, mean diameter 10 μm), which were covered with Rhodamine B dye, in order to separate the fluorescent emission and the elastically scattered light from the cavitation bubble at the laser wavelength, and hence, to distinguish between the gaseous and the liquid phase. The fluorescent light was transmitted through an optical filter, which was placed between the PIV camera and the long-distance microscope.

These results are linked to real injectors since appropriate dimensionless parameters are defined as relating to the flow conditions in the model and to the prototype. This could be achieved using Reynolds and Cavitation Numbers with, as suggested in [31], the Cavitation number being relevant as a dynamic similarity parameter only when the flow cavitated. The Reynolds number is defined as follows:

$$Re = \frac{\rho U_d d}{\mu} \quad (1)$$

where Re is the Reynolds number, ρ is the density of the working fluid, U_d is the bulk velocity at a reference section, and d is the characteristic length of the reference section, as illustrated in Figures 1 and 2 (which is taken as the diameter of the needle valve 'seat' at the reference section, namely 36 mm for Type A and Type B models, and 50 mm for Type C model) and μ is the dynamic viscosity of the working fluid.

The Cavitation number, formally derived, is as follows:

$$\sigma_v = \frac{P_d - P_v}{\frac{1}{2}\rho U_d^2} \quad (2)$$

where σ_v is the cavitation number, P_d is the static pressure at the reference section, P_v is the vapor pressure of the working fluid at 25 °C (466 Pa), ρ is the density of the working fluid, and U_d is the bulk velocity at the reference section. Note that other expressions for the cavitation number exist in the literature (as presented in [9]), which can replace the dynamic pressure with a pressure drop across the reference point: there are advantages to both definitions. In the present context, we prefer the formal definition above, which relates to changes in dynamic pressure. As seen in [12], the magnitudes of the Cavitation number in our flows were "large" in the sense that, on physical grounds, the Cavitation number compared the liquid static pressure to the dynamic pressure of the flow, and one expected the cavitation to arise when the value of σ_v was of order unity. This apparent discrepancy arose because of our choice of location for the reference section, which was remote from that of the location of cavitation, and, thus, the resultant values of σ_v were of the order of thousands. This unusual choice of location for the reference point was made because it was easier to measure the static pressure necessary for the calculation of the Cavitation number. Note that it was formally permissible to define the reference conditions and the location for the cavitation number at any convenient point in the model, provided that we conducted scaling with reference to the same location in the prototype.

The conditions for the visualization of cavitation with a high frame rate camera are shown in Table 1.

Table 1. Conditions for high frame rate visualization of the gasoline injector models.

	Type A	Type B	Type C
Reynolds Number	10,700	11,550	9200
Cavitation Number	2700	2300	6400

The conditions of the PIV measurements and the simultaneous (not at high frame rate) visualization of cavitation were the following: for the Type B model, the Reynolds Number was 11,450, and the Cavitation Number was 2650. For the Type C model, the Reynolds Number was 9200, and the Cavitation Number was 6400.

2.2. Proper Orthogonal Decomposition (POD)

Proper Orthogonal Decomposition, or POD, is a powerful method of data analysis. Based on the Karhunen–Loeve procedure of probability theory [32,33], POD aims at reducing the dimensionality of a dataset while retaining as much as possible the variations present in it [34,35]. The basic idea behind POD is to describe a given statistical ensemble with a minimum number of deterministic modes [36,37].

We considered an ensemble of instantaneous data $\Omega(t, x)$, with x and t as the spatial and temporal parameters, respectively. In the present work, Ω represents the two-dimensional velocity data from the PIV measurement of liquid velocity just upstream of the nozzles or the image intensity distribution in the shadowgraph images of the nozzles (see Figure 5). The mean velocity or mean intensity is subtracted from the instantaneous values so that the values of Ω represent fluctuations only. For the M number of flow realizations and N number of spatially located data points for each realization, POD decomposes $\Omega(t, x)$ into a sum of the product of spatial eigenvectors $\varphi_j(x)$ and temporal coefficients $a_{ij}(t)$; therefore,

$$\Omega(t, x) = \sum_{j=1}^{r} a_{ij}(t) \sqrt{\lambda_j} \varphi_j(x) \tag{3}$$

where i = 1 to M, j = 1 to N, λ_j represents the eigenvalue corresponding to each eigenvector $\varphi_j(x)$, and r is the rank of the matrix $[I]_{MN}$ so that $r = \min(M, N)$. Thus, the POD modes $\varphi_j(x)$ represent the average spatial features of the whole ensemble, while the corresponding coefficients $(\sqrt{\lambda_j} a_{1j}, \sqrt{\lambda_j} a_{2j} \ldots \ldots \sqrt{\lambda_j} a_{rj})$ signify their "weight" for the time instants i = 1, 2,..., M, respectively.

The eigenvalues λ_j are obtained by solving the eigenvalue equation, $R\varphi = \lambda \varphi$, Ra = λa, under the restriction that the norm of φ_j is 1, where R is the spatial cross-correlation matrix of size $N \times N$. However, when $M \ll N$, as in the present case, the calculation time could be dramatically reduced if the temporal cross-correlation matrix $[R_T]_{MM}$ is evaluated instead of $[R]_{NN}$; therefore:

$$R_T(t, t^*) = \frac{1}{N} \sum_{k=1}^{N} \Omega(t, x_k) \times \Omega(t^*, x_k) \tag{4}$$

This numerical procedure, as proposed by Sirovich [38], is popularly known as the "method of snapshots". The solution $R_T a = \lambda a$, $R_T \lambda = a\lambda$ leads to the orthonormal temporal coefficients $a_{ij}(t)$ corresponding to the eigenvalues λ_j. The symmetry and non-negative definiteness of R_T ensures $\lambda_j \geq 0$. The eigenvectors are obtained from the inverse relation $\varphi_j = \lambda_j^{-0.5} \sum_{i=1}^{M} a_{ij} I_i$. The eigenvalues are ordered as $\lambda_j > \lambda_{j+1}$, and the corresponding coefficients (a_{ij}) and modes (φ_j) are also arranged accordingly. Hence, the first mode φ_1 always represents the maximum spatial variations in the liquid velocity fluctuations upstream of the nozzle or the intensity fluctuations in the shadowgraph images of the nozzles. The significant advantage of POD is that due to its fast convergence property, the number of energetically significant modes is minimum. Hence, original intensity data can be reconstructed using only a few modes instead of considering all of them; therefore,

$$\Omega(t,x) = \sum_{j=1}^{r_{optimum}} a_{ij}(t)\sqrt{\lambda_j}\varphi_j(x), \; r_{optimum} < M \qquad (5)$$

In other words, only a few modes, $r_{optimum}$ (much less than the total number of modes, M), needed to be considered for the data analysis.

The present work used the method of the snapshot described above to obtain the POD modes. For the POD analysis of the PIV data, 1000 instantaneous two-dimensional liquid velocity vector fields (measured upstream of the nozzles) were considered, such that each sample contained a velocity measured at 42×42 grid points. In this case, the initial POD modes were considered to be synonymous with the dominant liquid flow structures upstream of the nozzles [39]. For the POD analysis of the shadowgraph images, 1000 instantaneous images were considered. Only certain sections of each shadowgraph image close to the nozzle inlet were considered for the POD calculation in order to optimize the computational time since cavitation was not observed too far downstream of the nozzle. In this case, the initial POD modes depicted the string or edge separation cavitation within the nozzle. The uncertainty in the amplitude of the spatial POD modes was found to be about 10%. However, the uncertainty for eigenvalues, which determined the significance of the modes, was about 1% or even less for all cases.

3. Results

3.1. Visualization of 'Bulk' Cavitation

Consecutively acquired cavitation visualization images of the Type A Injector Model are shown in Figure 6. Three of these images (for the instances t_0, $t_0 + 7/4500$ s and $t_0 + 14/4500$ s) are shown in [12]; however, here, the whole series acquired is shown in order for the reader to have a complete understanding of the cavitation evaluation. Note that the flow is from right to left. Two types of cavitation were identified in the flow. The first, which is commonly observed in the literature, is known as edge separation cavitation and is a result of flow separation due to the high associated streamline curvature that occurs at the edge of the nozzle inlet. The second is named 'bulk' cavitation and arises far from the walls and, thus, far from the streamline curvature, which is associated with salient points on a wall. The latter could be caused or affected by streamwise vortical structures that are present inside and just upstream of the injection nozzle; however, so far, there is no quantitative experimental flow field evidence for this. Both edge separation cavitation and 'bulk' cavitation were present in the flow. Edge separation cavitation is indicated by white dashed circles, and 'bulk' cavitation is indicated by white rectangles for the typical visualization images of each case. These were added to the images to assist the reader in identifying the regions that were occupied by 'bulk' cavitation or edge separation cavitation compared to the rest of the images. It can be noted that for the Type A model, there is a "scratch" at the surface of the material of the transparent model that is present in all the images of Figure 6, which is indicated by the red rectangle. Figure 6 shows that the edge separation cavitation was present at the bottom edge from t_0 to $t_0 + 3/4500$ s. From $t_0 + 4/4500$ s to $t_0 + 5/4500$ s, cavitation was visible only on the top corner of the nozzle inlet. The "top corner" cavitation could probably be related to the separation region, which is attached to the top corner of the nozzle, as shown in the PIV measurements of the Type A model presented in [25], while there is nothing comparable at the bottom edge. Therefore, the PIV results suggest that cavitation is unexpected at the bottom inlet corner.

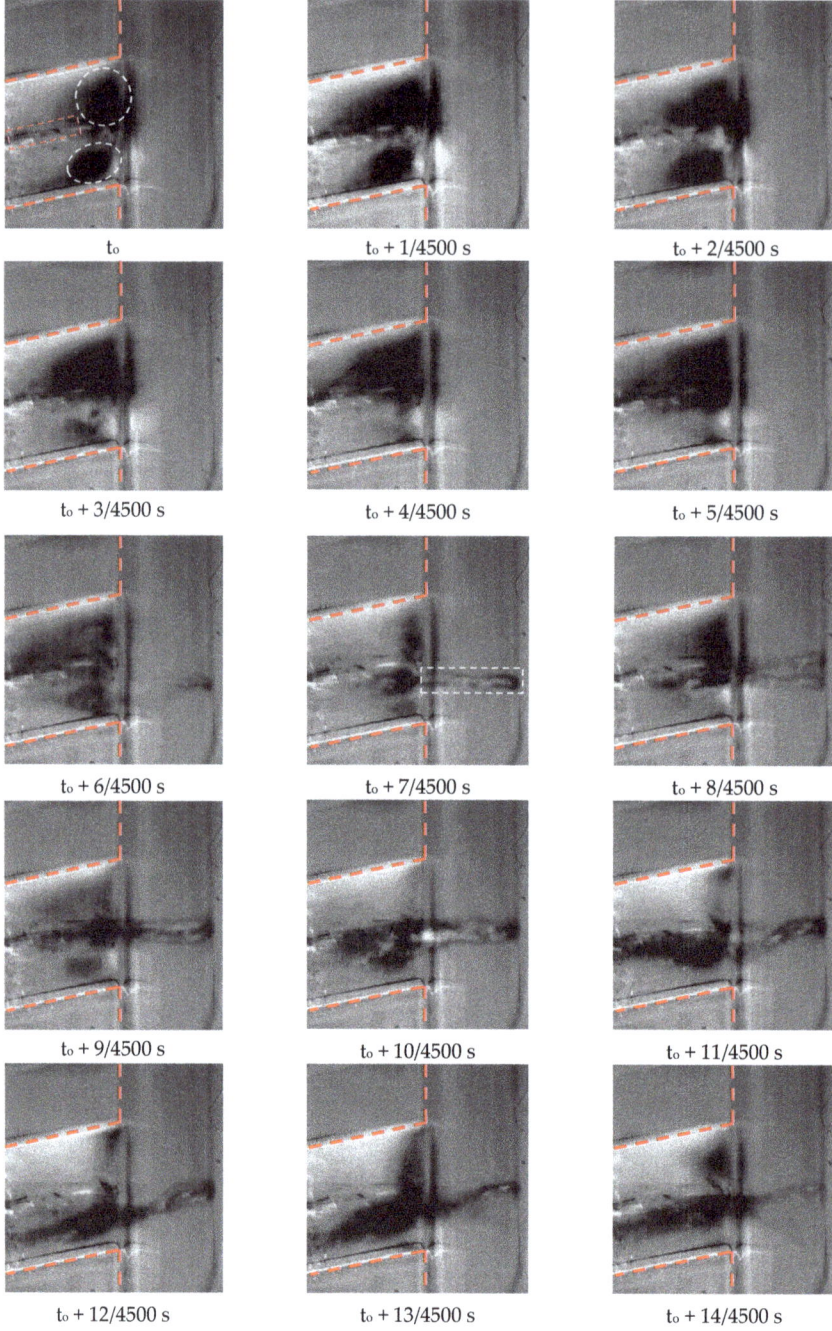

Figure 6. Cavitation visualization images of the Type A (Nozzle 4) model obtained with high speed camera. Flow is from right to left. White circles indicate edge separation cavitation, and white rectangles indicate string cavitation for typical images. Red rectangle indicates a "scratch" at the model's surface. Three of these images (for the instances t_o, $t_o + 7/4500$ s and $t_o + 14/4500$ s) are shown in [12]. Red dashed lines indicate nozzle boundaries.

At the instant of $t_0 + 6/4500$ s, the separation cavitation was weaker in comparison to the previous images, and a weak bubble string could be observed. From that moment on, we could see the development and existence of 'bulk' cavitation in this nozzle. At time $t_0 + 8/4500$ s, two strings of the bulk cavitation were visible, but it is unlikely that the second could be cavitation in the nozzle right behind the one of interest because of the small depth of field in the optics. Up to the image at time $t_0 + 8/4500$ s, the 'bulk' cavitation did not seem to extend inside the nozzle. After that time and until $t_0 + 11/4500$ s, the 'bulk' cavitation was also present inside the nozzle, and from that moment until time $t_0 + 13/4500$ s, the edge separation cavitation and 'bulk' cavitation coexisted in the nozzle. In the last image of this series, the two strings seemed to separate and become very thin, so after a few instants, 'bulk' cavitation was not present. The presence of the string at first only upstream of the nozzle inlet and then inside the nozzle led us to conclude that it started upstream of the nozzle, which suggested that it was caused by the streamwise vortical structures present at that region which was also the motivation for the PIV measurements appearing normal to the notional axis of the injector, as presented in the next section.

The cavitation visualization images of the Type B model (Figure 7) show that both edge separation and 'bulk' cavitation were present again. From the image acquired at time t_0 until the image at time $t_0 + 3/4500$ s, only edge separation cavitation could be observed, which was located both at the top and bottom corners of the nozzle inlet. In the picture, at time $t_0 + 4/4500$ s, a weak string appeared, and from that moment on until time $t_0 + 8/4500$ s, a clear string was present in the examined nozzle. From time $t_0 + 9/4500$ s, the image of the string became very weak, and then only edge separation cavitation was present at both inlet corners. Therefore, edge separation cavitation was present during the time that 'bulk' cavitation occurred.

It is noted that the Type B model had a larger nozzle plenum just upstream of the nozzle in comparison to the Type A model, as can be concluded by the geometry of the injector models presented in the earlier text. Although from purely geometrical considerations, $r_{streamline}$, where $r_{streamline}$ is the radius of the curvature for the streamlines that enter the nozzle, might be different between the two models and it was hard to draw any conclusions since the static pressure field was related to $\rho U^2/r_{streamline}$. The presence of 'bulk' cavitation appeared to affect the existence of edge separation cavitation, at least at the initial development stages of the string, since at the initial development stages of the string for the Type A model, edge separation cavitation was not present. When a string appeared, the liquid that entered the nozzle met an abrupt turn in the flow direction near the inlet corner of the Type A model, restricted edge separation, and, as a consequence, the edge separation cavitation was not present. During these stages, it might be that the liquid flow at the region where edge separation usually happens was settling without forming a recirculation, which could reduce the local pressure below the boiling point (466 Pa as explained in [40]) since the flow around that section was occupied by the string. In the case of the Type B model, the streamline curvature as the liquid entered the nozzle might be smaller than Type A; therefore, smaller recirculation zones were formed at the inlet edges, which were not disrupted by the presence of the strings. This could be the reason why, for the type B model, 'bulk' cavitation and edge separation cavitation coexisted at all times. This might be a useful conclusion for the design of the injector. Although we refer to gasoline injectors here, this could also be applicable to diesel and other types of injectors.

The cavitation visualization images of the Type C Injector are illustrated in Figure 8. Again, three of these images (for the instances t_0, $t_0 + 7/4500$ s and $t_0 + 14/4500$ s) are shown in [12], and for the same reason, as Type A is also presented here. In this case, three main conclusions can be drawn. First, in all the images, there was a continuous presence of 'bulk' cavitation, with the string "precessing", which at least suggested that the lifetime of 'bulk' cavitation was longer in this case than in the other types. Secondly, the diameter of the string in these images was significantly larger than for Type A and Type B models; therefore, it could be said that, in this case, whatever the flow structure that gave rise to cavitation and was a result of the geometry of the injector, this allowed the longer presence of cavitation. Thirdly, bubbles seemed to cover the whole nozzle region since this was all shadowed, although it was not possible to decide from the images if this was also a result of edge separation cavitation.

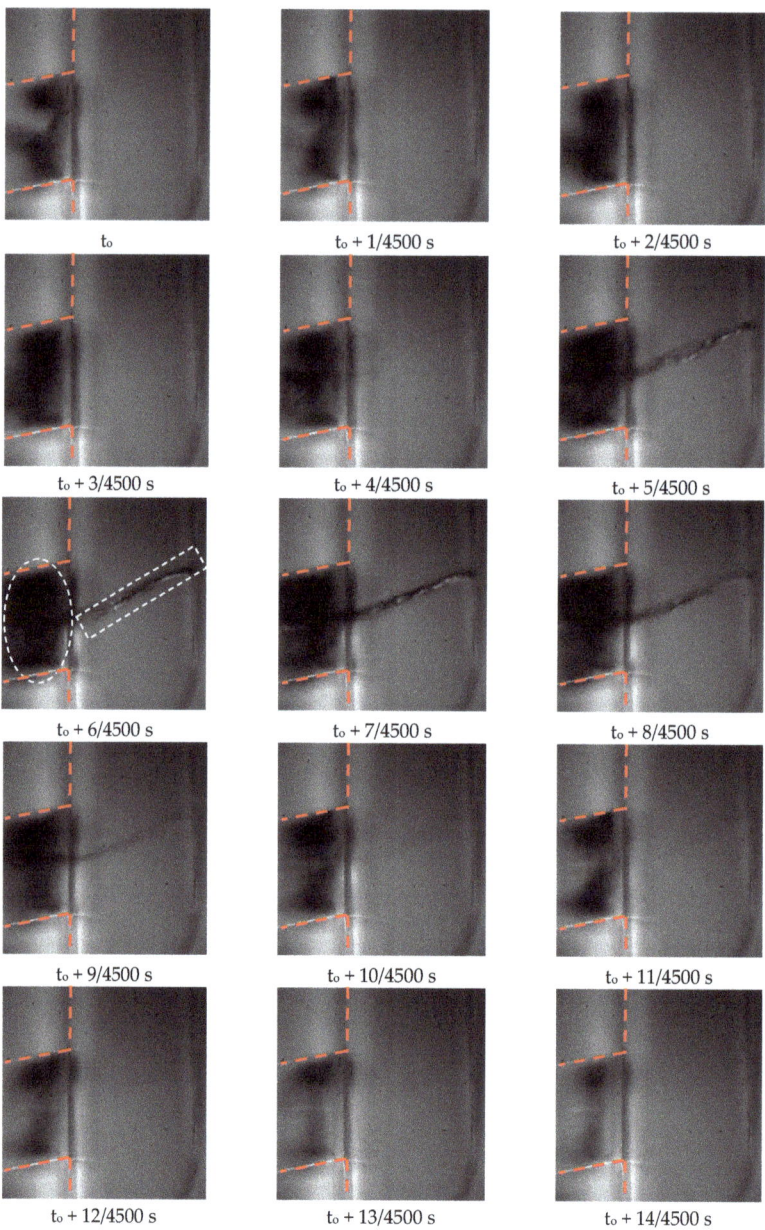

Figure 7. Cavitation visualization images of Type B (Nozzle 4) model obtained with high speed camera. White circles indicate edge separation cavitation and white rectangles represent string cavitation for typical images. Flow is from right to left. Red dashed lines indicate nozzle boundaries.

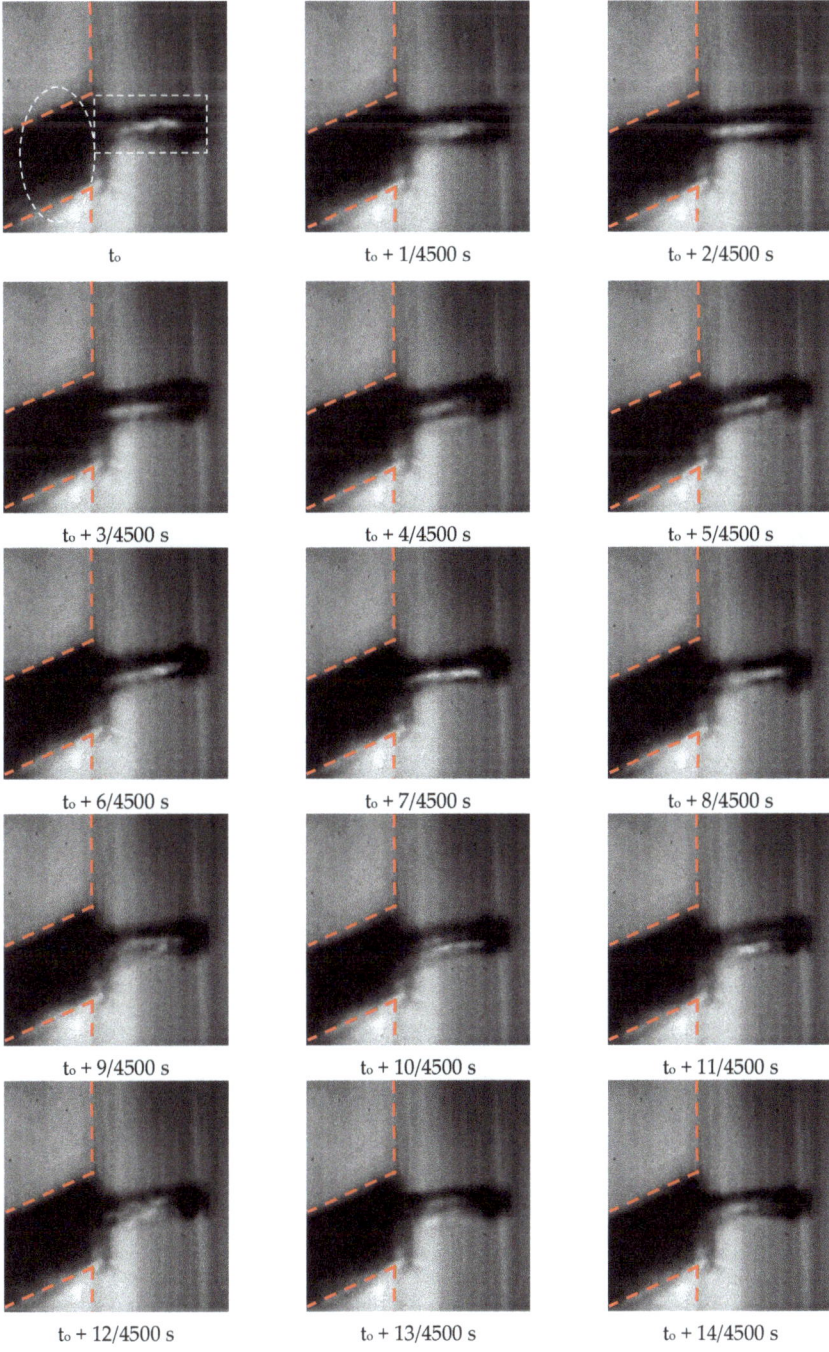

Figure 8. Cavitation visualization images of the Type C model (Nozzle 5) obtained with high-speed camera. Flow is from right to left. White circles indicate edge separation cavitation, and white rectangles indicate string cavitation for typical images. Three of these images (for the instances t_o, $t_o + 7/4500$ s and $t_o + 14/4500$ s) are shown in [12]. Red dashed lines indicate nozzle boundaries.

3.2. Mean Flow Field and Cavitation Visualization

Velocity measurements at the planes just upstream of the nozzles were conducted, and the results are illustrated for the cases where 'bulk' cavitation was present. More specifically, for the Type B injector model, the results are illustrated for nozzle 1 and nozzle 4 (present at the edge and the interior locations, respectively, as shown in Figure 1). Simultaneously with the fluorescent PIV liquid velocity measurements, the cavitation was visualized in order to see if 'bulk' cavitation was present or not. The PIV results were averaged over 1000 images, and the cavitation visualization images were typical shadowgraph images. For nozzle 1 (Figure 9), only corner-separation cavitation was present, and the average, at least for the flow field just upstream of the nozzle, had only a weak clockwise swirling motion. In the case of nozzle 4 (Figure 10), where 'bulk' cavitation was present, the mean flow field just upstream had no mean swirl but was similar to a "potential flow sink with cross flow" (specifically, the internal flow *inside* the half body solution).

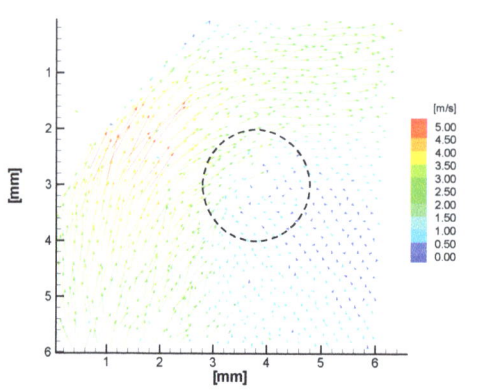

Figure 9. Left hand side (LHS): time-averaged liquid velocity measurements in the "plane of measurement", just upstream of the entrance to nozzle 1 (please refer to Figure 5) for Type B model. **Right-hand** side (RHS): visualization of cavitation (image plane is parallel to that of Figure 1). The dashed circle refers to the nozzle location.

These results are surprising for two reasons. First, we would reasonably expect 'bulk' cavitation to be associated with the swirling flow centered on the nozzle; however, this did not seem to be the case. Secondly, we might reasonably expect the swirling flow to inhibit edge cavitation; however, this was also not the case. Quite why the initiation of 'bulk' cavitation was promoted by this sink-like flow is not clear. One possible explanation is that the flow measurements were time-averaged values and not representative of the instantaneous flow, which could give rise to cavitation at specific times in the flow.

Flow velocity measurements were also conducted for the Type C model (see Figure 2). Referring to the same figure, we considered nozzle 1, nozzle 6 (which are located at the edge), and nozzle 5 (interior nozzle) to demonstrate different cavitation types. Figures 11–13 presents the mean velocity field and cavitation visualization images for different nozzles of the Type C injector. It was observed that for nozzle 5 (Figure 12), which showed 'bulk' cavitation, an approximately sink-like flow was present. For nozzles 1 and 6 (Figures 11 and 13, respectively), edge separation cavitation was present, and the flow formed two vortices just upstream of the nozzle inlet, which was different for the sink-like flow that was observed just upstream of nozzle 5. It is very important for the designer that, in both injector models, the flow field just upstream of nozzles shows a 'bulk' cavitation that is qualitatively similar since the designer can induce geometry features that promote or inhibit this kind of flow.

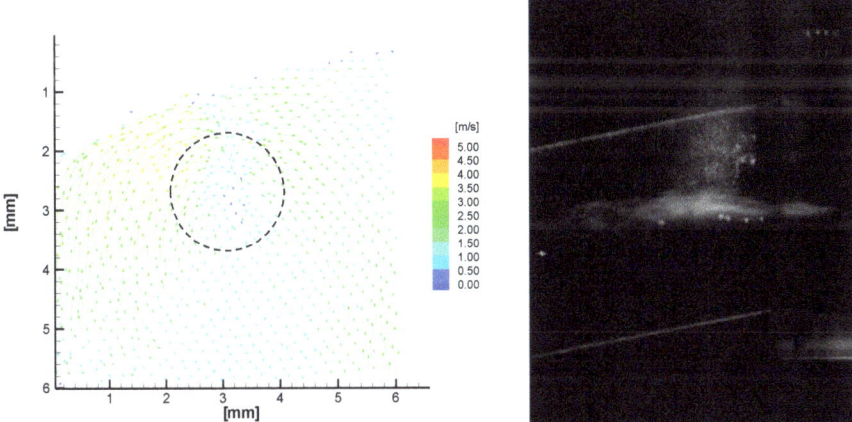

Figure 10. **Left hand** side (LHS): time-averaged liquid velocity measurements in the "plane of measurement", just upstream of the entrance to nozzle 4 (please refer to Figure 5) for Type B model. **Right-hand** side (RHS): visualization of cavitation (image plane is parallel to that of Figure 1). The dashed circle refers to the nozzle location.

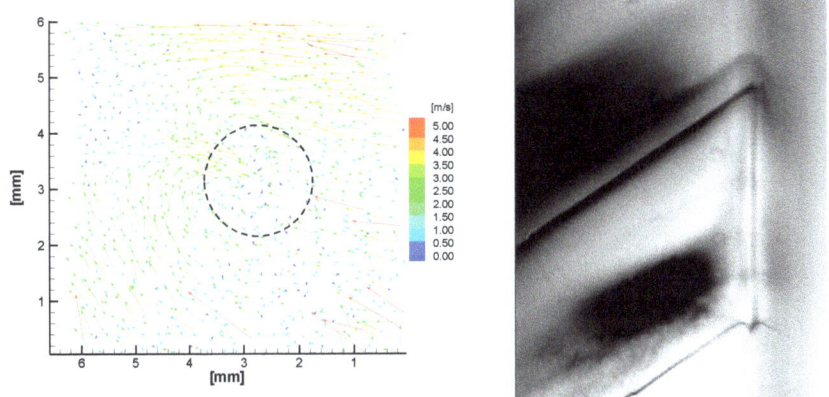

Figure 11. **Left hand** side (LHS): time-averaged liquid velocity measurements in the "plane of measurement", as defined in Figure 2, just upstream of the entrance to nozzle 1 (please refer to Figure 5) for Type C model. **Right-hand** side: visualization of cavitation (image plane is parallel to that of Figure 2). The dashed circle refers to the nozzle location.

Obtaining PIV measurements simultaneously with cavitation visualization images at nozzles 1, 5, and 6 (refer to Figure 5 as mentioned above) was conducted because the remaining nozzles were positioned symmetrically with respect to the examined ones and were expected to show the same results. This is the reason why the flows in the three nozzles were examined for the Type C model and in two nozzles for Type B (with Type A needle valve).

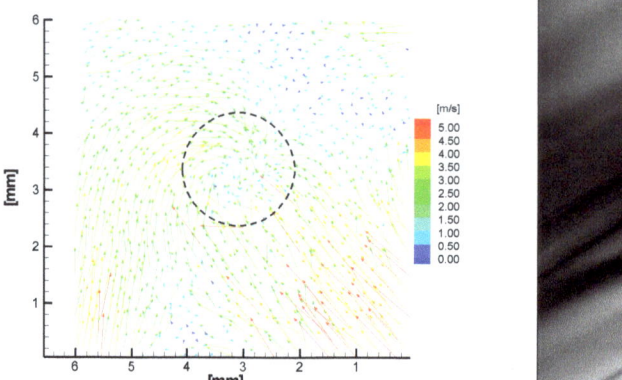

Figure 12. **Left hand** side (LHS): time-averaged liquid velocity measurements in the "plane of measurement", as defined in Figure 2, just upstream of the entrance to the nozzle 5 (please refer to Figure 5) for Type C model. **Right-hand** side: visualization of cavitation (image plane is parallel to that of Figure 2). The dashed circle refers to the nozzle location.

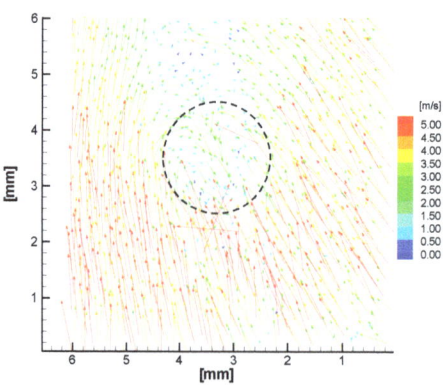
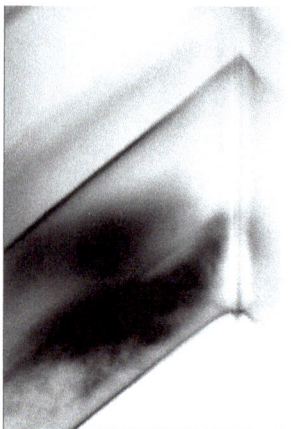

Figure 13. **Left hand** side (LHS): time-averaged liquid velocity measurements in the "plane of measurement", as defined in Figure 2, just upstream of the entrance to the nozzle 6 (please refer to Figure 5) for Type C model. **Right-hand** side: visualization of cavitation (image plane is parallel to that of Figure 2). The dashed circle refers to the nozzle location.

3.3. Proper Orthogonal Decomposition (POD) of Liquid Velocity Field in Cavitating Flow Conditions

POD was applied to the instantaneous liquid velocity fields just upstream of the nozzles to captivate the flow conditions, as shown in the previous section. The sum of all the eigenvalues represents the total turbulent kinetic energy of the flow since the decomposition occurred over the fluctuations of the liquid velocity from the mean value. The distribution of the eigenvalues for the nozzle Type B injector model with respect to the mode number is shown in Figure 14. These decreased rapidly after about the first 10 initial modes. Therefore, the eigenvectors corresponding to the first few eigenvalues were expected to correspond to the dominant turbulent structures of the liquid flow.

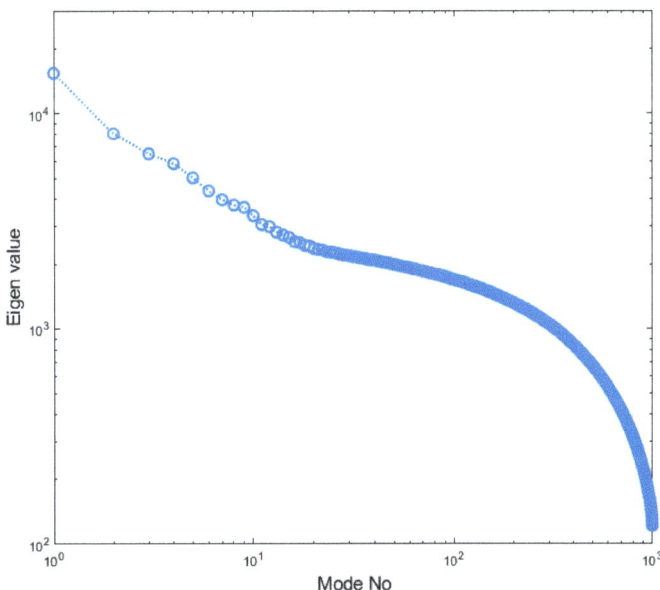

Figure 14. Eigenvalue spectrum for the fluctuating velocity components u and v for the Type B injector model.

The first two POD modes are presented in Figure 15 and in Figure 16. We can observe that the very first mode (Figure 15) depicted the presence of a vortical structure. The first mode had the maximum average correlation with all the instantaneous velocity fields, and hence, it represents the most common flow structure. Thus, in the present case, it could be correlated with the presence of 'bulk' cavitation since the flow conditions were selected so that 'bulk' cavitation was present. It is worth noting that this vortical structure was not present in the mean flow field results. The second mode (Figure 16) was qualitatively different and not obviously related to a flow structure that could promote 'bulk' cavitation.

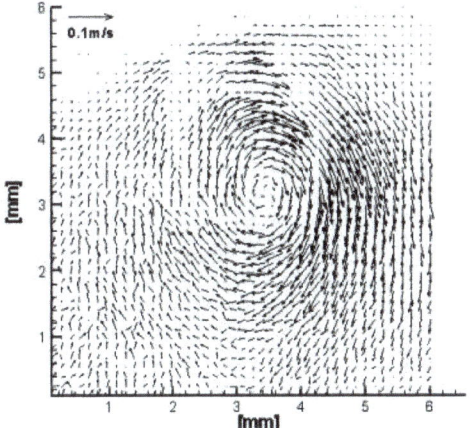

Figure 15. POD Mode 1 of the liquid velocity field for the Type B injector model.

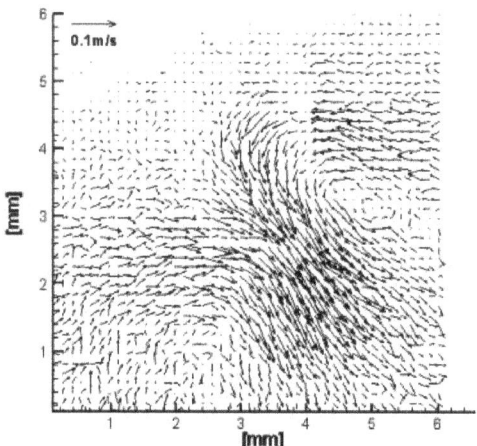

Figure 16. POD Mode 2 of the liquid velocity field for the Type B injector model.

In order to estimate if the pressure drop in relation to this vortical structure could lead to 'bulk' cavitation, the instantaneous velocity field of Type B was reconstructed by considering the first mode only. Figure 17 below shows a typical radial profile of absolute velocity, where point '0' corresponds to the position of the nozzle center. The POD mode 1 signifies the existence of a free vortex or, strictly speaking, a Rankine vortex since the fluid possesses finite viscosity. Therefore, away from the nozzle axis, the tangential velocity first increased and then decreased close to the outer wall. At the radius of the order of the nozzle radius (2 mm), the tangential fluid velocity was about 1 m/s. Since the angular momentum of the vortical structure had to be conserved, the tangential velocity was inversely proportional to the radius or $v \times r = const$. Hence, when the vortical structure entered the nozzle, the fluid velocity increased close to its axis. Therefore, at $r = R_{nozzle}/10$ and $r = R_{nozzle}/20$, the fluid velocity accelerated to about 10 m/s and 20 m/s, respectively. Assuming that far away from the axis of the vortex, the pressure was atmospheric, and the tangential velocity was zero, $\Delta P_{critical} = P_{atm} - P_{vap.pressure}$ represent the minimum pressure drop necessary to initiate cavitation. The actual pressure drop was $\Delta P_{actual} = \rho v^2/2$, and therefore, the ratio $\Delta P_{critical}/\Delta P_{actual}$ was found to be equal to about **2** and **0.5** for $r = R_{nozzle}/10$ and $r = R_{nozzle}/20$. This means 'bulk' cavitation must occur close to the nozzle axis. Since both pressure drop and vorticity proportionally increase with the inverse of the square from the distance from the nozzle, a small initial rotation of the fluid upstream of the nozzle is sufficient to create a concentrated vortex inside the nozzle, which can cavitate the liquid. It should be noted that this is the first time that proof has been provided that the local flow characteristics can cause cavitation. In previous studies, the emphasis has been on the gas entrained inside the nozzle from the environment outside the nozzle or by gas phase components that remain in the injector sac volume by previous injection events. Due to the experimental arrangement of the current study, the previously proposed mechanisms of explaining cavitation in the literature, according to the previous sentence, cannot occur. In this way, only specific instantaneous flow structures can induce cavitation.

The eigenvalue spectrum for the corresponding nozzle of the Type C injector model is shown in Figure 18. The trend of the eigenvalue spectrum for the fluctuating liquid velocity components was similar to the corresponding nozzle of the Type B model, with a sharp decrease in the initial eigenvalues. Again, this means that the dominant liquid flow structures are represented by the first few modes. The first mode for this case is illustrated in Figure 19, which again shows a vortical structure, which might be responsible for the formation of 'bulk' cavitation. Higher modes are associated with flow structures that have smaller length scales.

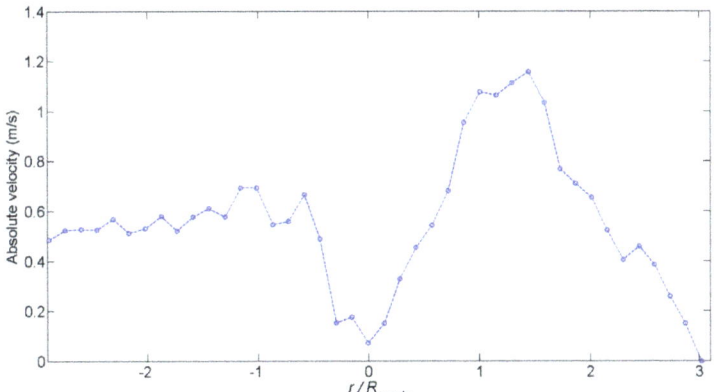

Figure 17. Typical reconstructed radial profile of the absolute velocity of the Type B injector model using POD mode 1 at the plane of the measurement just upstream of Nozzle 4 (see Figure 1). Point '0' corresponds to the position of the nozzle axis.

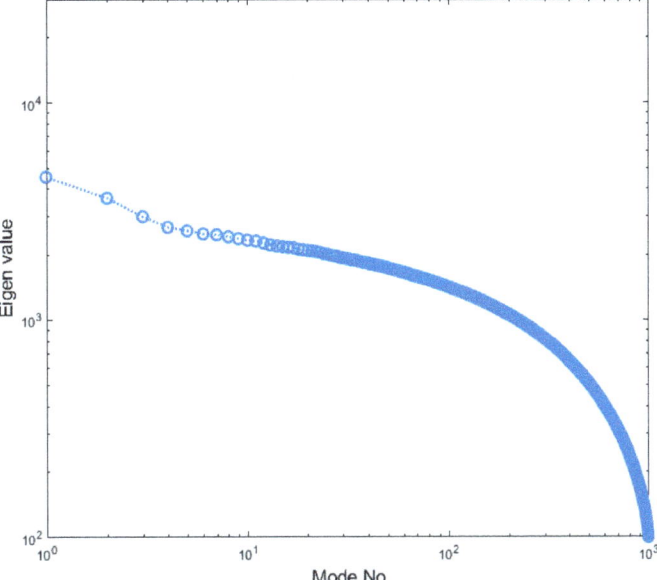

Figure 18. Eigenvalue spectrum for the fluctuating liquid velocity components u and v for the Type C injector model.

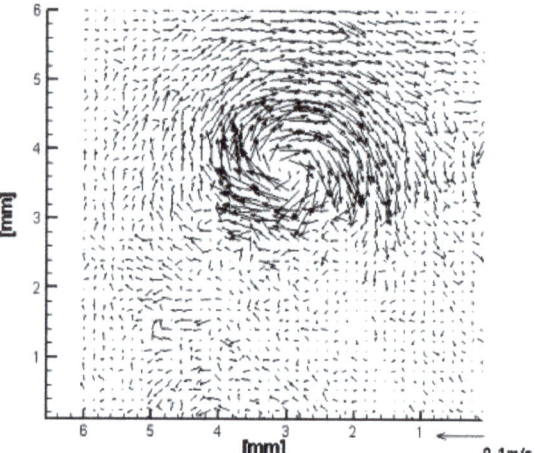

Figure 19. POD Mode 1 of the liquid velocity field for the Type C injector model.

3.4. POD of Shadowgraphic Cavitation Images of Type B Injector Model

POD was applied to the intensity of the shadow graphic cavitation images (those acquired simultaneously with PIV, so 1000 images for each nozzle) for nozzle 1 and nozzle 4 (see Figure 1) of the Type B injector model. This technique was not applied to the Type C model because the multiple nozzles were densely located, and there was noise generated on the shadow graphic images by the presence of cavitation in the nozzles located behind nozzles 1, 5, and 6 (Figure 2) for which PIV data have been acquired and examined. Figure 20 shows the first two modes of POD's application to the shadow graphic cavitation images for the Type B injector model.

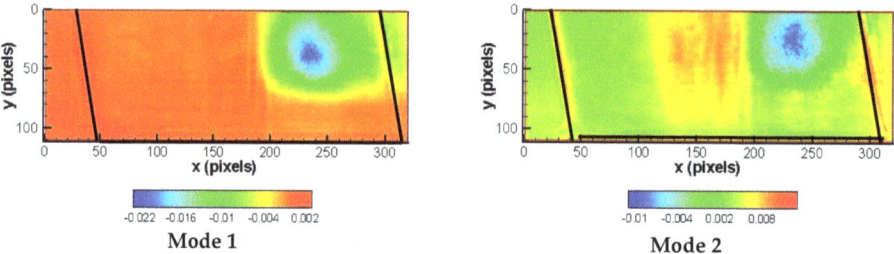

Mode 1   **Mode 2**

Figure 20. POD modes 1 and 2 for the shadow graphic cavitation images of Nozzle 4 (see Figure 1) of the Type B injector model. Flow is upwards (different orientation from that of Figure 1). The color of the contour plots refers to the fluctuations in pixel intensity values for the shadow graphic images. Nozzle borders are marked with black lines.

It should be noted that the liquid flow in the images of Figure 20 has an upward direction. The original shadow graphic images were contaminated due to unwanted reflected light. Therefore, modes 1 and 2 show a bright spot (blue color in the contour plots of Figure 20) near the right edge of the nozzle, which does not correspond to edge separation cavitation, but the intensity of reflected light. To avoid these effects, POD was applied to the original images after cropping the right side, and the first four POD modes are shown in Figure 21.

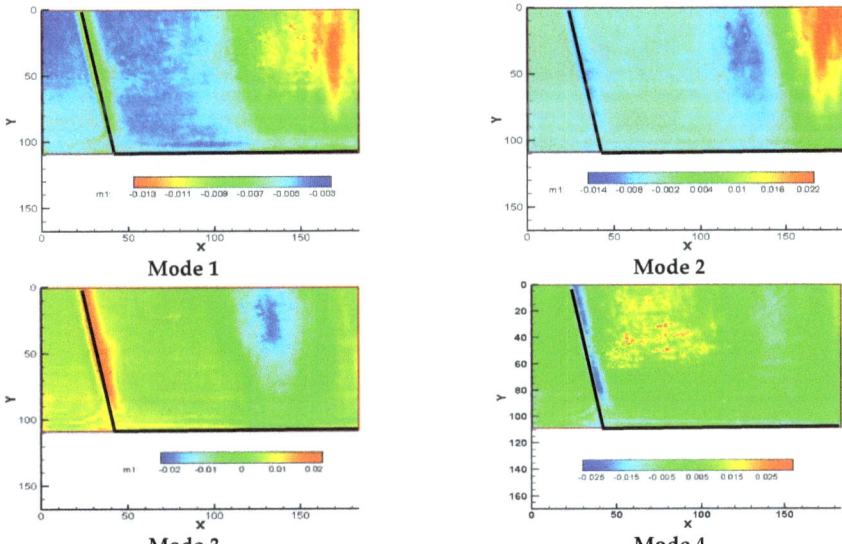

Figure 21. POD modes 1, 2, 3 and 4 of the shadow graphic cavitation images for Nozzle 4 (see Figure 1) of the Type B injector model after cropping the right-hand side of the original shadow graphic images. Flow is upwards (different orientation than that of Figure 1). The color of the contour plots refers to the fluctuations in pixel intensity values for the shadow graphic images. Nozzle borders are marked with black lines.

The first three modes of Figure 21 clearly depict the presence of 'bulk' cavitation. It should be noted that the right-hand side of Figure 21 corresponds nearly to the axis of the nozzle flow, as the reader can notice by comparing the coordinates between the images of Figures 20 and 21. Edge separation cavitation appeared for mode 4. In addition, the first four POD modes of the shadowgraphs in nozzle 1 (see Figure 1) are the same as the injector model presented in Figure 22.

Figure 22 shows that edge-separation cavitation was present for all the modes shown. If we examine the eigenvalue contribution of each mode (Figure 23), we can observe that for nozzle 4 (see Figure 1), the contribution of the first three modes where 'bulk' cavitation was observed was approximately 45% of the total value, while, for nozzle 1 (see Figure 1), the contribution of the first four modes, where we only had edge separation cavitation, was 80% of the total value. These percentages represent the probabilities of having a 'bulk' or edge separation cavitation in the flow of the examined nozzles. Therefore, in nozzle 4 (see Figure 1) of the Type B injector model, the probability of having 'bulk' cavitation was about 45%, while, in nozzle 1 of the same model, the probability of having edge separation cavitation was about 80%. So, the probability of having 'bulk' cavitation in nozzle 1 was low, which could be verified by observing the shadow graphic cavitation images of nozzle 1, 'bulk' cavitation was not present at all.

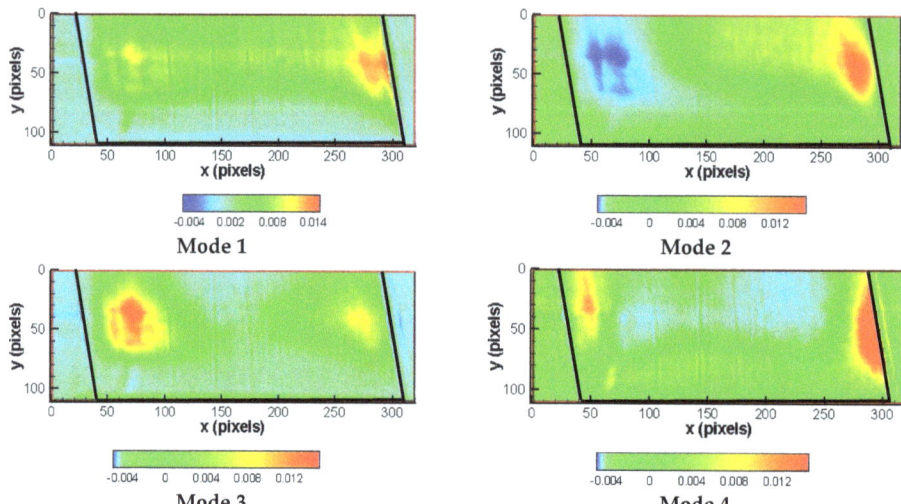

Figure 22. POD modes 1, 2, 3 and 4 of the shadow graphic cavitation images of Nozzle 1 (see Figure 1) of the Type B injector model. Flow is upwards (different orientation than that of Figure 1). The color of the contour plots refers to the fluctuations in pixel intensity values of the shadow graphic images. Nozzle borders are marked with black lines.

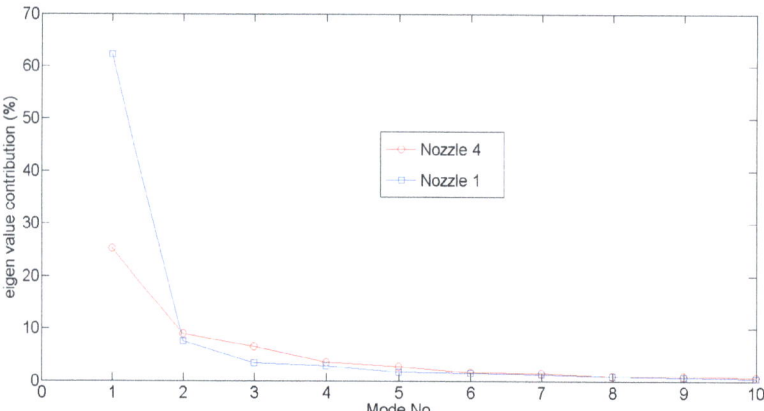

Figure 23. Eigenvalue contribution of each of the first 10 modes of the shadow graphic cavitation images of Nozzles 1 and 4 in Figure 1 of the Type B injector model.

4. Conclusions

In this work, 'bulk' cavitation was studied in three gasoline multi-hole injectors. Two-dimensional micron resolution Particle Imaging Velocimetry was employed to measure the internal flow field of the 10:1 super-scale transparent models of multi-hole injectors just upstream of the entrance to the holes of the injector plates in the vicinity of an operating regime just after the onset of cavitation. Our motivation was to understand the physics behind the formation of bulk cavitation and its correlation with the injector flow field. 'Bulk' cavitation was found to be present in the specific nozzles of three geometrically different injector models by means of fast camera visualization, where the time evolution of cavitation was also recorded. For Type A and Type B injector models, 'bulk' cavitation was not always present, while for the case of the Type C injector model, 'bulk' cavitation was present for all the images of cavitation visualization, which indicated that the residence time of 'bulk' cavitation in this type of injector was longer compared to the other injector models.

The liquid flow field at the nozzles of the two injector models (Type B and Type C) was quantified, and it was found that the mean liquid flow velocity just upstream of the exit holes resembled, as expected, the internal flow inside the half body corresponding to the classical potential flow solution for a sink with cross flow. We expected, a priori, 'bulk' cavitation to be associated with the existence of the swirling flow centered on the nozzle axis at a *given instant*. However, we found no such swirling flow structure in the *mean* flow field results. We thus applied Proper Orthogonal Decomposition (POD) to the instantaneous velocity data in order to identify the dominant liquid flow structures, which could be related to cavitation. It was found that "mode 1" eigenvalues indeed corresponded to swirling flow structures and were dominant for the cases when 'bulk' cavitation was present for both injector models where the flow was quantified. This might be related to the origin of 'bulk' cavitation. It is the first time that quantitative flow field experimental evidence has been presented, identifying that the above local flow structures could be related to local 'bulk' cavitation.

Finally, we applied POD to the shadow graphic cavitation images of the Type B injector model, and the eigenvalue contribution of each mode depicting 'bulk' cavitation was found to be representative of the probability of having this kind of cavitation.

In summary, this paper has demonstrated for the first time in a quantitative way the importance of instantaneous liquid flow structures on the initiation of different types of cavitation in injector nozzles. It highlights the importance of describing the instantaneous flow structures in computations of such flows in order to predict the different types of cavitation that can occur and the associated probability of their appearance and time-dependent behavior.

Author Contributions: Conceptualization, Y.H., A.M.K.P.T. and A.A.; methodology, D.K. and S.S.; software, D.K. and S.S.; validation, D.K. and S.S.; formal analysis, D.K. and S.S.; investigation, D.K. and S.S.; resources, Y.H., A.M.K.P.T. and A.A.; data curation, D.K. and S.S.; writing—original draft preparation, D.K. and S.S.; writing—review and editing, Y.H. and A.T; visualization, D.K.; supervision, Y.H. and A.T; project administration, Y.H. and A.T; funding acquisition, Y.H., A.T. and A.A. All authors have read and agreed to the published version of the manuscript.

Funding: This research was funded by Keihin Corporation, Japan.

Data Availability Statement: The data presented in this study are available on request from the corresponding author.

Acknowledgments: We acknowledge Keihin Corporation Japan for providing financial support and the optical injector models.

Conflicts of Interest: The authors declare no conflict of interest.

References

1. Payri, F.; Bermudez, V.; Payri, R.; Salvador, F.J. The influence of cavitation on the internal flow and the spray characteristics in diesel injection nozzles. *Fuel* **2004**, *83*, 419. [CrossRef]
2. Ulrich, H.; Lehnert, B.; Guénot, D.; Svendsen, K.; Lundh, O.; Wensing, M.; Berrocal, E.; Zigan, L. Effects of liquid properties on atomization and spray characteristics studied by planar two-photon fluorescence. *Phys. Fluids* **2022**, *34*, 083305. [CrossRef]
3. Oda, T.; Hiratsuka, M.; Goda, Y.; Kanaike, S.; Ohsawa, K. *Experimental and Numerical Investigation about Internal Cavitating Flow and Primary Atomization of a Large-Scaled VCO Diesel Injector with Eccentric Needle*; ILASS-Europe: Brno, Czech Republic, 2010.
4. Watanabe, H.; Nishikori, M.; Hayashi, T.; Suzuki, M.; Kakehashi, N.; Ikemoto, M. Visualization analysis of relationship between vortex flow and cavitation behaviour in diesel nozzle. *Int. J. Engine Res.* **2015**, *16*, 5. [CrossRef]
5. Zhong, W.; He, Z.; Wang, Q.; Shao, Z.; Tao, X. Experimental study of flow regime characteristics in diesel multi-hole nozzles with different structures and enlarged scales. *Int. Commun. Heat Mass Transf.* **2014**, *59*, 1. [CrossRef]
6. McGinn, P.; Tretola, G.; Vogiatzaki, K. Unified modeling of cavitating sprays using a three-component volume of fluid method accounting for phase change and phase miscibility. *Phys. Fluids* **2022**, *34*, 082108. [CrossRef]

7. Kim, J.H.; Nishida, K.; Yoshizaki, T.; Hiroyasu, H.H. Characterization of Flows in the Sac Chamber and the Discharge Hole of a D.I. Diesel Injection Nozzle by Using a Transparent Model Nozzle. *SAE Tech. Pap.* **1997**, *972942*, 11–23.
8. Arcoumanis, C.; Flora, H.; Gavaises, M.; Badami, M.M. *Cavitation in Real–Size Multi–Hole Diesel Injector Nozzles*; SAE International: Diego County, CA, USA, 2000; Volume 109, pp. 1485–1500.
9. Arcoumanis, C.; Flora, H.; Gavaises, M.; Kampanis, N.; Horrocks, R. *Investigation of Cavitation in a Vertical Multi-Hole Injector*; SAE International: Diego County, CA, USA, 1999; Volume 108, pp. 661–678.
10. Gilles-Birth, I.; Bernhardt, S.; Spicher, U.; Rechs, M. A Study of the In-Nozzle Flow Characteristics of Valve Covered Orifice Nozzles for Gasoline Direct Injection. *SAE Tech. Pap.* **2005**, *1*, 3684.
11. Nouri, J.M.; Mitroglou, N.; Yan, Y.; Arcoumanis, C. Internal Flow and Cavitation in a Multi-hole Injector for Gasoline Direct-Injection Engines. *SAE Tech. Pap.* **2007**, *1*, 1405.
12. Kolokotronis, D.; Hardalupas, Y.; Taylor, A.M.K.P.; Aleiferis, P.G.; Arioka, A.; Saito, M. Experimental Investigation of Cavitation in Gasoline Injectors. *SAE Tech. Pap.* **2010**, *1*, 1500.
13. Reid, B.A.; Hargrave, G.K.; Garner, C.P.; Wigley, G. An investigation of string cavitation in a true-scale fuel injector flow geometry at high pressure. *Phys. Fluids* **2010**, *22*, 031703. [CrossRef]
14. Mitroglou, N.; McLorn, M.; Gavaises, M.; Soteriou, C.; Winterbourne, M. Instantaneous and ensemble average cavitation structures in Diesel micro-channel flow orifices. *Fuel* **2014**, *116*, 736. [CrossRef]
15. Reid, B.A.; Gavaises, M.; Mitroglou, N.; Hargrave, G.K.; Garner, C.P.; Long, E.J.; McDavid, R.M. On the formation of string cavitation inside fuel injectors. *Exp. Fluids* **2014**, *55*, 1662. [CrossRef]
16. Giannadakis, E.; Gavaises, M.; Arcoumanis, C. Modelling of cavitation in diesel injector nozzles. *J. Fluid Mech.* **2008**, *616*, 153. [CrossRef]
17. Gavaises, A.M.; Arcoumanis, C.C. Vortex flow and cavitation in diesel injector nozzles. *J. Fluid Mech.* **2008**, *610*, 195.
18. Gavaises, M.; Andriotis, A.; Papoulias, D.; Mitroglou, N.; Theodorakakos, A. Characterization of string cavitation in large-scale Diesel nozzles with tapered holes. *Phys. Fluids* **2009**, *21*, 52. [CrossRef]
19. Salvador, F.J.; Romero, J.V.; Rosello, M.D.; Martinez-Lopez, J. Validation of a code for modelling cavitation phenomena in Diesel injector nozzles. *Math. Comput. Model.* **2010**, *52*, 1123. [CrossRef]
20. Chang, N.A.; Yakushiji, R.; Dowling, D.R.; Ceccio, S.L. Cavitation visualization of vorticity bridging during the merger of co-rotating line vortices. *Phys. Fluids* **2007**, *19*, 058106. [CrossRef]
21. Choi, J.; Ceccio, S.L. Dynamics and noise emission of vortex cavitation bubbles. *J. Fluid Mech.* **2007**, *575*, 1. [CrossRef]
22. Walther, J.; Schaller, J.K.; Wirth, R.; Tropea, C. Characterization of Cavitating Flow Fields in Transparent Diesel Injection Nozzles Using Fluorescent Particle Image Velocimetry (FPIV). In Proceedings of the ILASS 2000, Darmstadt, Germany, 11–13 September 2000.
23. Roth, M.H.; Gavaises, M.; Arcoumanis, C. *Cavitation Initiation, Its Development and Link with Flow Turbulence in Diesel Injector Nozzles*; SAE International: Diego County, CA, USA, 1999; Volume 111, pp. 561–580.
24. Allen, J.; Hargrave, G.; Khoo, Y. In-Nozzle and Spray Diagnostic Techniques for Real-Sized Pressure Swirl and Plain Orifice Gasoline Direct Injectors. *SAE Tech. Pap.* **2003**, *1*, 3151.
25. Aleiferis, P.G.; Hardalupas, Y.; Kolokotronis, D.; Taylor, A.M.K.P.; Arioka, A.; Saito, M. Experimental Investigation of the Internal Flow Field of a Model Gasoline Injector Using Micro-Particle Image Velocimetry. *SAE Trans. J. Fuels Lubr.* **2006**, *115*, 597.
26. Aleiferis, P.G.; Hardalupas, Y.; Kolokotronis, D.; Taylor, A.M.K.P.; Kimura, T. Investigation of the Internal Flow Field of a Diesel Model Injector Using Particle Image Velocimetry and CFD. *SAE Tech. Pap.* **2007**, *1*, 1897.
27. Mauger, C.; Méès, L.; Michard, M.; Azouzi, A.; Valette, S. Shadowgraph, Schlieren and interferometry in a 2D cavitating channel flow. *Exp. Fluids* **2012**, *53*, 1895. [CrossRef]
28. Mauger, C.; Méès, L.; Michard, M.; Lance, M. Velocity measurements based on shadowgraph-like image correlations in a cavitating micro-channel flow. *Int. J. Multiph. Flow* **2014**, *58*, 301. [CrossRef]
29. Mauger, C. Cavitation in a Diesel Injector Model Micro-Channel: Visualization Methods and Influence of Surface Condition. Ph.D. Thesis, École Centrale Lyon, Lyon, France, 2012.
30. Charalambides, G. Charge Stratified HCCI Engine. Ph.D. Thesis, Department of Mechanical Engineering, Imperial College London, London, UK, 2006.
31. Franc, J.P.; Michel, J.M. *Fundamentals of Cavitation*; Kluwer Academic Publishers: Dordrecht, The Netherlands, 2004; ISBN 1-4020-2232-8.
32. Loeve, M.M. *Probability Theory*; van Nostrand: Princeton, NJ, USA, 1955.
33. Golub, G.; Loan, C.V. *Matrix Computations*; North Oxford Academic: Oxford, UK, 1983.
34. Kumar, A.; Sahu, S. Liquid jet disintegration memory effect on downstream spray fluctuations in a coaxial twin-fluid injector. *Phys. Fluids* **2020**, *32*, 073302. [CrossRef]
35. Charalampous, G.; Hadjiyiannis, C.; Hardalupas, Y. Proper orthogonal decomposition of primary breakup and spray in co-axial airblast atomizers. *Phys. Fluids* **2019**, *31*, 043304. [CrossRef]
36. Charalampous, G.; Hardalupas, Y. Application of Proper Orthogonal Decomposition to the morphological analysis of confined co-axial jets of immiscible liquids with comparable densities. *Phys. Fluids* **2014**, *26*, 113301. [CrossRef]
37. Kumar, A.; Sahu, S. Large scale instabilities in coaxial air-water jets with annular air swirl. *Phys. Fluids* **2019**, *31*, 124103. [CrossRef]
38. Sirovich, L. Turbulence and the dynamics of coherent structures. *Q. Appl. Math.* **1987**, *45*, 561–571. [CrossRef]

39. Lumley, J.L. The structure of inhomogeneous turbulent flows. In Proceedings of the International Colloqium on the Fine Scale Structure of the Atmosphere and its Influence on Radio Wave Propagation, Nauka, Moscow, 1967.
40. Kolokotronis, D. Experimental Investigation of the Internal Flow Field of Model Fuel Injectors. Ph.D. Thesis, Department of Mechanical Engineering, Imperial College London, London, UK, 2007.

Disclaimer/Publisher's Note: The statements, opinions and data contained in all publications are solely those of the individual author(s) and contributor(s) and not of MDPI and/or the editor(s). MDPI and/or the editor(s) disclaim responsibility for any injury to people or property resulting from any ideas, methods, instructions or products referred to in the content.

Article

Shedding of Cavitation Clouds in an Orifice Nozzle

Taihei Onishi [1,2,*], Kaizheng Li [1], Hong Ji [3] and Guoyi Peng [1,*]

[1] College of Engineering, Nihon University, Koriyama 963-8643, Fukushima, Japan
[2] SU Endoscope Development Department, Olympus Corporation, Hachioji 192-8507, Tokyo, Japan
[3] College of Energy and Power Engineering, Lanzhou University of Technology, Lanzhou 730050, China
* Correspondence: taihei.ohnishi531@gmail.com (T.O.); peng@mech.ce.nihon-u.ac.jp (G.P.)

Abstract: Focused on the unsteady property of a cavitating water jet issuing from an orifice nozzle in a submerged condition, this paper presents a fundamental investigation of the periodicity of cloud shedding and the mechanism of cavitation cloud formation and release by combining the use of high-speed camera observation and flow simulation methods. The pattern of cavitation cloud shedding is evaluated by analyzing sequence images from a high-speed camera, and the mechanism of cloud formation and release is further examined by comparing the results of flow visualization and numerical simulation. It is revealed that one pair of ring-like clouds consisting of a leading cloud and a subsequent cloud is successively shed downstream, and this process is periodically repeated. The leading cloud is principally split by a shear vortex flow along the nozzle exit wall, and the subsequent cloud is detached by a re-entrant jet generated while a fully extended cavity breaks off. The subsequent cavitation cloud catches the leading one, and they coalesce over the range of $x/d \approx 1.8 \sim 2.5$. Cavitation clouds shed downstream from the nozzle at two dominant frequencies. The Strouhal number of the leading cavitation cloud shedding varies from 0.21 to 0.29, corresponding to the injection pressure. The mass flow rate coefficient fluctuates within the range of $0.59 \sim 0.66$ at the same frequency as the leading cloud shedding under the effect of cavitation.

Keywords: cavitation; bubble cloud; orifice nozzle; water jet; flow visualization

---

Citation: Onishi, T.; Li, K.; Ji, H.; Peng, G. Shedding of Cavitation Clouds in an Orifice Nozzle. *Fluids* **2024**, *9*, 156. https://doi.org/10.3390/fluids9070156

Academic Editors: Nguyen Van-Tu and Ricardo Ruiz Baier

Received: 30 April 2024
Revised: 1 June 2024
Accepted: 28 June 2024
Published: 5 July 2024

Copyright: © 2024 by the authors. Licensee MDPI, Basel, Switzerland. This article is an open access article distributed under the terms and conditions of the Creative Commons Attribution (CC BY) license (https://creativecommons.org/licenses/by/4.0/).

1. Introduction

High-speed water jets, where pressurized water or liquid mixture is issued from a small nozzle at high speed, have been developed and applied to many fields of industry [1–3]. Among them, submerged water jets injected into still water have received much attention for their capacity to continually cause intensive cavitation impact with the collapse of cavitation bubbles [4–6]. For this particular property, submerged water jets are often used in various industry fields such as the cleaning of complex mechanical products, peening of metal materials, and decomposing and sterilizing of sewage waters [1,2,6,7]. However, their processing performance is closely dependent upon the unsteady behavior of cavitation clouds related to the nozzle system and operating conditions [8–10]. Although some experimental studies [11–13],B14-fluids-3013913,B15-fluids-3013913 were made on water jets concerning the effects of driven pressure, nozzle geometry, and standoff distance as well as temperature, etc., the inner structure of cavitating jets and the interaction between cavitation bubbles and liquid flow are still unclear with respect to the complexity of turbulent cavitating flow, especially in the case of high-pressure submerged water jets accompanying intensive cavitation. Hutli et al. [16] reported an experimental study on the frequency of cavitation clouds discharging in a high-pressure submerged water jet using image analysis of high-speed camera observations. There are few works on the inner structure of cavitating flow and the mechanism of cavitation cloud releasing within a narrow nozzle for the case of high-speed submerged water jets accompanying intensive cavitation [3,17,18].

Cavitation usually occurs in the high-velocity region of the nozzle throat once the local pressure decreases to a critical level. The occurrence of cavitation induces strong pressure fluctuations, noise, vibrations, and the erosion of nozzles, especially in intensively cavitating flows, where cavitation cavities break off and multi-scale cavitation clouds shed and collapse periodically [17,18]. The large pressure fluctuation is a major source of flow instability, resulting in liquid/vapor density variances. The sharp density variations, i.e., pure liquid, pure vapor, and liquid/vapor mixture, significantly alter the flow field distribution and cavity dynamics. Moreover, the transient multi-scale cavity behavior from small vapor bubbles to large-scale cloud cavities produces strong pressure loads, such as high-frequency pressure fluctuations and impulsive pressure peaks. Although numerical simulation has become a useful way to perform flow investigations, with great progress in computational resources, the modeling of unsteady cavitation flow requires careful consideration of cavitation dynamics and the interaction between bubble cavities and liquid flow. Due to the strong coherent interactions between the cavitation dynamics and the flow structure, numerical simulations of intensively cavitating water jets remain a challenge [19–21].

With the purpose of clarifying the unsteady cavitating flow structure of high-speed submerged water jets used for industry, such as water jet cleaning and peening, this paper presents a fundamental investigation of the flow pattern of cavitation cloud releasing in a simplified orifice nozzle by combined utilization of flow visualization and numerical simulation methods. High-speed camera observations of cavitation jets were conducted, and the periodicity of cloud shedding under different injection pressures was evaluated using image analysis. Concerned with the two-phase flow structure and the mechanism of cloud generation and release with the development of jet flow, a numerical analysis was performed by using a compressible gas-vapor/liquid mixture cavitation model in consideration of the effect of sharp density variation caused by cavitation [21]. The assumption of a homogenous gas–liquid two-phase fluid was adopted, and the gas phase contained in the cavitation bubbles was assumed to consist of vapor and non-condensable components. The compressibility of the vapor component was treated semi-empirically as a constant, and the growth rate of the gas void fraction was logically evaluated by using the sonic speeds in both gas and liquid fluid media. The model was embedded in an in-house unsteady Reynolds-averaged Navier–Stokes (URANS) solver for compressible fluids by employing the realizable k-ε turbulence model [22]. The relation between the flow structure and the unsteady behavior of cavitation cloud shedding, as well as the effect of a re-entrant jet, were then investigated.

The main findings of this work are summarized as follows: (1) One pair of ring-like clouds, consisting of a leading cloud and a subsequent cloud, occurs at the nozzle throat and shed downstream successively when the cavitation number decreased to below 0.5. (2) The leading cloud is principally split from the nozzle exit by the shear vortex flow, and the subsequent cloud is detached from the throat wall by the re-entrant jet generated while a fully extended cavity breaks off. The subsequent cavitation cloud catches the leading one, and they coalesce over the range of $x/d \approx 1.8$~2.5. (3) Cavitation clouds shed from the nozzle at two dominant frequencies. The Strouhal number of the leading cavitation cloud shedding varies from 0.21 to 0.29, corresponding to the injection pressure. The mass flow rate coefficient fluctuates periodically in the range of $0.59 \sim 0.66$ at the same frequency as the leading cloud shedding under the effect of cavitation. The above points are expected to be referential to understanding the unsteady flow structure of cavitating water jet and then improving the performance of jet nozzle.

2. Experimental Apparatus and Method

Figure 1 shows the schematic diagram of the experimental device. An open-type rectangular water tank made of acrylic acid resin is set horizontally and an orifice nozzle is installed at the center of the left side wall. The mean diameter of the nozzle $d = 5.0$ mm at its throat and the length of the throat $L_d = 0.6d$. The inlet diameter of the nozzle $D = 2.6d$, and the length of inlet pipe is $6.0d$. To observe its inner flow behavior, the nozzle is also made of transparent acrylic material. The nozzle is connected to a closed pressure tank via a high-pressure hose. The lower half of the pressure tank is filled with tap water and the upper half is filled with pressurized air. The transparent observation water tank is also filled with tap water and the water depth is kept to the level of 450 mm by using an overflow pipe. Pressurized water supplied from the pressure tank is injected from the nozzle into the transparent observation tank and then a submerged water jet is generated. The absolute injection pressure is adjustable from 0.3 MPa to 0.8 MPa by adjusting the air pressure within the pressure tank with a compressor. For monitoring of the injection pressure, a high frequency pressure sensor (whose measuring error equals ±0.4 kPa) is installed at the inflow pipe just in front of the nozzle inlet and its output is recoded in real-time via a universal recorder. Also, the timely averaged mean flow rate of the jet is measured by a turbine flow meter (FT200-030, Japan Flow Controls Co. Ltd., Tokyo, Joan), whose uncertainty is estimated as 2.0%. As an index of a cavitating water jet, cavitation number σ is defined as

$$\sigma = \frac{p_o - p_v(T_\infty)}{P_{in} - p_o} \quad (1)$$

where P_{in} denotes the injection pressure (absolute), p_o represents the surrounding static pressure (absolute) at the nozzle exit, and p_v denotes the saturated vapor pressure under the reference temperature T_∞. Similarly, the flow rate coefficient c_q of the nozzle is defined in terms of the mass flow rate q_m as follows.

$$c_q = \frac{q_m}{0.25\pi d^2 V_{th} \rho_w} \quad (2)$$

in which ρ_w denotes the density of water under the given condition, and V_{th} represents the theoretical injection velocity defined as follows by neglecting all the hydraulic losses.

$$V_{th} = \sqrt{\frac{2(P_{in} - p_o)/\rho_w}{1 - (d/D)^4}} \quad (3)$$

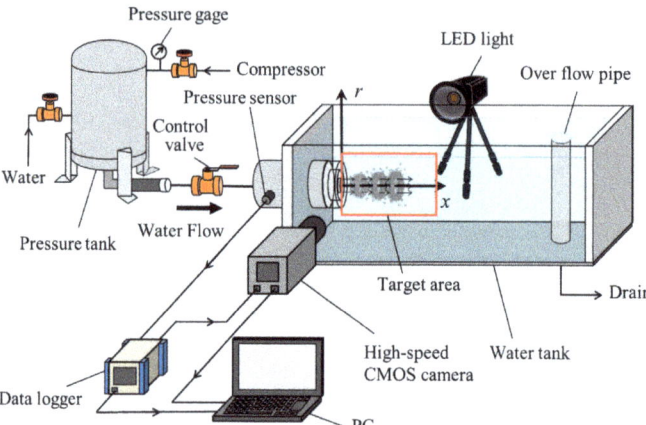

Figure 1. Scheme of experimental device.

High-speed camera observation of unsteady cavitating water jet was performed and the unsteady behavior is evaluated via image analysis [23]. Instantaneous images of cavitation clouds, where cavitation bubbles are used as flow tracers, were captured and recorded by using a high-speed CMOS camera (Photron FASTCAM SA-NX2, 1024 × 1024 pixels with 12-bit gray level, Photron, Tokyo, Japan). The observation area was axisymmetrically fitted to the nozzle central axis and its size was adjusted from the inlet of the nozzle throat to the downstream of the nozzle exit ($\sim$ 60 mm(in raial direction)× $\sim$ 90 mm (in axial direction) according to the injection pressure. The image resolution of the camera was set to be 1024 × 512 pixels. Photographs were taken under transmission light conditions by setting a panel-type high-intensity LED lamp on the opposite side of the camera. The shooting frame rate was set to be 30,000 fps. For comparison, fluorescent nylon microparticles (Kanomax, Andover, NJ, USA, ORGASOL 0457, Light wave length $\lambda = 590 \sim 610$ nm) were also used as flow tracers and a long-wave pass polarizer filter ($\lambda \geq 560$ nm) was fitted to the camera lens for the purpose of decreasing scattered light reflection on bubble surfaces.

In order to describe the flow field, a cylindrical coordinate system (x, r) was adopted, where the origin was located at the nozzle exit and the coordinates, x and r, were, respectively, set along the streamwise direction and the radial direction. The components of the velocity vector in the x and r directions are denoted as u, v, respectively. Then, the compound velocity is defined by $V = (u^2 + v^2)^{1/2}$.

3. Results and Discussions

3.1. Instantaneous Flow Visualization

Figure 2 shows a sequence of visualization images demonstrating the behavior of cavitation cloud shedding from the nozzle at a well-developed stage when the injection absolute pressure $P_{in} = 0.6$ MPa. These photos were taken using transmission light, and cavitation clouds appear to be dark gray, and the background water appears to be bright white. Figure 2(1) shows an instantaneous flow distribution when two relatively large cavitation clouds, A (middle) and A' (downstream), are released from the nozzle. The upstream cloud connected to the nozzle exit is denoted as B. Figure 2(2–4) show that cloud B is entirely detached from the nozzle and then runs after cloud A. Cloud A' contracts and collapses while cloud A expands. Figure 2(5) indicates that a parent cavity is newly generated at the nozzle throat. Figure 2(6) shows that the parent cavity, which is denoted as A'', grows up and expands nearly to its maximum. Then, it detaches from the nozzle wall as shown in Figure 2(7), whereupon a new cavitation cloud denoted by the blue solid line is released. Figure 2(6–9) demonstrate that cloud B catches up with cloud A, and then they coalesce. Thus, two new relatively large cavitation clouds are generated as shown in Figure 2(9). The solid red line with arrows denotes the motion of cloud A, and the dashed line with an arrow of cloud B. Figure 2(10,11) show the flowing of clouds newly released from the nozzle. Then, a new cloud is generated at the nozzle exit as shown in Figure 2(12), which is quite similar to Figure 2(1). Just the same as in Figure 2(1), the upstream three cavitation clouds near the nozzle exit are denoted as B, A, and A', and Figure 2(13–18) show a repeat of the above releasing and coalescing process. Because clouds A and B are always released in order, cloud A is called to be the leading cloud cavity and B is the succeeding cloud. The solid lines with arrows denote the shedding of the leading cloud A and the dashed lines with arrows indicate the motion of the succeeding cloud B. The periodicity of cloud releasing and coalescing is demonstrated. According to the figure, we note that the ring-like cavitation clouds release from the nozzle throat and then coalesce consequently within the range $x/d \leq 3$. The large well-developed clouds collapse in the approximate range of $4 < x/d < 7$.

3.2. Periodicity of Cavitation Cloud Shedding

In order to evaluate the periodic characteristics of bubble cloud shedding, image analysis of high-speed camera photographs was performed by investigating the temporal variation of the gray level [23]. Sample images of one-pixel width in the axial direction were cut from a series of photographs taken by a high-speed camera at the positions of $x/d = 0.5, 1.0, 2.0, 3.0, 4.0$, and 6.0, respectively. Then, the temporal variations of the gray level were investigated, and the waveform of the average gray level variation was analyzed. The periodic spectrum was calculated by fast Fourier transform analysis (FFT) [24]. Figure 3 shows the distribution of the power spectral density (PSD) of the average gray level oscillations, where the prominent large values are denoted in green to red colors for visibility. Two dominant frequency components, $f_1 = 1483$ Hz and

$f_2 = 2791$ Hz, are demonstrated by the figure. That is to say, cavitation clouds release and coalesce at multiple frequencies. The first frequency f_1 appears in the range $0 \leq x/d < 6$, which corresponds with the release of the leading cloud A. The second one appears in the range $0 \leq x/d < 3$, which corresponds to the release of both the leading and the succeeding clouds A and B. They coalesce near $x/d \approx 1.8 \sim 2.5$, and $f_2/f_1 \approx 2$.

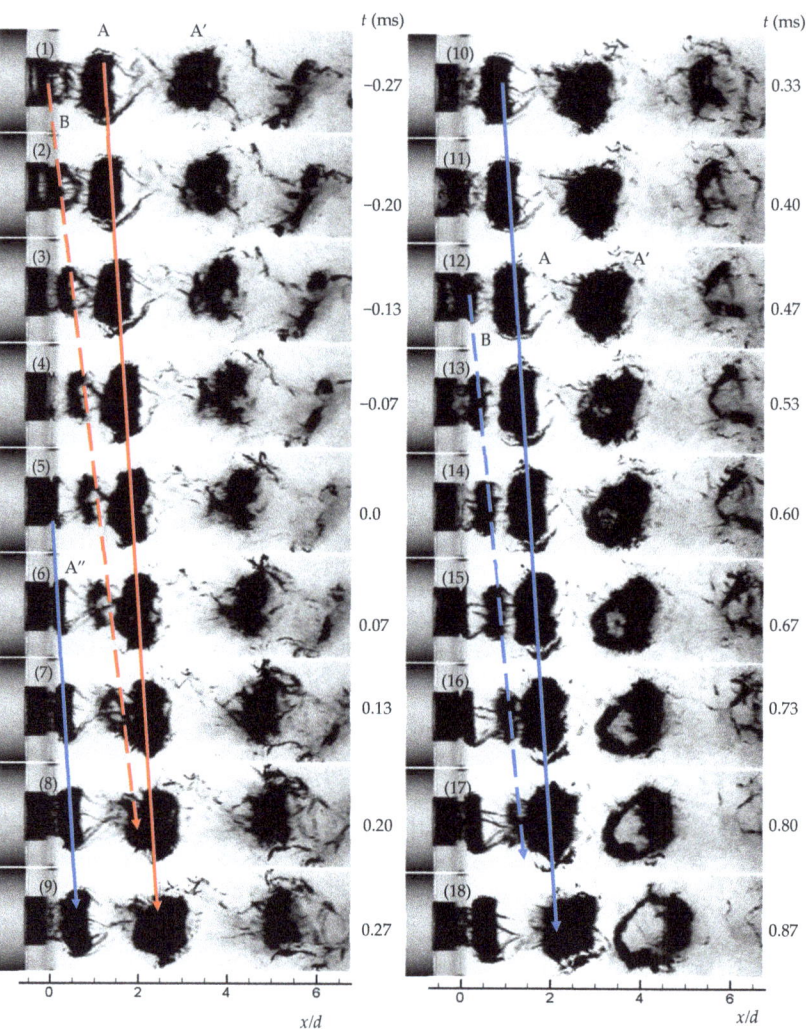

Figure 2. Periodic shedding of cavitation clouds ($P_{in} = 0.6$ MPa, $\sigma \cong 0.18$).

As a dimensionless index for such a periodic phenomenon, the Strouhal number, St, is defined as follows by using the dominant frequency f of cavitation cloud shedding and the nozzle diameter d.

$$St = fd/V_{th} \qquad (4)$$

Then, two Strouhal numbers corresponding to f_1 and f_2 are calculated to be $St_1 \cong 0.23$ and $St_2 \cong 0.46$.

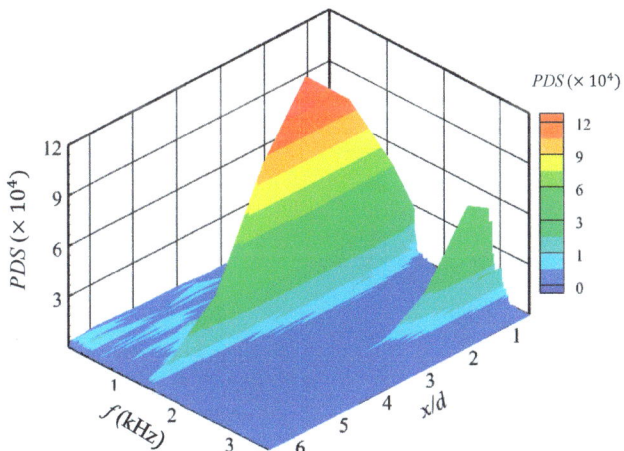

Figure 3. Dominant frequencies of cavitation cloud shedding ($P_{in} = 0.6$ MPa, $\sigma \cong 0.18$).

Table 1 shows the experimental results under different injection pressures, where the dominant frequencies and Strouhal numbers of cavitation cloud shedding are presented. As shown in the table, the dominant frequency of the leading cavitation cloud increases gradually with the increase of injection pressure and the Strouhal number St_1 varies in the range of 0.21–0.29. Similar results were reported by Nishimura et al. [25] The frequency f_2 denoting the motion of both the leading and the succeeding clouds appears within the range of $x/d < 3$.

Table 1. The dominant frequencies and Strouhal numbers of cloud shedding.

P_{in} (MPa)	V_{th} (m/s)	Re ($\times 10^5$)	f_1 (s^{-1})	St_1	f_2 (s^{-1})	St_2
				$x/d < 6$		$x/d < 3$
0.3	20.3	0.7	1161	0.29	2322	0.57
0.4	24.9	0.8	1443	0.29	2856	0.57
0.5	28.6	0.9	1373	0.24	2725	0.48
0.6	32.0	1.0	1483	0.23	2791	0.44
0.7	35.0	1.1	1545	0.22	3076	0.44
0.8	37.8	1.2	1571	0.21	3124	0.41

3.3. Mechanism of Cavitation Cloud Shedding

Regarding the flow structure as well as the interaction of cavitation cloud and flow field [26,27] numerical simulations were further conducted to clarify the interior of the intensively cavitating water jet. To capture the unsteady fluid dynamic effect of cavitation, a practical compressible gas-vapor/liquid mixture cavitation model based on the homogeneous multiphase flow approach was adopted, which allows for the practical treatment of the problem of high-speed cavitating water jets. The gas phase contained in the cavitation bubbles is assumed to consist of vapor and non-condensable components, and the compressibility of the vapor component is treated semi-empirically as a constant. The growth rate of the gas void fraction caused by cavitation is estimated by using the sonic speeds in both the gas and the liquid media. The model is embedded in an in-house unsteady Reynolds-averaged Navier–Stokes (URANS) solver for compressible fluids by employing the realizable k-ε turbulence model to evaluate the effect of turbulence. The details may be referred to in [21]. Numerical simulations were performed, and the relation between the flow structure and the unsteady behavior of cavitation cloud shedding was investigated. Then, the characteristics of cavitation cloud shedding, especially the process of cloud shedding and the effect of a re-entrant jet are analyzed.

Figure 4 shows a comparison of experimental results and numerical ones, where a sequence of images demonstrating the periodic release and coalescence of ring-like cavitation clouds is presented. The right-hand side shows the computational results, where the instantaneous velocity vector distributions and contour maps of gas volumetric fraction α_G in the $x - r$ section are presented in time sequence. The red color denotes the gaseous phase caused by cavitation and the blue color the liquid phase. The black vectors show the magnitude and the direction of local dimensionless velocities. The bright regions of $\alpha_G \geq 0.01$ represent cavitation clouds. The right-hand side shows the experimental data of the flow visualization taken by the high-speed video camera under similar working conditions.

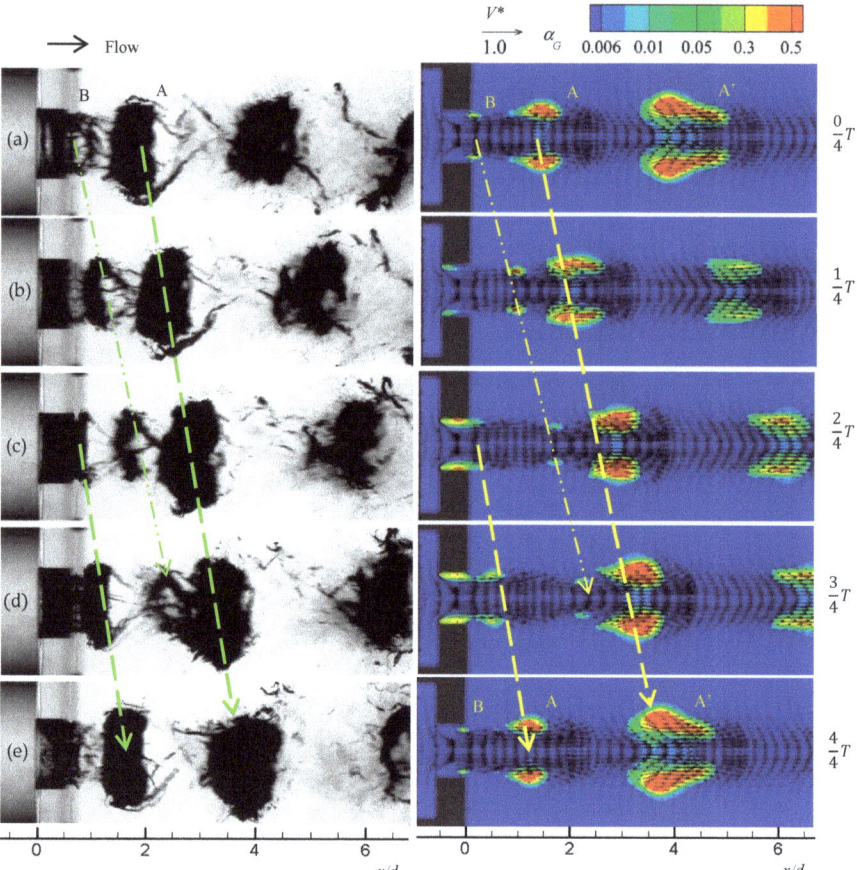

Figure 4. Periodic release and coalescence of cavitation clouds at the developed stage: (**a–e**, **left**) experimental high-speed video camera observations in a time sequence and (**a–e**, **right**) contour map of gas void fraction obtained by numerical simulation.

Figure 4a shows an instantaneous flow distribution when a succeeding cloud denoted as B is nearly detached from the nozzle exit, where two relatively large cavitation clouds denoted as A (middle) and A' (downstream) are already released from the nozzle. Figure 4b,c show that cloud B becomes entirely detached from the nozzle and runs after cloud A while cloud A expands and cloud A' contracts. Figure 4d indicates that a new parent cavity formed at the nozzle throat is fully extended and almost breaks into two parts under the effect of shear vortex flow formed at the nozzle exit, whereupon a new cavitation cloud is going to be released. The coalescence of clouds B and A is demonstrated. Figure 4e shows a repeat of the scenario in Figure 4a, where clouds A and B are combined as cloud A' and a new cycle of cloud shedding is starting. Computational results show a good agreement with the experimental ones, and the periodic release and coalescence of cavitation clouds within the range of $x/d \leq 3$ is predicted reasonably well, except the collapse pattern of cloud A' ($x/d > 4$). The reason may be concluded to be that the assumption of axisymmetric flow was adopted in the numerical simulations to reduce the computation cost.

Figure 5 shows the temporal pulsation of the mass flow rate coefficient at the well-developed stage. The solid blue line denotes the computational result of mass flow rate coefficient c_q. The dashed blue line with a solid circle shows the results of experimental measurement by a turbine flowmeter (10 Hz response) under the same working condition, where the high-frequency fluctuation of the flow rate is not detected for the limitation of the flowmeter response. The black solid line denotes the temporal variation of gas volumetric fraction α_G at a given scanning position ($x/d = 2.0, r/d = 0.48$). The figure demonstrates that α_G pulsatively varies from 0.002 to 0.6 at the scanning position, reflecting the periodic release of cavitation clouds during the developed stage. The flow rate coefficient pulsates from 0.59 to 0.65 approximately, where the effect of cavitation cloud shedding is demonstrated. The frequency of the flow rate coefficient pulsation agrees to the value of f_1 evaluated by image analysis of high-speed camera observation. The average value of c_q evaluated from the numerical simulation results coincides with the experimental ones [28], and the reliability of present simulations is further confirmed.

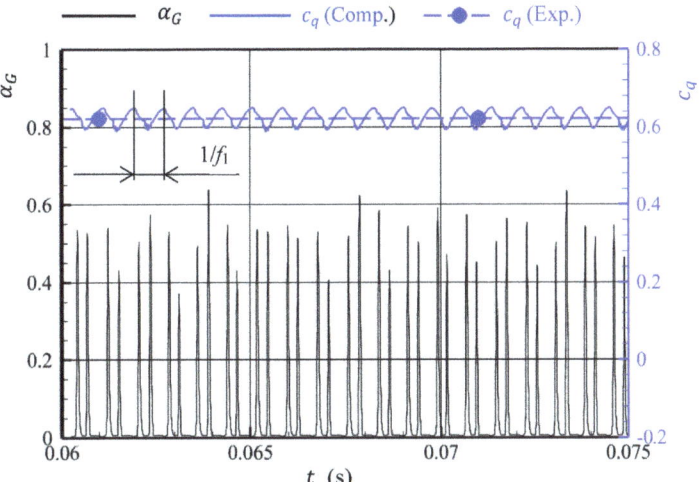

Figure 5. Temporal pulsation of the mass flow rate coefficient c_q and the variation of gas void fraction α_G at a given sampling position ($x/d = 2.0$, $r/d = 0.48$).

Concerning the inner flow structure of the cavitating jet and the mechanism of cloud shedding, the unsteady flow behavior in the local region of the nozzle throat was investigated. Figure 6 shows an instantaneous flow distribution of the cavitating jet and the succeeding cloud-releasing, coalescing, and generating process. The top panel presents a contour map of the gas void fraction α_G and velocity vector distribution in the $x - r$ section when a ring-like cavitation cloud attached to the nozzle throat expands to the outside of the nozzle exit. The red color denotes the gaseous phase and the blue color the liquid phase. The figure demonstrates that the cloud attached to the nozzle throat extends nearly to its maximum while previously released ring-like clouds travel downstream.

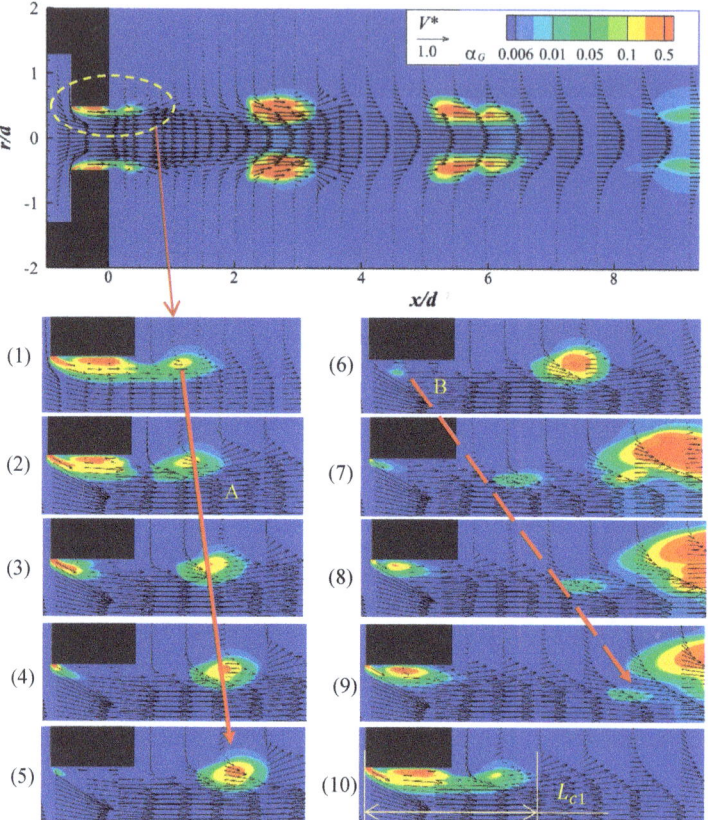

Figure 6. Instantaneous flow distributions of cavitating water jet, where (1)–(10) show a circle of cavitation cloud shedding from the nozzle throat in time sequence.

The lower panels of Figure 6 illustrate one circle of cavitation cloud shedding in a time sequence. Figure 6(1) shows an enlargement of the local flow field near the nozzle throat, where the fully extended cavity (cavitation cloud) begins to break under the effect of shear vortex flow. At step (2), the extended cavity splits into a leading part, which is marked as A, and a subsequent part attached to the nozzle wall. Corresponding to the breakdown of the large cavity, a re-entrant jet is formed along the throat wall as shown in (3). At step (3), the leading cavity A travels downstream and the subsequent cavity begins to contract, while the re-entrant jet detaches the cavity from the adjacent wall. In steps (4) and (5), a reverse flow region forms in the nozzle throat near the wall under the action of the re-entrant jet, and most parts of the wall-attached subsequent cavity become separated from the wall. At step (6), the subsequent cavity detaches from the parent cavity attached to the leading edge of the nozzle throat, and it is marked as B. In steps (7) and (8), cavity B runs after the leading cloud A while the leading edge attached parent cavity gradually extends. At step (9), the subsequent cavity combines with the leading cavity, while the wall-attached cavity expands quickly. At step (10), the wall-attached cavity is nearly fully extended to begin a new cycle as shown in step (1). The solid red line denotes the release of the leading cloud A, and the dashed line that of the subsequent cloud B. The length of the fully extended cavity, which is one of important parameters indicating the intensity of cavitation, is estimated to be $L_{c1}/d \approx 0.8 - 1.1$. Summarizing the above, we know that the leading cloud A is principally split by the shear flow, and the subsequent cloud B is principally detached by the re-entrant jet. They successively shed downstream, and this process is periodically repeated.

4. Conclusions

The property of cavitation cloud shedding in a sharp-edged orifice nozzle has been investigated by combined utilization of high-speed camera observation and flow simulation. The inner structure of the cavitating jet and the mechanism of cloud shedding are clarified. The results demonstrate that:

(1) One pair of ring-like clouds, consisting of a leading cloud A and a subsequent cloud B, is successively shed downstream, and this process is periodically repeated in the well-developed stage.

(2) The leading cloud is principally split by the shear vortex flow, and the subsequent cloud is detached by the re-entrant jet generated while a fully extended cavity breaks down. The subsequent cavitation cloud B catches the leading cloud A, and they coalesce over the range of $x/d \approx 1.8 \sim 2.5$.

(3) The Strouhal number of the leading cavitation cloud shedding varies from 0.21 to 0.29, corresponding to the injection pressure. The mass flow rate coefficient fluctuates from $0.59 \sim 0.66$ at the same frequency as the shedding of the leading cloud, and its average equals approximately 0.63 under the given condition.

Author Contributions: Conceptualization, T.O. and G.P.; methodology, T.O., H.J. and G.P.; investigation, T.O. and K.L.; writing—original draft preparation, T.O.; writing—review and editing, H.J. and G.P.; supervision, G.P.; project administration, G.P.; funding acquisition, G.P. All authors have read and agreed to the published version of the manuscript.

Funding: This research was partly supported by JSPS KAKENHI (C) 17K06169.

Data Availability Statement: The data presented in this study are available from the corresponding author upon reasonable request.

Acknowledgments: The authors would like to thank K. Yoshida, R. Ishizuka, and K. Shirota, former students of Nihon University, for their assistance in the experiments and computations.

Conflicts of Interest: Author Taihei Oishi was employed by the company Olympus Corporation. All authors declare that the research was conducted in the absence of any commercial or financial relationships that could be construed as a potential conflict of interest.

References

1. Howell, J.; Ham, E.; Jung, S. Ultrasonic Bubble Cleaner as a Sustainable Solution. *Fluids* **2023**, *8*, 291. [CrossRef]
2. Kalumuck, K.M.; Chahine, G.L. The use of cavitating jets to oxidize organic compounds in water. *J. Fluids Eng.* **2000**, *122*, 465–470. [CrossRef]
3. Hutli, E.A.F.; Nedeljkovic, M.S.; Radovic, N.A.; Bonyár, A. The relation between the high speed submerged cavitating jet behaviour and the cavitation erosion process. *Int. J. Multiph. Flow* **2016**, *83*, 27–38. [CrossRef]
4. Safaei, S.; Mehring, C. Effect of Dissolved Carbon Dioxide on Cavitation in a Circular Orifice. *Fluids* **2024**, *9*, 41. [CrossRef]
5. Cui, Y.; Zhao, M.; Ding, Q.; Cheng, B. Study on Dynamic Evolution and Erosion Characteristics of Cavitation Clouds in Submerged Cavitating Water Jets. *J. Mar. Sci. Eng.* **2024**, *12*, 641. [CrossRef]
6. Soyama, H. Cavitating jet: A review. *Appl. Sci.* **2020**, *10*, 7280. [CrossRef]
7. Mohod, A.V.; Teixeira, A.C.S.C.; Bagal, M.V.; Gogate, P.R.; Giudici, R. Degradation of organic pollutants from wastewater using hydrodynamic cavitation: A review. *J. Environ. Chem. Eng.* **2023**, *11*, 109773. [CrossRef]
8. Michael, M.; Wright, B.E.; Dropkin, A.; Truscott, T.T. Cavitation of a submerged jet. *Exp. Fluids* **2013**, *54*, 1541–1543.
9. Peng, K.; Tian, S.; Li, G.; Huang, Z.; Zhang, Z. Cavitation in water jet under high ambient pressure conditions. *Exp. Therm. Fluid Sci.* **2017**, *89*, 9–18. [CrossRef]
10. Peng, G.; Itou, T.; Oguma, Y.; Shimizu, S. Effect of ventilation on the velocity decay of cavitating submerged water jet. In *Fluid-Structure-Sound Interactions and Control*; Zhou, Y., Lucey, A.D., Liu, Y., Huang, L., Eds.; Springer: London, UK, 2018; Volume 9, pp. 93–98.
11. Ullas, P.K.; Dhiman Chatterjee, S. Vengadesan; Experimental study on the effect of throat length in the dynamics of internal unsteady cavitating flow. *Phys. Fluids* **2023**, *35*, 023332. [CrossRef]
12. Liu, Y.; Huang, B.; Zhang, H.; Wu, Q.; Wang, G. Experimental investigation into fluid–structure interaction of cavitating flow. *Phys. Fluids* **2021**, *33*, 093307. [CrossRef]

13. Huang, S.; Huang, J.; He, K. Research on Erosion Effect of Various Submerged Cavitating Jet Nozzles and Design of Self-Rotating Cleaning Device. *Appl. Sci.* **2024**, *14*, 1433. [CrossRef]
14. Ge, M.; Manikkam, P.; Ghossein, J.; Subramanian, R.; Coutier-Delgosha, O.; Zhang, G. Dynamic mode decomposition to classify cavitating flow regimes induced by thermodynamic effects. *Energy* **2022**, *254*, 124426. [CrossRef]
15. Ge, M.; Zhang, G.; Petkovšek, M.; Kunpeng Long, K.; Coutier-Delgosha, O. Intensity and regimes changing of hydrodynamic cavitation considering temperature effects. *J. Clean. Prod.* **2022**, *338*, 130470. [CrossRef]
16. Hutli, E.A.F.; Nedeljkovic, M.S. Frequency in shedding/discharging cavitation clouds determined by visualization of a submerged cavitating jet. *J. Fluids Eng.* **2008**, *130*, 021304. [CrossRef]
17. Franc, J.P.; Michel, J.M. *Fundamentals of Cavitation*; Kluwer Academic Publishers: Dordrecht, The Netherlands, 2004; pp. 1–55.
18. Podbevšek, D.; Petkovšek, M.; Ohl, C.D.; Dular, M. Kelvin-Helmholtz instability governs the cavitation cloud shedding in Venturi microchannel. *Int. J. Multiph. Flow* **2021**, *142*, 103700. [CrossRef]
19. Peng, G.; Yang, C.; Oguma, Y.; Shimizu, S. Numerical analysis of cavitation cloud shedding in a submerged water. *J. Hydrodyn.* **2016**, *28*, 986–993. [CrossRef]
20. Cruz-Ávila, M.; León-Ruiz, J.; Carvajal-Mariscal, I.; Klapp, J. CFD turbulence models assessment for the cavitation phenomenon in a rectangular profile Venturi tube. *Fluids* **2024**, *9*, 71. [CrossRef]
21. Oishi, T.; Peng, Y.; Ji, H.; Peng, G. Numerical simulations of cavitating water jet by an improved cavitation model of compressible mixture flow with an emphasis on phase change effects. *Phys. Fluids* **2023**, *35*, 073333. [CrossRef]
22. Shih, T.H.; Liou, W.W.; Shabbir, A.; Yang, Z.; Zhu, J. A new k-ε eddy-viscosity model for high Reynolds number turbulent flows—Model development and validation. *Comput. Fluids* **1995**, *24*, 227–238. [CrossRef]
23. Peng, G.; Wakui, A.; Oguma, Y.; Shimizu, S.; Ji, H. Periodic behavior of cavitation cloud shedding in submerged water jets issuing from a sheathed pipe nozzle. *J. Flow Control Meas. Vis.* **2018**, *6*, 15–26. [CrossRef]
24. Lin, Y.; Kadivar, E.; Moctar, O. Experimental study of the cavitation effects on hydrodynamic behavior of a circular cylinder at different cavitation regimes. *Fluids* **2023**, *8*, 162. [CrossRef]
25. Nishimura, A.; Takakuwa, O.; Soyama, H. Similarity law on shedding frequency of cavitation cloud induced by a cavitating jet. *J. Fluid Sci. Technol.* **2012**, *7*, 405–420. [CrossRef]
26. Liu, H.; Kang, C.; Zhang, W.; Zhang, T. Flow structures and cavitation in submerged waterjet at high jet pressure. *Exp. Therm. Fluid Sci.* **2017**, *88*, 504–512. [CrossRef]
27. Liu, B.; Pan, Y.; Ma, F. Pulse pressure loading and erosion pattern of cavitating jet. *Eng. Appl. Comput. Fluid Mech.* **2019**, *14*, 136–150. [CrossRef]
28. Nurick, W.H. Orifice cavitation and its effect on spray mixing. *J. Fluids Eng.* **1976**, *98*, 681–687. [CrossRef]

Disclaimer/Publisher's Note: The statements, opinions and data contained in all publications are solely those of the individual author(s) and contributor(s) and not of MDPI and/or the editor(s). MDPI and/or the editor(s) disclaim responsibility for any injury to people or property resulting from any ideas, methods, instructions or products referred to in the content.

Article

Experimental Study of the Cavitation Effects on Hydrodynamic Behavior of a Circular Cylinder at Different Cavitation Regimes

Yuxing Lin [†], Ebrahim Kadivar [*,†] and Ould el Moctar [*]

Institute of Ship Technology, Ocean Engineering and Transport Systems, University of Duisburg-Essen, 47057 Duisburg, Germany; yuxing.lin@uni-due.de
* Correspondence: ebrahim.kadivar@uni-due.de (E.K.); ould.el-moctar@uni-due.de (O.e.M.)
† These authors contributed equally to this work.

Abstract: In this work, we experimentally investigated the cavitation effects on the hydrodynamic behavior of a circular cylinder at different cavitating flows. We analyzed the cavitation dynamics behind the circular cylinder using a high-speed camera and also measured the associated hydrodynamic forces on the circular cylinder using a load cell. We studied the cavitation dynamics around the cylinder at various types of the cavitating regimes such as cloud cavitation, partial cavitation and cavitation inception. In addition, we analyzed the cavitation dynamics at three different Reynolds numbers: 1×10^5, 1.25×10^5 and 1.5×10^5. The results showed that the hydrodynamics force on the circular cylinder can be increased with the formation of the cavitation behind the cylinder compared with the cylinder at cavitation inception regime. The three-dimensional flow caused complex cavitation behavior behind the cylinder and a strong interaction between vortex structures and cavity shedding mechanism. In addition, the results revealed that the effects of the Reynolds number on the cavitation dynamics and amplitude of the shedding frequency is significant. However the effects of the cavitation number on the enhancement of the amplitude of the shedding frequency in the cavitating flow with a constant velocity is slightly higher than the effects of Reynolds number on the enhancement of the amplitude of the shedding frequency at a constant cavitation number.

Keywords: cavitation; hydrodynamic loading; vortex induced vibration

Citation: Lin, Y.; Kadivar, E.; el Moctar, O. Experimental Study of the Cavitation Effects on Hydrodynamic Behavior of a Circular Cylinder at Different Cavitation Regimes. *Fluids* **2023**, *8*, 162. https://doi.org/10.3390/fluids8060162

Academic Editors: Manolis Gavaises and D. Andrew S. Rees

Received: 13 April 2023
Revised: 16 May 2023
Accepted: 19 May 2023
Published: 23 May 2023

Copyright: © 2023 by the authors. Licensee MDPI, Basel, Switzerland. This article is an open access article distributed under the terms and conditions of the Creative Commons Attribution (CC BY) license (https://creativecommons.org/licenses/by/4.0/).

1. Introduction

With the development of the marine applications and hydraulic machinery components, such as high-speed underwater vehicles, propeller-rudder systems and central flow pumps, the negative effects of cavitation and control of this phenomenon on the solid structure have became a matter of attention. The high velocity of flow around an immersible body can induce a low pressure region below the vapor pressure of the liquid and generate a phenomenon known as cavitation. The cavitation can grow on the surface of the immersible bodies and ultimately the extended cavity can be collapsed near the solid boundaries in the high pressure region and induce noise, vibration, erosion and mitigation of the performance of the systems [1–5]. As a non-linear system, the cavitation dynamics formed on an immersible body can be affected by different factors, such as surface roughness [6,7], material properties [8], cavitation nuclei density and nuclei radius [9], liquid temperature [10–12], and etc. Based on how the cavitation is created and oscillation mechanism of the cavitation dynamics, the cavitation can be classified into different types such as sheet cavitation, partial cavitation, cloud cavitation and super-cavitation on the marine operating systems and hydraulic components. The cavity structures around the immersible bodies have different volumes which can induce various hydrodynamic forces or vibration on the system. The sheet cavity can appear directly after the cavitation inception close to the leading edge of a hydrofoil with decreasing of the cavitation number. The cloud cavity structure can be formed after the detachment of the attached cavity from the solid surface and may collapse near the trailing edge of the hydrofoil.

Coutier-Delgosha et al. [13] investigated the internal structure of the attached cavity on a two dimensional hydrofoil by using the novel endoscopic and X-ray technology. They found the void fraction distribution inside the sheet cavitation and presented the vapor/liquid morphology for different cavitation shapes. Furthermore, Barre et al. [14] used the optical probe to measure the void ratio of a 'quasi-stable' sheet cavitation in a venturi-geometry and compared their results with their numerical simulation. They presented that the sheet cavitation dynamics is quasi stable, however small cavity structures are detached from the surface at the aft region of the sheet cavity due to the effects of the re-entrant jet. Pelz et al. [15] developed an analytical model to investigate the transition from the sheet to cloud cavitation by describing the growth of the attached sheet cavitation and the extension of the re-entrant jet. They modelled the re-entrant jet as a spreading film under the cavity and their work helped us to understand the transition between sheet and cloud cavitation phenomenon. Their results showed that sheet to cloud cavitation can be affected by the cavitation nuclei density, the viscous effect of flow and the Reynolds number. The cloud cavitation has been identified as the most dangerous type of cavitation because of the collapse of large scale of cloud cavity on the surface of immersible bodies at different cavitating regimes with lower cavitation numbers [16].

In the various previous studies, the fundamental of the shedding mechanism of different cavitation dynamics have been investigated experimentally and numerically. To understand the behavior of the partial cavitation, Le et al. [17] observed the cavity structure on a plano-convex hydrofoil placed in the cavitation tunnel and measured the wall pressure distribution around the hydrofoil. Their findings showed that the re-entrant jet is the main reason of the shedding mechanism of the cloud cavitation and the cavitation instability which may induce a periodic cavitation structure. Stutz and Reboud [18] studied experimentally the shedding mechanism of the cavitation and the phase transition in a venturi-type test section. They pointed out that in addition to the re-entrant jet, a local pressure reduction can be formed by the large turbulence fluctuation in the partial cavitation regime. The visualization of the re-entrant jet and the cavitation periodic behaviour around a hydrofoil was obtained by Callenaere et al. [19]. In their work, some essential factors for the generation of re-entrant jet have been discussed: 1. the large adverse pressure at the cavity closure region; 2. the thickness of the cavity structure should be large enough, which can also present the attached cavity length. They showed that the re-entrant jet can play as the primary shedding mechanism in the partial and cloud cavitation regimes.

Leroux et al. [20] studied experimentally and numerically the partial cavitation shedding on a two-dimensional hydrofoil to understand the periodic shedding mechanism of the cavity structure. Their results revealed that a shock wave can be appeared by the cloud cavity collapse at low cavitation numbers. They found that the extension of the shock wave can also change the cavity shedding dynamics near the leading-edge of the hydrofoil. The role of the shock wave on the cavity oscillation was investigated by different researchers Genesh et al. [21] and Wu et al. [22]. They studied the shock wave propagation on a wedge-apex geometry and around a two dimensional hydrofoil by using the high speed X-ray densitometry. They both observed that the large scale of the shedding cavity can be collapsed subsequently near the solid surface and a type of pressure pulsation can be generated at the low cavitation numbers. In addition, they found that the pressure pulsation can affect the growth of the attached partial cavitation on the solid boundary. Their results showed that the shock wave front on the downstream region has been formed which was the symbol of the shock wave in the cavity shedding mechanism. Furthermore, they indicated that the pressure pulsation inside the cavitation structure can play as a dominant role for the cavity shedding mechanism. Karathanassis et al. [23] performed high-speed X-ray phase-contrast imaging of the cavitating flow developing within an axisymmetric throttle orifice using high-flux synchrotron radiation. Karathanassis et al. [24] studied X-ray phase contrast and absorption imaging for the quantification of transient cavitation in high-speed nozzle flows. Their results revealed that the X-ray phase-contrast imaging is suitable for capturing fine morphological fluctuations of transient cavitation structures.

However, the technique may not provide information on the quantity of vapor within the orifice. Kadivar et al. [25,26] studied the effects of the pressure fluctuations in the cavitation surge regime around a two dimensional hydrofoil and a flat plate with semi-circular leading edge. They presented that the cavitation dynamics on the hydrofoil was driven mainly by the re-entrant jet in the cloud cavitation regime. However, the pressure pulsations formed around the flat plate can generate pressure waves inside the cavitation structure which may play as a dominant role for the cavity shedding mechanism.

The relationship between cavitation and the turbulent flows on different geometries such as wedge-type geometries, circular cylinders and propellers were investigated. The experimental work from Kermen & Parkin [27] showed that the formation of the cavitation inception phenomenon on a circular disk depend on the Reynolds number. In addition, Arndt [28] extended the experimental studies around a disk and confirmed the relationship between cavitation inception number and the Reynolds number. The work from Kermen & Parkin and Arndt presented the counting near-wake vortices and the visualization of coherent structures behind the circular disk. However, the size and the rotation rate of the vortices haven't been clarified. In study of Belahadji et al. [29] with a two dimensional wedge, the near-wake vortices and Benard-Karman vortices were observed. One model between the cavitation inception number and Reynolds number was corroborated with the experimental data and they explained the transition of vortices from near-wake region to the far-wake region. Recently Wu et al. [30] used an advance void-fraction method, a high-speed X-ray densitometry, to observe the cavitation vortices behind a wedge geometry. They examined the cavitation structure and corresponded void fraction behind the wedge and found out that the cavitation void and the local Mach number increased with reducing the cavitation numbers. They found that the sound speed in the mixture region can be reduced by increasing the void fraction. Besides the investigations of cavitating flows behind a standard bluff body such as wedge, the investigations of cavitation behind circular cylinders play also an important roll in the hydraulic and ship industries.

The single phase flows behind a circular cylinder has been studied extensively by different researchers (Roshko [31]; Bearman [32]; Wei & Smith [33]; Williamson [34]; Szepessy & Bearman [35]; Norberg [36]). However, only few studies are considered the cavitating flows around the circular cylinder and the hydrodynamics effect of cavitation regimes. Fry [37] performed an experiment of cavitation behind a cylinder in a free stream flow and captured the cavitation-induced noise spectra together with the cavitation dynamic observation. His results showed that the noise peaks was related to the cavity collapse intensity and the cavity/wake dynamics. Matsudaira et al. [38] measured the bubble collapse pressure in the wake dynamics and studied the relationship between the bubble collapse-induced pressure and the separation of Karmen-vortex cavity from a cylinder with Reynolds number region from 4.5×10^5 to 6.0×10^5 and cavitation numbers from 0.9 to 1.6. Their work showed that by the reduction of the cavitation number and increasing of the Reynolds number, the maximum bubble collapse induced pressure can be increased. In addition, they showed that more severe cavitation-induced erosion can be observed on the solid boundaries at higher Reynolds numbers. The effects of the erosion from the bubble collapse-induced pressure on the circular cylinder were investigated by Saito & Sato [39] mounting an aluminum plate behind the cylinder in the cavitating flows. They observed that the cavity shedding structures near the aluminum plate and measured the erode pits to get a ratio between the passing cavity shedding and the pits formed on the aluminum plate.

Recently some investigations around the cylinder in the cavitating flows were performed experimentally and numerically to study the relationship between cavitation regime, vortex shedding behind cylinder and the relationship between cavitation and transition from laminar to turbulence flows. Franc & Michel [40] developed a semi-empirical approach for the prediction of the location of cavity detachment in laminar separation for circular and elliptical cylinder and validated their approach with the experimental data. Gnanaskandan & Mehesh [41] studied the cavitating flows over a circular cylinder numerically and compared the results of the cavitating regimes with the results of the single phase flows.

Their results showed that the cavitation suppressed the turbulence in the near-wake region and modified the vortex shedding mechanism significantly by the vorticity dilatation due to the phase changing inside the laminar separation. They investigated also the Reynolds number effect on the cavity length and found out that the cavity volume in vortex shedding process can be increased at higher Reynolds numbers. Kumar et al. [42] investigated the cavitation dynamics around a circular cylinder at the Reynolds number of 64,000 and the relationship between cavity life time and the cavitation number. The results from their work showed that the shedding frequency can be reduced and the pressure fluctuation due to the cavitation around the cylinder can be increased when the cavitation number decreased. Ghahramani et al. [43] performed experimental and numerical investigations about three-dimensional cavitating flow around a semi-cylinder at high Reynolds numbers and low cavitation numbers. They found that different vortices pattern formed and started to grow behind the semi-cylinder with the reduction of the cavitation numbers. In addition, their results revealed that at the lowest cavitation number, the cavitation dynamics affected the vorticity and vertical vortex structure significantly.

Brandao et al. [44] performed a numerical study on the cavitation around a circular cylinder based on the work done by Gnanaskandan & Mehesh [41] and extended the water-vapor model to include non-condensable gas effects. They considered the shock wave propagation inside the wake flow after the cavity collapse and presented the influence of non-condensable gas on delaying the transition to the low cavity shedding frequency. In addition, their results showed that the growth of two-dimensional cavity reduced the vortex stretching and baroclinic torque. Dobroselsky [45] investigated cavitation dynamics around a circular cylinder with the Reynolds number effect by using the Particle Image Velocimetry (PIV) measurement technology. His results presented the details of the transition from laminar to turbulent flow and the dependence of kinematic characteristics and the separation angle of the boundary layer on the Reynolds number at non-cavitating regime. In addition, different wake patterns and vortex shedding over a long period have been observed behind the cylinder. Sadri and Kadivar [46] performed a numerical simulation using a high-order compact finite-difference scheme to study cavitating flow and the cavitation-induced noise around one and two circular cylinders. They studied the cavitating flow for different gaps of two side-by-side cylinders. Their results revealed that the wakes behind the side-by-side cylinders are merged together and a single vortex street is generated by the gap reduction between two cylinders.

Previous works on cavitation dynamics around cylinder are mostly investigated the cavitating flows behind two dimensional cylinder with a very small gap between the cylinder and the test section. In this work, we focused on the highly three-dimensional cavitating flow around a stainless steel circular cylinder. In addition, we considered the Reynolds number effect on the cavitation dynamics and visualized two different cavitation regimes as cloud—and partial cavitations at high sampling frequency. We measured the synchronized hydrodynamics forces on the cylinder, which presented the effects of Reynolds numbers on the cavitation and vortex structures behind the cylinder. The remainder of this paper is arranged as follows: In Section 2, the experimental setup, experimental conditions and the test cases will be described. The main results of the cavitation around circular cylinders at different cavitating regimes and various Reynolds numbers are then discussed in Section 3. Finally, the main conclusions will be listed in Section 4.

2. Experimental Set-Up

The experimental investigation was carried out in the cavitation tunnel K23 of Institute of Ship Technology, Ocean Engineering and Transport Systems at the University of Duisburg-Essen. In this cavitation tunnel, the absolute pressure range from 0.1 to 2 bar and the maximum velocity of 9 m/s can be achieved. Figure 1 showed the experimental setup of the present work. The test case is a circular cylinder which was mounted on one of the Plexiglas wall of the cavitation tunnel test section. The test section of the tunnel has a cross-section of 0.3×0.3 m^2 and a length of 1.1 m. The velocity of the inlet section was measured using a differential pressure sensor and the inlet-/outlet absolute pressures are measured using the absolute pressure sensors mounted on the test section of the tunnel. The initial turbulence intensity for the inlet flow of was about 2%. The high-speed camera Phantom V9.1 was placed on the side of cavitation tunnel test section to capture the side view of the cavitation dynamics around the circular cylinder. For the visualization, the sampling frequency was adjusted 1000 Hz and the exposure time for each image was 30 ms. One user trigger signal was generated at the begin of the visualization and the force measurement was started to recording the data to achieve the synchronization between visualization and force measurement. The force measurement was carried out using a force sensor mounted between the tunnel test section and the cylinder. The recording frequency of the force measurement was adjusted to 4800 Hz.

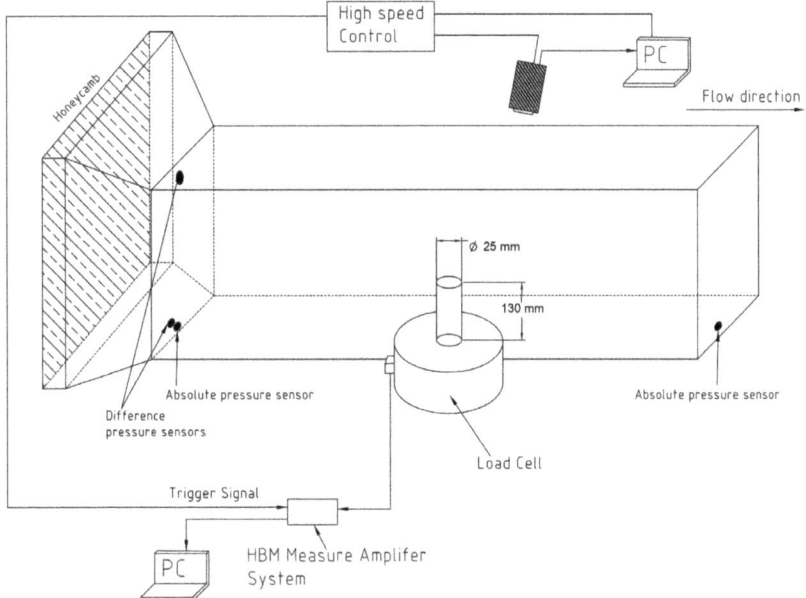

Figure 1. Schematic sketch of the experimental set-up for the experiment of cavitation dynamics behind a circular cylinder.

In this investigation, we visualized cavitation dynamics around the circular cylinder and measured the hydrodynamics forces on the cylinder. The circular cylinder has a diameter of 25 mm and the length of 130 mm. We performed the experiments for three different Reynolds numbers ranging from 1.0 to 1.5×10^5 in this work. The experimental environment such as water temperature was maintained in a constant level for different Reynolds numbers to adjust the same situation for capturing the cavitation dynamics. The air content was between 1.2 and 1.4 mg/L and the water temperature was about 16 °C. The Reynolds number and cavitation number are defined as follows:

$$\sigma = \frac{p_{ref} - p_v}{1/2 \rho_{ref} V_{ref}^2} \quad (1)$$

$$Re = \frac{V_{ref} l_{ref}}{\nu} \quad (2)$$

where p_{ref} and p_v are the reference static pressure at the middle of the inlet section and the saturation vapor pressure of the operating liquid, respectively, V_{ref} is the incoming flow velocity, and ρ_{ref} and ν are the reference density and kinematic viscosity of the operating liquid, respectively.

The absolute error of measurement of saturation vapor pressure depends on the water temperature during the experimental testing in the cavitation tunnel which can be estimated indirectly using a temperature measurement system. The water temperature was 16 °C with an uncertainty of 0.1 °C during the experiment. The measurement error of vapor pressure was estimated about 24 Pa for temperature variation of 0.1 °C. The uncertainty of the inlet pressure was in the range of 100–200 Pa. The uncertainty of the measured cavitation number for different testings was in the range of 0.05–0.1 during the experiment. The uncertainty of the inlet velocity was obtained for different experiments and at different flow conditions. The measurement uncertainty of the Reynolds number based on the cylinder diameter and the mean flow velocity is proportional to the absolute error of the inlet velocity. The uncertainties of the inlet velocity were about 0.07 at the inlet velocity of 4 m/s, 0.09 at inlet velocity of 5 m/s and 0.1 at inlet velocity of 6 m/s. Therefore, the maximum measurement error of the inlet velocity was about 1% for different experiments which led to an uncertainty of the measured Reynolds number of about 1–2%.

3. Results

In this section, the results of two different cavitating regimes around the circular cylinder are presented. Figures 2 and 3 illustrate one cyclic behavior of the cloud cavitation and partial cavitation dynamics around the circular cylinder with inlet flow velocity of 4 m/s, respectively. The Reynolds number for this velocity is 1.0×10^5. Table 1 shows a summary of initial conditions such as inlet velocity, cavitation number, Reynolds number and outlet pressure for different experiments of the present work.

Table 1. The initial conditions of each experiment of the present work.

Inlet Velocity (m/s)	Cavitation Number (-)	Reynolds Number (-)	Outlet Pressure (bar)
4.0	1.1	1×10^5	0.104
	1.6	1×10^5	0.121
5.0	1.02	1.25×10^5	0.135
	1.55	1.25×10^5	0.199
6.0	1.05	1.49×10^5	0.189
	1.55	1.49×10^5	0.277

3.1. Cavitation Dynamics behind the Cylinder

As the results of cloud cavitation dynamics show, relative large vortex structures behind the cylinder were formed and developed downstream of the cylinder with the flow during a cyclic behavior of cavitation in the cloud cavitation regime. In addition, a quasi strong tip vortex cavitation can be seen at the free-end of the cylinder. The attached cavity on the cylinder was detached from the cylinder surface only after a short distance of about 4–6 mm on the surface. The flow moving downstream of the cylinder can interact with the cavity structures behind the circular cylinder and affect the cavitation dynamics. The interaction between the tip vortex cavitation and detached cavitation occurred at the position approximately 50 mm vertically from the tip and 25–30 mm horizontally behind the cylinder. This interaction can modify the cavitation shedding mechanism and affect on the pressure distribution of the cavitation and leading to a change of the cavity shedding frequency. The pressure wave propagation inside the cavity resulting from the cavity collapse may be the main reason of the cavity shedding in this case.

In Figure 3, the side-view of the partial cavitation behind the cylinder was presented. As it can be seen from the results, the cavity volume was significantly reduced with increasing of the cavitation number compared with the case of cloud cavitation. In addition, the images show that the cavity structures appeared behind the cylinder mostly initiated from the tip vortex cavitation and has a tendency from the top side to the bottom. In other words, a substantial amount of the cavitation can be induced in the region close to the tip of the cylinder. The main reason can be the formation of the low pressure region near the cylinder tip the where the flow with higher velocity passing from the cylinder tip. Therefore, in the partial cavitation regime, the cavity structures which can be started from the middle of the cylinder surface was much lower compared with the cavity structures in the cloud cavitation regime.

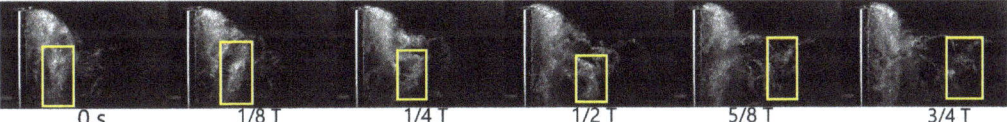

Figure 2. Cavitation dynamics on the circular cylinder at the velocity of 4 m/s and cavitation number of 1.0. The flow is from left to the right direction. Cavity shedding period (T) is 38.28 ms. Yellow squares show the formation of strip-shape cavity structures in the images.

Figure 3. Cavitation dynamics on the circular cylinder at the velocity of 4 m/s and cavitation number of 1.5. The flow is from left to the right direction. Cavity shedding period (T) is 40.58 ms. Yellow squares show the formation of strip-shape cavity structures in the images.

Figure 4 illustrates the cavitation dynamics around the cylinder at velocity of 5 m/s with the Reynolds number of 1.25×10^5 and the cavitation number of 1.0. As the results show, a larger volume of the cavity structure behind the cylinder can be seen compared to the cavity structure at a velocity of 4 m/s in the cloud cavitating flow. The reason of larger cavity structure could be due to the larger low pressure region formed behind the cylinder due to the higher flow velocity compared to the case with lower Reynolds number. The tip vortex cavitation dynamics in the experiment with the velocity 5 m/s was remained mostly unchanged and the attached cavity on the upper side of cylinder appeared persisted for the entire period of cloud cavitation. The mean width of the cavity structure from the cylinder tip was about 50–55 mm. In addition, the attached cavity on the front view of the cylinder has a 'finger' shape. In this regime, both the cavity from the cylinder tip and from the cylinder surface crossed behind the cavitation vortex structure and created a strip-shape cavity (see the yellow squares in the images). These strip-shape cavitation structures can be observed in most of the time steps in the horizontal direction. Considering the vortex shedding mechanism behind the cylinder, the opposite direction of strip-shape cavities could be explained by detachment of the vortex structures from top and bottom of the cylinder which generated the low pressure regions at two parallel vortex streets. In the wake region of the cylinder, most cavitation structure collapse after a distance of about 5 cylinder diameters behind the cylinder. The reason could be due to the pressure enhancement in the region far from the cylinder. However, the details of flow pressure distribution needs a further investigation to understand this physical phenomenon inside the cavitation dynamics. The partial cavitation regime around the cylinder was presented in Figure 5 and the cavity structure exhibited a similar movement pattern as cavity dynamics seen in Figure 3. The main cavity structure could be generated by the flow passing over the cylinder tip. Simultaneously, the results reveal the detachment of the some cavities from the bottom section of the cylinder near the wall. These cavity structures have a movement diagonally upwards to the region near the middle of the cylinder.

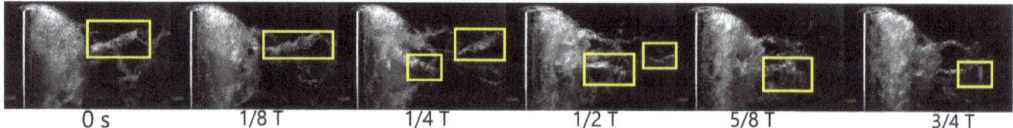

Figure 4. Cavitation dynamics on the circular cylinder at the velocity of 5 m/s and cavitation number of 1.0. The flow is from left to the right direction. Cavity shedding period (T) is 31.60 ms. Yellow squares show the formation of strip-shape cavity structures in the images.

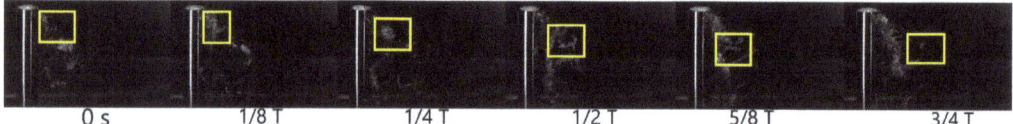

Figure 5. Cavitation dynamics on the circular cylinder at the velocity of 5 m/s and cavitation number of 1.5. The flow is from left to the right direction. Cavity shedding period (T) is 30.04 ms. Yellow squares show the formation of strip-shape cavity structures in the images.

In this work, we also studied the cavity dynamics on the circular cylinder at a velocity of 6 m/s, which presented in Figures 6 and 7. The results show that the cloud cavitation structures on the cylinder at this velocity have the largest volume of cavity among all cases. Therefore, larger vortical structures can be generated behind the cylinder. In addition, the near wake region behind the cylinder extended over a larger range from about 4 cylinder diameters. The cavity structure condensed quickly at a distance of approximately 6 cylinder diameters. Compared to the cavity at the velocity of 5 m/s, the detached cavity formed behind the cylinder showed a parallel shedding structure from the whole cylinder surface in the downstream direction. This aspect could be attributed to the weaker influence of the tip vortex cavitation on the larger cavity structure which can increase the cavitation stability behind the cylinder.

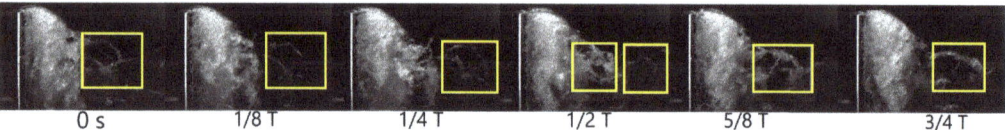

Figure 6. Cavitation dynamics on the circular cylinder at the velocity of 6 m/s and cavitation number of 1.0. The flow is from left to the right direction. Cavity shedding period (T) is 26.0 ms. Yellow squares show the formation of strip-shape cavity structures in the images.

Figure 7. Cavitation dynamics on the circular cylinder at the velocity of 6 m/s and cavitation number of 1.5. The flow is from left to the right direction. Cavity shedding period (T) is 26.54 ms. Yellow squares show the formation of strip-shape cavity structures in the images.

In the Figure 7, the partial cavitation dynamics behind the circular cylinder at Reynolds number of 1.5×10^5 was presented. In this regime, the effects of the tip vortex cavitation on the dynamics of the cavitation formed behind the cylinder is dominant. As it can be seen from the images, the main part of the cavity volume is from the tip vortex cavitation which can be formed during the shedding process. In addition, the detached cavity from cylinder in the downstream has a similar parallel structure to the cloud cavitation at this velocity.

The reason for this parallel structure is the large vortex shedding process generated by the higher velocity field.

3.2. Hydrodynamics Forces on the Cylinder

The time histories of lift coefficients amplitude on the circular cylinder at three different velocities and in the cloud cavitation regime are presented in Figure 8. The lift coefficient is defined as follows:

$$c_l = \frac{F_{lift}}{0.5 \cdot A \cdot \rho \cdot v^2} \quad (3)$$

In the definition of the lift coefficient, the parameter is lift force and "A" is the relevant surface and calculated by the multiplication of the cylinder diameter and the cylinder height. The fluctuation of the lift coefficient at the Reynolds number 1.0×10^5 (velocity 4 m/s) demonstrates a strong repeatability with the peaks and troughs remaining at the same level during the recording period with no significant differences between each fluctuation. Based on the results, it can be deduced that the hydrodynamics force on the circular cylinder at the velocity of 4 m/s is continually repeated in the cloud cavitation regime. The lift coefficient on the cylinder at the velocity of 5 m/s presented a similar repeatability in its fluctuations, however with higher frequency vibration.

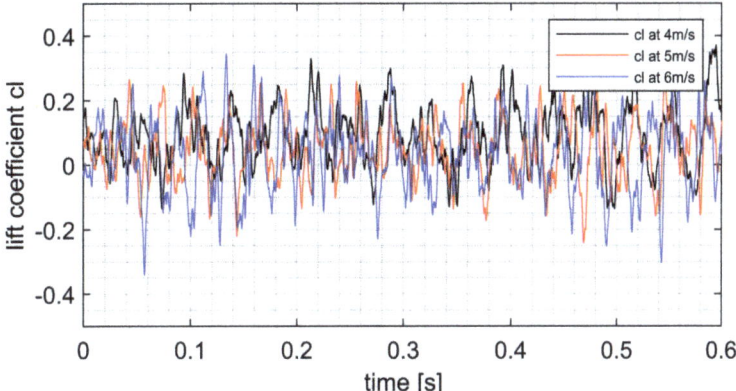

Figure 8. Time history of hydrodynamics lift coefficient on the circular cylinder at three different Reynolds numbers with the cavitation number = 1.0. The mean lift coefficient at the velocities of 4 m/s, 5 m/s and 6 m/s are 0.091, 0.053 and 0.023, respectively. The root mean square of the lift coefficient at the velocities of 4 m/s, 5 m/s and 6 m/s are 0.103, 0.104 and 0.112, respectively.

The time histories of the lift coefficients at the velocities of 4 m/s and 5 m/s have almost the same root mean square value which shows that the combined hydrodynamic force on the circular cylinder at these two velocities is at the same level. However, the lift coefficient on the cylinder at the velocity of 6 m/s shows a strong non-linear behavior characterized by a low frequency signal and multiple high frequency signals. The peak values exhibit multi similar fluctuations in high frequency region. This could be due to the highly instability behavior and the interaction between cloud cavitation collapse and tip vortex shedding behind the circular cylinder.

Figure 9 presents the time history of lift coefficient on the circular cylinder at three different velocities in the partial cavitation regime. In this regime, the tip vortex cavitation plays a dominant roll and the lift coefficient on the cylinder was affected by the tip vortex cavitation periodically. The cavity volume was increased by the increasing of the velocity. This enhancement of cavity volume can be observed by the increasing of the root mean square value of the amplitude of the lift coefficient fluctuations. Generally, the lift coefficients of the cylinder in three different velocities present a non-linearity behavior due to the influence of the interaction between tip vortex generation and cavity shedding. Further discuss about the hydrodynamics effect would be carried out in the frequency domain.

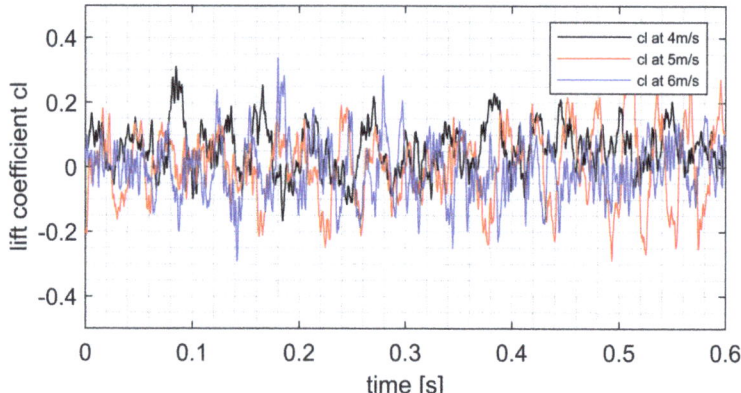

Figure 9. Time history of hydrodynamics lift coefficient on the circular cylinder at three different Reynolds numbers with the cavitation number = 1.5. The mean lift coefficient at the velocities of 4 m/s, 5 m/s and 6 m/s are 0.069, 0.012 and 0.021, respectively. The root mean square of the lift coefficient at the velocities of 4 m/s, 5 m/s and 6 m/s are 0.081, 0.091 and 0.097, respectively.

Figures 10 and 11 present comparisons between lift forces in the frequency domain at cloud cavitation, partial cavitation and cavitation inception regimes at the velocity of 4 m/s. The cavitation inception dynamics on the circular cylinder appeared at the cavitation number of 2.4 but no visible cavitation structure was seen in this regime. Therefore, we ignored to present the results of the cavitation on the cylinder in the cavitation inception regime and focused on the cavitation effect on the hydrodynamic behavior of the partial- and cloud cavitation dynamics. Frequencies f1 to f6 and the corresponding amplitudes in the figures represented the peaks of the frequency amplitude in the fast Fourier transform (FFT) calculation.

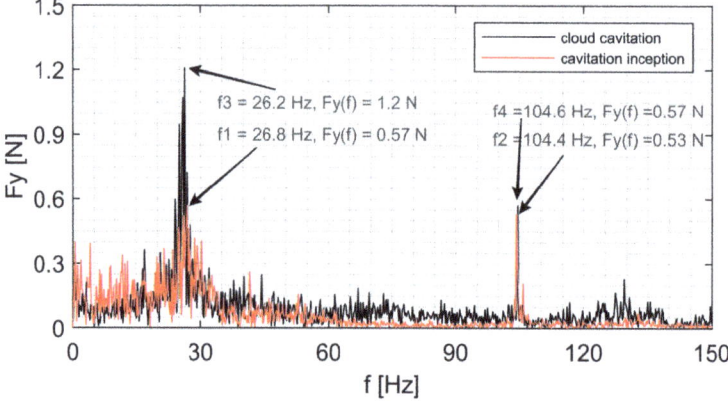

Figure 10. Comparison between lift forces on the circular cylinder at cloud cavitation regime (cavitation number = 1.0) and cavitation inception regime in the frequency domain at the velocity of 4 m/s.

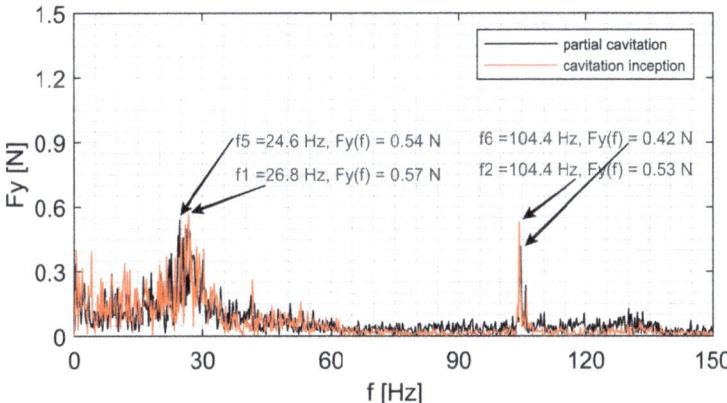

Figure 11. Comparison between lift forces on the circular cylinder at partial cavitation regime (cavitation number = 1.5) and cavitation inception in the frequency domain at the velocity of 4 m/s.

The peaks at relative low frequency ranges can be corresponded to the shedding frequency of the large-scale cavity structures. The peaks at the high frequency range remained constant for all cases at different cavitation regimes. In addition, the peaks at relative low frequency ranges for each cases can be the reason of the flow induced hydrodynamics forces on the cylinder. The tip vortex shedding and small cavities shedding can affect the hydrodynamic forces in addition to the effects of the collapse of the large cavity structures behind the cylinder. The peaks at higher frequency range show not significant shift in the frequency range. From the Figure 10 can be seen that the peaks f1 and f3 represent the hydrodynamics lift forces at the low frequency range for the cavitation inception and cloud cavitation, respectively. The peaks f2 and f4 reveal the hydrodynamic lift forces at the high frequency domain for the cavitation inception and cloud cavitation, respectively. It can be deduced that the amplitude of the frequency f3 is about twice of the frequency f1 at a similar frequency. This means that the presence of the cavitation can affect the vibration behavior on the circular cylinder. In other words, the vibration amplitude of the cylinder can be increased at the cavitation cloud regime compared to the vibration amplitude of the cylinder at the cavitation inception regime. It can be deduced that the cavitation cloud can merged with Karman vortex street formed behind the cylinder at non-cavitating regime and induce much higher vibration amplitude acted on the cylinder. In this case, the frequency value and amplitude of frequency at relatively high frequency range show no significant difference at the cavitation inception and cloud cavitation regimes. Therefore, it can be included that the presence of the cavitation has less affect on the vibration amplitude of the cylinder at high frequency range in the cloud cavitation regime compared to the cavitation inception regime.

Figure 11 presents the comparison of the hydrodynamic lift forces on the cylinder in the frequency domain at partial cavitation regime (cavitation number = 1.5). The peaks at the low frequency domain, f1 and f5 have mostly the same amplitudes however an increasing of the frequency value about 2 Hz in the cavitation inception regime can be observed. The reason can be the low cavitation volume formed on the cylinder at cavitation inception regime and the large distance between the cavity collapse position and the cylinder surface. As it can be seen from the results, the cavitation has a influence on the vibration amplitude at the high frequency range in the partial cavitation regime. A reduction of about 20% for the vibration amplitude can be deduced due to the effects of the cavitation on the hydrodynamic force on the cylinder. One of the reason for this amplitude reduction could be the effects of the tip vortex cavitation shedding on the cavitation dynamics behind the cylinder.

Figure 12 and 13 show the hydrodynamic lift forces on the circular cylinder in the cloud cavitation regime ($\sigma = 1.0$), partial cavitation regime ($\sigma = 1.5$) and cavitation inception ($\sigma = 2.4$) at velocity of 5 m/s. The lift force on the cylinder at this velocity shows a strong noise in the frequency range from 23 to 40 Hz. The vortex shedding at this range was affected by the small cavitation structures formed behind the cylinder. The peaks f7 and f9 represent the amplitude of the lift force on the cylinder in the cavitation inception and cloud cavitation regimes, respectively. In other words, the vibration amplitude in the cloud cavitation regime is about 1.5 times of the vibration amplitude on the cylinder in the cavitation inception regime. However, the vibration amplitude increasing rate at the flow velocity of 5 m/s is lower than the amplitude increasing rate at the flow velocity of 4 m/s in the low frequency range. In addition, the vibration amplitude on the cylinder at lower Reynolds number is higher than the vibration amplitude at higher Reynolds number. In relatively high frequency range from 45 to 135 Hz in the Figure 12, the lift force on the circular cylinder in the cloud cavitation regime has several small peaks with amplitudes over 0.3 N. These peaks were probably caused by the small cavitation structure shedding, which have also relatively higher frequency. Furthermore, the lift force peak of the frequency in the cloud cavitation regime (f10) is about 60% of the lift force peak in the cavitation inception regime (f8). In other words, the vibration amplitude for the cavitation inception regime at higher frequency is higher than the vibration amplitude for the case with the cloud cavitation. In Figure 13, the results show that the lift forces have similar vibration magnitude at the cavitation inception regime in the low frequency range compared to the partial cavitation regime at the same condition. In addition, the value and amplitude of the frequency for both cavitating regimes are similar to each other.

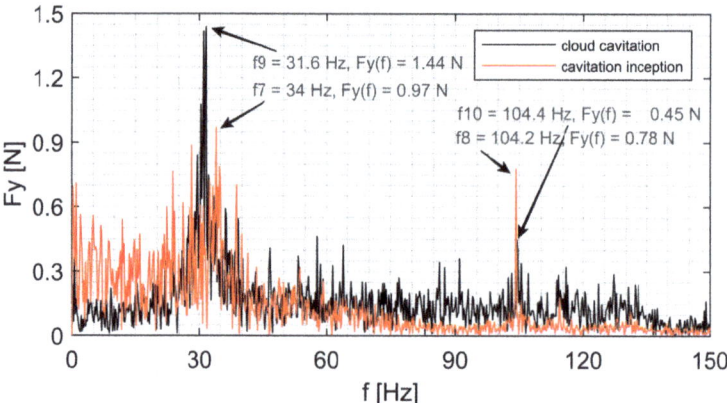

Figure 12. Comparison between hydrodynamic lift forces on the circular cylinder at cloud cavitation regime (cavitation number = 1.0) and cavitation inception in the frequency domain at the velocity of 5 m/s.

The results revealed, at the velocity of 4 m/s, the amplitude of the lift force at cloud cavitation regime (cavitation number of 1.0) is 2.2 times of the amplitude of the lift force at partial cavitation regime (cavitation number of 1.5). At the velocity of 5 m/s, the vibration amplitude at cloud cavitation regime (cavitation number of 1.0) is 1.46 times of the vibration amplitude at partial cavitation regime (cavitation number of 1.5). However, at a constant cavitation number of 1.0, the vibration amplitude at the velocity of 5 m/s is 1.2 times of the vibration amplitude at the velocity of 4 m/s. In addition, at a constant cavitation number of 1.5, the vibration amplitude at the velocity of 5 m/s is 1.8 times of the vibration amplitude at the velocity of 4 m/s.

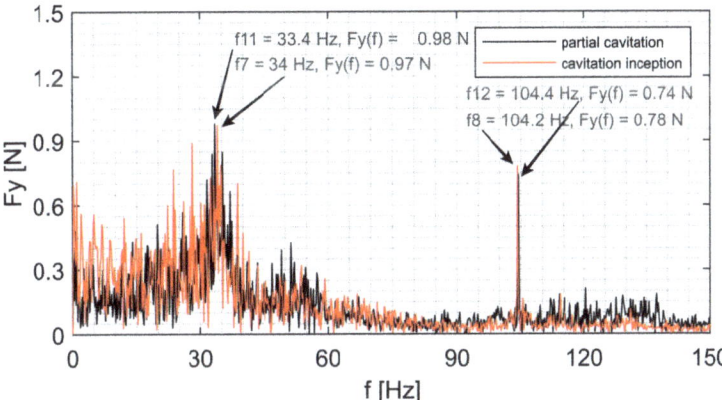

Figure 13. Comparison between hydrodynamic lift forces on the circular cylinder at partial cavitation regime (cavitation number = 1.5) and cavitation inception in the frequency domain at the velocity of 5 m/s.

4. Conclusions

We performed an experimental study on the cavitation structure and the corresponding hydrodynamic force on a circular cylinder at three Reynolds numbers and two different cavitating regimes. The cavity shedding mechanism, the time history of force coefficient and the lift force in frequency domain have been analyzed to understand the cavitation dynamics and the effects of cavitation on the hydrodynamic force on the circular cylinder. The results revealed that by increasing the flow velocity, larger cavity volume on the circular cylinder can be observed which lead to a stronger hydrodynamic forces. With increasing of the Reynolds number, the frequency of lift force on the cylinder shifted to a higher frequency position, indicating an increase in vortex and cavity shedding frequencies. At the same Reynolds number, changing the cavitation number affected the cavitation dynamics and the amplitude of the hydrodynamics forces on the circular cylinder. However, the cavitation regime had only a small influence on the frequency value of the lift force in the low frequency range. It can be concluded that the effects of the Reynolds number on the cavitation dynamics and amplitude of the shedding frequency is significant. However the effects of the cavitation number on the enhancement of the amplitude of the shedding frequency in a cavitating flow with a constant velocity is slightly higher than the effects of Reynolds number on the enhancement of the amplitude of the shedding frequency at a constant cavitation number.

Author Contributions: Conceptualization, E.K.; methodology, E.K.; validation, E.K. and Y.L.; formal analysis, E.K. and Y.L.; investigation, E.K. and Y.L.; resources, E.K. and Y.L.; data curation, E.K. and Y.L.; writing—original draft preparation, E.K. and Y.L.; writing—review and editing, E.K. and Y.L.; visualization, E.K. and Y.L.; supervision, O.e.M.; project administration, O.e.M.; funding acquisition, E.K. and O.e.M. All authors have read and agreed to the published version of the manuscript.

Funding: This work was supported by Deutsche Forschungsgemeinschaft (DFG) with the project number 469042952.

Data Availability Statement: The data presented in this study are available in the article.

Acknowledgments: Authors acknowledge the support provided by students; Minoo Ataei and Khashayar Ardalan for the design of the test samples.

Conflicts of Interest: The authors declare no conflict of interest.

References

1. Reisman, G.; Wang, Y.; Brennen, C.E. Observations of shock waves in cloud cavitation. *J. Fluid Mech.* **1998**, *355*, 255–283. [CrossRef]
2. Patella, R.F.; Choffat, T.; Reboud, J.L.; Archer, A. Mass loss simulation in cavitation erosion: Fatigue criterion approach. *Wear* **2013**, *300*, 25–215.
3. Kadivar, E.; el Moctar, O.; Skoda, R.; Löschner, U. Experimental study of the control of cavitation-induced erosion created by collapse of single bubbles using a micro structured riblet. *Wear* **2021**, *486–487*, 204087. [CrossRef]
4. Dular, M.; Bachert, B.; Stoffel, B.; Sirok, B. Relationship between cavitation structures and cavitation damage. *Wear* **1998**, *355*, 255–283. [CrossRef]
5. Lin, Y.; Kadivar, E.; el Moctar, O.; Neugebauer, J.; Schellin, T.E. Experimental investigation on the effect of fluid–structure interaction on unsteady cavitating flows around flexible and stiff hydrofoils. *Phys. Fluids* **2022**, *34*, 083308. [CrossRef]
6. Arndt, R.E.; Ippen, A.T. *Cavitation near Surfaces of Distributed Roughness*; Massachusetts Institute of Technology: Cambridge, MA, USA, 1967.
7. Kadivar, E.; el Moctar, O.; Sagar, H. Experimental study of the influence of mesoscale surface structuring on single bubble dynamics. *J. Ocean Eng.* **2022**, *260*, 111892. [CrossRef]
8. Young, Y.L. Fluid–structure interaction analysis of flexible composite marine propellers. *J. Fluids Struct.* **2008**, *24*, 255–283. [CrossRef]
9. Venning, J.A.; Pearce, B.W.; Brandner, P.A. Nucleation effects on cloud cavitation about a hydrofoil. *J. Fluid Mech.* **2022**, *947*, A1. [CrossRef]
10. Phan, T.-H.; Kadivar, E.; Nguyen, V.-T.; el Moctar, O.; Park, W.-G. Thermodynamic effects on single cavitation bubble dynamics under various ambient temperature conditions. *Phys. Fluids* **2022**, *34*, 023318. [CrossRef]
11. Ge, M.; Petkovšek, M.; Zhang, G.; Jacobs, D.; Coutier-Delgosha, O. Cavitation dynamics and thermodynamic effects at elevated temperatures in a small Venturi channel. *Int. J. Heat Mass Transf.* **2021**, *170*, 120970. [CrossRef]
12. Ge, M.; Manikkam, P.; Ghossein, J.; Subramanian, R.K.; Coutier-Delgosha, O.; Zhang, G. Dynamic mode decomposition to classify cavitating flow regimes induced by thermodynamic effects. *Energy* **2022**, *254*, 124426. [CrossRef]
13. Coutier-Delgosha, O.; Devillers, J.F.; Pichon, T.; Vabre, A.; Woo, R.; Legoupil, S. Internal structure and dynamics of sheet cavitation. *Phys. Fluids* **2006**, *18*, 017103. [CrossRef]
14. Barre, S.; Rolland, J.; Boitel, G.; Goncalves, E.; Patella, R.F. Experiments and modeling of cavitating flows in venturi: Attached sheet cavitation. *Eur. J. Mech.-B/Fluids* **2009**, *28*, 444–464. [CrossRef]
15. Pelz, P.F.; Keil, T.; Groß, T.F. The transition from sheet to cloud cavitation. *J. Fluid Mech.* **2017**, *817*, 439–454. [CrossRef]
16. Kadivar, E. Experimental and Numerical Investigations of Cavitation Control Using Cavitating-Bubble Generators. Ph.D. Thesis, University of Duisburg-Essen, Duisburg, Germany, 2020.
17. Le, Q.; Franc, J.P.; Michel, J.M. Partial cavities: Global behavior and mean pressure distribution. *J. Fluids Eng.* **1993**, *115*, 243–248. [CrossRef]
18. Stutz, B.; Reboud, J.-L. Two-phase flow structure of sheet cavitation. *Phys. Fluids* **1997**, *9*, 3678–3686. [CrossRef]
19. Callenaere, M.; Franc, J.-P.; Michel, J.-M.; Rionde, M. The cavitation instability induced by the development of a re-entrant jet. *J. Fluid Mech.* **2001**, *444*, 223–256. [CrossRef]
20. Leroux, J.-B.; Astolfi, J.A.; Billard, J.Y. An Experimental Study of Unsteady Partial Cavitation. *J. Fluids Eng.* **2004**, *126*, 94–101. [CrossRef]
21. Ganesh, H.; Mäkiharju, S.A.; Ceccio, S.L. Bubbly shock propagation as a mechanism for sheet-to-cloud transition of partial cavities. *J. Fluid Mech.* **2016**, *802*, 37–78. [CrossRef]
22. Wu, J.; Ganesh, H.; Ceccio, S. Multimodal partial cavity shedding on a two-dimensional hydrofoil and its relation to the presence of bubbly shocks. *Exp. Fluids* **2019**, *60*, 1–17. [CrossRef]
23. Karathanassis, I.; Koukouvinis, P.; Kontolatis, E.; Lee, Z.; Wang, J.; Mitroglou, N.; Gavaises, M. High-speed visualization of vortical cavitation using synchrotron radiation. *J. Fluid Mech.* **2018**, *838*, 148–164. [CrossRef]
24. Karathanassis, I.; Heidari-Koochi, M.; Zhang, Q.; Hwang, J.; Koukouvinis, P.; Wang, J.; Gavaises, M. X-ray phase contrast and absorption imaging for the quantification of transient cavitation in high-speed nozzle flows. *Phys. Fluids* **2021**, *33*, 032102. [CrossRef]
25. Kadivar, E.; Timoshevskiy, M.V.; Nichik, M.Y.; el Moctar, O.; Schellin, T.E.; Pervunin, K.S. Control of unsteady partial cavitation and cloud cavitation in marine engineering and hydraulic systems. *Phys. Fluids* **2020**, *32*, 052108. [CrossRef]
26. Kadivar, E.; Ochiai, T.; Iga, Y.; el Moctar, O. An experimental investigation of transient cavitation control on a hydrofoil using hemispherical vortex generators. *J. Hydrodyn.* **2020**, *33*, 1139–1147. [CrossRef]
27. Kermeen, R.W.; Parkin, B.R. Incipient Cavitation and Wake Flow behind Sharp-Edged Disks. 1957. Hydrodynamics Lab Report No. 85-4. Available online: https://apps.dtic.mil/sti/citations/AD0144749 (accessed on 3 April 2023).
28. Arndt, R.E. Semiempirical Analysis of Cavitation in the Wake of a Sharp-Edged Disk. *J. Fluids Eng.* **1976**, *98*, 560–562. [CrossRef]
29. Belahadji, B.; Franc, J.P.; Michel, J.M. Cavitation in the rotational structures of a turbulent wake. *J. Fluid Mech.* **1995**, *287*, 383–403. [CrossRef]
30. Wu, J.; Deijlen, L.; Bhatt, A.; Ganesh, H.; Ceccio, S.L. Cavitation dynamics and vortex shedding in the wake of a bluff body. *J. Fluid Mech.* **2021**, *917*, A26. [CrossRef]

31. Roshko, A. Experiments on the flow past a circular cylinder at very high Reynolds number. *J. Fluid Mech.* **1961**, *10*, 345–356. [CrossRef]
32. Bearman, P.W. On vortex shedding from a circular cylinder in the critical Reynolds number regime. *J. Fluids. Mech.* **1969**, *37*, 577–585. [CrossRef]
33. Wei, T.; Smith, C.R. Reboud. Secondary vortices in the wake of circular cylinders. *J. Fluids. Mech.* **1986**, *169*, 513–533. [CrossRef]
34. Williamson, C.H.K. Three-dimensional wake transition. *J. Fluid Mech.* **1996**, *328*, 345–407. [CrossRef]
35. Szepessy, S.; Bearman, P. Aspect ratio and end plate effects on vortex shedding from a circular cylinder. *J. Fluid Mech.* **1992**, *234*, 191–217. [CrossRef]
36. Norberg, C.; Reboud, J.-L. An experimental investigation of the flow around a circular cylinder: Influence of aspect ratio. *J. Fluid Mech.* **1994**, *258*, 287–316. [CrossRef]
37. Fry, S.A. Investigating cavity/wake dynamics for a circular cylinder by measuring noise spectra. *J. Fluid Mech.* **1997**, *42*, 187–200. [CrossRef]
38. Matsudaira, Y.; Gomi, Y.; Oba, R. Characteristics of bubble-collapse pressures in a karman-vortex cavity. *JSME Int. J.* **1992**, *35*, 179–185. [CrossRef]
39. Saito, Y.; Sato, K. Cavitation bubble collapse and impact in the wake of a circular cylinder. In Proceedings of the Fifth International Symposium on Cavitation (CAV2003), Osaka, Japan, 1–4 November 2003; pp. 3–8.
40. Franc, J.P.; Michel, J.M. Fundamentals of Cavitation; Springer Science & Business Media: Berlin/Heidelberg, Germany, 2006; Volume 76.
41. Gnanaskandan, A.; Mahesh, K. Numerical investigation of near-wake characteristics of cavitating flow over a circular cylinder. *J. Fluid Mech.* **2016**, *790*, 453–491. [CrossRef]
42. Ghahramani, E.; Jahangir, S.; Neuhauser, M.; Bourgeois, S.; Poelma, C.; Bensow, R.E. Experimental and numerical study of cavitating flow around a surface mounted semi-circular cylinder. *Int. J. Multiph. Flow* **2020**, *124*, 103191. [CrossRef]
43. Kumar, P.; Chatterjee, D.; Bakshi, S. Experimental investigation of cavitating structures in the near wake of a cylinder. *Int. J. Multiph. Flow* **2017**, *89*, 207–217. [CrossRef]
44. Brandao, F.L.; Bhatt, M.; Mahesh, K. Numerical study of cavitation regimes in flow over a circular cylinder. *J. Fluid Mech.* **2019**, *885*, A19. [CrossRef]
45. Dobroselsky, K. Cavitation streamlining of a round cylinder in the critical range. *J. Phys. Conf. Ser.* **2020**, *1677*, 012056. [CrossRef]
46. Sadri, M.; Kadivar, E. Numerical investigation of the cavitating flow and the cavitation-induced noise around one and two circular cylinders. *Ocean Eng.* **2023**, *277*, 114178. [CrossRef]

Disclaimer/Publisher's Note: The statements, opinions and data contained in all publications are solely those of the individual author(s) and contributor(s) and not of MDPI and/or the editor(s). MDPI and/or the editor(s) disclaim responsibility for any injury to people or property resulting from any ideas, methods, instructions or products referred to in the content.

Article

Prediction of Critical Heat Flux during Downflow in Fully Heated Vertical Channels

Mirza M. Shah

Engineering Research Associates, 10 Dahlia Lane, Redding, CT 06896, USA; mshah.erc@gmail.com

Abstract: Boiling with downflow in vertical channels is involved in many applications such as boilers, nuclear reactors, chemical processing, etc. Accurate prediction of CHF (Critical Heat Flux) is important to ensure their safe design. While numerous experimental studies have been done on CHF during upflow and reliable methods for predicting it have been developed, there have been only a few experimental studies on CHF during downflow. Some researchers have reported no difference in CHF between up- and downflow, while some have reported that CHF in downflow is lower or higher than that in upflow. Only a few correlations have been published that are stated to be applicable to CHF during downflow. No comprehensive comparison of correlations with test data has been published. In the present research, literature on CHF during downflow in fully heated channels was reviewed. A database for CHF in downflow was compiled. The data included round tubes and rectangular channels, hydraulic diameters 2.4 mm to 15.9 mm, reduced pressure 0.0045 to 0.6251, flow rates from 15 to 21,761 kg/m^{2}s, and several fluids with diverse properties (water, nitrogen, refrigerants). This database was compared to a number of correlations for upflow and downflow CHF. The results of this comparison are presented and discussed. Design recommendations are provided.

Keywords: critical heat flux; downflow; tubes; rectangular channels; correlations; prediction

Citation: Shah, M.M. Prediction of Critical Heat Flux during Downflow in Fully Heated Vertical Channels. *Fluids* **2024**, *9*, 79. https://doi.org/10.3390/fluids9030079

Academic Editors: Yi-Tung Chen and Van Tu Nguyen

Received: 26 January 2024
Revised: 10 March 2024
Accepted: 15 March 2024
Published: 20 March 2024

Copyright: © 2024 by the author. Licensee MDPI, Basel, Switzerland. This article is an open access article distributed under the terms and conditions of the Creative Commons Attribution (CC BY) license (https://creativecommons.org/licenses/by/4.0/).

1. Introduction

Boiling with downflow in vertical channels is involved in many applications such as boilers, nuclear reactors, chemical processing, etc. Accurate prediction of CHF (Critical Heat Flux) is important to ensure their safe design. Many experimental studies have been done on CHF during upflow and reliable methods for predicting it have been developed. There have been comparatively few experimental studies on CHF during downflow. There are differences in the results reported by various researchers. Some have stated that they found no difference between the CHF during upflow and downflow; for example, Barnett (1963) [1]. Some have reported that CHF during downflow is higher or lower than that during upflow under various conditions; for example, Chen (1993) [2]. During upflow, buoyancy force is in the direction of flow. During downflow, buoyancy force is against the flow direction. Hence, some differences in the CHF in these two directions may be expected. Only a few correlations have been published that are stated to be applicable to CHF during downflow. No comprehensive comparison of correlations with test data has been published. There is a lack of well-verified methods to predict CHF during downflow.

The objective of this research was to determine whether, in fact, there is a significant difference between CHF in upflow and downflow, and to develop a reliable prediction method for downflow CHF if a significant difference was found. To achieve this objective, literature was surveyed to identify experimental studies, data sources, and prediction methods. Of special interest were experimental studies in which CHF was measured with flow in both upward and downward directions. A comprehensive database was developed and compared to the best available correlations for upflow and downflow CHF. The results of this research are presented and discussed. It is to be noted that this research was confined to fully heated channels; partially heated channels are not included.

2. Previous Work

2.1. Experimental Work

Gambill and Bundy (1961) [3] measured CHF during downflow of water in rectangular channels. They compared their data with correlations based on upflow CHF. The agreement was fairly good.

Barnett (1963) [1] conducted tests at pressures of 38 bar and 138 bar with water flowing in a vertical tube. He found no effect of flow direction on the boiling crisis.

Pappel et al., (1966) [4] performed tests with nitrogen in a vertical tube. Nitrogen was subcooled at the entrance to the tube. They found that CHF in downflow was lower than that in upflow at low flow rates. The difference disappeared at high flow rates. Pappel (1972) [5] performed similar tests with zero inlet quality and the results were similar.

Kirby et al., (1967) [6] performed tests with up- and downflow with water at 1.7 bar in an annulus. They report that CHF in downflow was 10 to 30 percent lower in downflow, the larger difference being at the lowest flow rate.

Cumo et al., (1977) [7] performed tests with R-12 flowing up and down a vertical tube. They concluded that CHF during downflow is 10 to 30% lower than that in upflow, especially at low inlet qualities. They attributed this difference to the effect of buoyancy.

Lazarek and Black (1982) [8] performed tests with R-113 in a vertical tube. They found no difference between CHF during upflow and downflow.

Mishima et al., (1985) [9] performed upflow and downflow CHF tests on a 6 mm diameter tube with water at atmospheric pressure as the test fluid. Tests were done alternatively with a stiff system and a soft system. In the stiff system, precautions were taken to prevent instability, such as by applying strong throttling at tube inlet, while such precautions were not taken in the soft system. CHF in the stiff system was considerably higher than that in the soft system. They found no difference between the CHF in upflow and downflow.

Remizov et al., (1985) [10] did tests on a vertical tube in which critical quality was measured in upflow and downflow at identical inlet subcooling, flow rate, and heat flux. They found that at the lowest flow rate, critical quality was always lower for upflow, though the difference decreased with increasing heat flux. At the highest flow rate, critical quality was lower for upflow at low heat flux but higher at high heat flux.

Deqiang et al., (1987) [11] performed tests with R-12 in an 8 mm diameter vertical tube. They found the downflow CHF to be lower than the upflow CHF at low flow rates, but equal at high flow rates.

Chang et al., (1991) [12] performed tests with atmospheric pressure in vertical tubes. Their tests showed that CHF in up- and downflow was essentially the same at low flow rates. At higher flow rates, CHF in upflow was higher, though the difference was small. They found that it was more difficult to maintain stability in downflow. They proposed a correlation for CHF applicable to both upflow and downflow without any factor for the effect of flow direction.

Chen (1993) [2] analyzed experimental data for upflow and downflow critical heat flux of water and freon in a vertical tube. It was found that the total rms (root-mean-square) of the comparison of upflow and downflow data and predicting downflow data using upflow CHF correlation are in the range of 6–14%. The CHF for upflow was regularly greater than that for downflow, but was smaller than that in downflow in the range of low critical quality. The downflow CHF was 80% of the upflow value at the point of the maximum difference between the two. (This description is based on the abstract of this report.)

Ruan et al., (1993) [13] performed tests on downflow of water in a vertical tube. Tests were done with different amounts of instability. They found that, in a stable system, downflow CHF approached that for upflow. In very unstable systems, CHF value corresponded to flooding CHF.

Ami et al., (2015) [14] performed tests with water in a vertical tube. For the data in which the location of CHF was known, CHF in upflow and downflow was about equal at lower flow rates. At the highest flow rate, CHF in downflow was about 15% higher than in upflow.

Sripada et al., (2021) [15] measured CHF with water flowing downwards in a 6 mm diameter vertical tube. Their measured CHF was very low, even much lower than that by

Mishima et al., (1985) [9] under unstable conditions. They had not done any throttling at the tube inlet. These data are clearly for unstable conditions. No conclusions can be drawn from such unstable CHF data.

2.2. Prediction Methods

While there are many correlations for CHF during upflow, only a few correlations have been proposed which are stated to be applicable to CHF during downflow. The more verified among them are discussed below.

Sudo et al., (1985) [16] have given the following correlation based on data for tubes and rectangular channels which is applicable to both upflow and downflow. It is given below.

$$q^* = 0.005 G^{*0.611} \tag{1}$$

$$q^* = \left(\frac{A_F}{A_H}\right) x_{in} G^* \tag{2}$$

At very low flow rates, CHF was considered to be due to flooding and the following equation was given for it:

$$q^* = C^2 \left(\frac{A_F}{A_H}\right) \frac{(D/\lambda)^{0.5}}{\left(1 + (\rho_G/\rho_L)^{1/4}\right)^2} \tag{3}$$

For rectangular channels, D is replaced by the channel width W. The constant C^2 is 0.71.

G^* and q^* are defined as:

$$q^* = \frac{q_c}{i_{LG}[\lambda \rho_G (\rho_L - \rho_G) g]^{0.5}} \tag{4}$$

$$G^* = \frac{G}{[\lambda \rho_G (\rho_L - \rho_G) g]^{0.5}} \tag{5}$$

$$\lambda = \frac{\sigma^{0.5}}{[(\rho_L - \rho_G) g]^{0.5}} \tag{6}$$

For upflow, q^* is the larger of those given by Equations (1) and (3). For downflow, Equation (1) applies when $G^* > 10^4$. For $G^* < 10^4$, q^* is the larger of those from Equations (2) and (3).

Hirose et al., (2024) [17] have given the following correlation for downflow based on data from several sources:

$$q^* = 0.422 G^{*0.564} (L_c/D)^{-0.902} \tag{7}$$

$$q^* = C^2 \left(\frac{A_F}{A_H}\right) \frac{i_{fg}(\rho_G g D(\rho_L - \rho_G))}{\left(1 + (\rho_G/\rho_L)^{1/4}\right)^2} \tag{8}$$

The higher of the q^* given by Equations (7) and (8) is to be used. Equation (8) is for CHF due to flooding. The constant C is to be determined from experimental data. They used $C = 1.18$.

Darges et al., (2022) [18] have given the following correlation, which is intended to be applicable to all flow directions:

$$Bo = 0.353 We_D^{-0.314} \left(\frac{L_c}{D_{HP}}\right)^{-0.226} \left(\frac{\rho_L}{\rho_G}\right)^{-0.481} \left[1 - x_{in}\left(\frac{\rho_L}{\rho_G}\right)^{-0.094}\right]$$
$$\times \left(1 + Fr_\theta^{-1}\right) \left(1 + 0.008 \frac{Bd_\theta}{We_D^{0.543}}\right) \tag{9}$$

where,

$$We_D = \frac{G^2 D_{HP}}{\rho_L \sigma} \quad (10)$$

$$Fr_\theta = \frac{G^2}{\rho_L^2 \cdot D_{HP} Sin\theta g} \quad (11)$$

$$Bd_\theta = \frac{gCos\theta(\rho_{L-}\rho_G)D_{HP}^2}{\sigma} \quad (12)$$

This correlation was based on data obtained by a team at Purdue University through tests on partially heated channels using FC-72 and nPFH fluids for many years. Tests were done in earth gravity, as well as in micro gravity. All flow directions were included in those tests. All of these tests were done on channels 2.5 mm × 5 mm made of plastic with heaters inserted in their sides.

Chang et al., (1991) [12] have given a correlation based on their own data as well as some data for low pressure water. Its predictions are the same for both up- and downflow. The reported accuracy is not very good.

There are many correlations for CHF during upflow. The best known among them are Shah (1987) [19] and Katto and Ohno (1984) [20]. Both of these were verified with wide ranging databases. Shah (2017) [21] had compared these correlations as well as several other correlations to data for CHF in small diameter channels. Shah's correlation was found to be the most accurate, followed by the correlations of Katto and Ohno and Zhang et al., (2006). The correlation of Wojtan et al., (2006) [22] was found to give fairly good agreement with refrigerant data.

3. Data Analysis

Efforts were made to collect data for downflow CHF. As noted by Rohsenow (1973) [13], only the data taken under stable conditions can be correlated and interpreted. Hence, data which showed instability were not considered. The data of Sripada et al., (2021) [23] were not considered as they were clearly obtained under unstable conditions, as discussed in Section 2.1. Ruan et al., (1993) [10] and Mishima et al., (1985) [9] have pointed out which of their data were taken under unstable conditions. Those data were not included in the present data analysis.

The figures in Mishima et al., (1985) [24] show no difference in CHF between upflow and downflow. These figures show CHF to initially increase linearly with mass velocity but show little or no effect of mass velocity at higher flow rates. The behavior at higher flow rates is against the trend shown by most data and these data are greatly overpredicted by all correlations. Hence, these were not included in the present study.

Some of the papers did not provide sufficient details to enable the analysis of data in them. For example, Deqiang et al., (1987) [11] have not given the length of the test tube without which their data cannot be analyzed.

In the paper by Ami et al., (2015) [14], CHF location is not given for most of the data and was therefore not analyzable. Some data are given for a 10 mm tube for which CHF location is stated. These were analyzed and the results are discussed in Section 4.2.

All data were read from figures in the publications except those of DeBortoli et al., which were read from tables.

The data for downward flow CHF that were analyzed are listed in Table 1. These were compared to the correlations of Shah, Katto and Ohno, Zhang et al., and Wojtan et al., which are based on upflow data, as well as the correlations of Sudo et al., Darges et al., and Hirose et al., which are stated to be applicable to downflow CHF.

Table 1. Range of data for downflow in vertical fully heated channels and the results of their comparison with some correlations.

Source	Channel Shape	D (D_{HYD}), mm	L_c/D	Fluid	p_r	G Kg/m²s	$Y^* \times 10^{-4}$	x_{in}	x_c	N	Katto–Ohno	Zhang et al.	Wojtan et al.	Darges et al.	Sudo et al.	Hirose et al.	Shah
Dougherty et al., (1994) [25]	Round	15.9	153	water	0.0209	1706 8010	320 3200	−0.28 −0.15	0.00 0.26	28	17.3 17.3	21.1 21.1	37.3 −37.3	0.7 8.6	9.8 −5.5	36.7 −36.7	15.8 15.8
Mishima et al., (1985) [9]	Round	6.0	57.3	water	0.0045	20 239	0.067 6.0	−0.13 −0.04	0.55 0.84	13	20.7 20.7	8.6 −5.1	64.0 64.0	26.7 16.9	56.9 −56.9	26.8 −18.1	6.1 0.2
Lazarek & Black (1982) [8]	Round	3.1	81.9	R-113	0.0383	235 498	9.1 35	−0.22 −0.02	0.72 0.89	9	4.3 −0.5	12.3 −12.3	19.1 −19.1	88.7 88.7	51.5 −51.5	34.0 −34.0	26.1 −26.1
Chang et al., (1991) [12]	Round	9.0	76	water	0.0045	15 25	0.05 0.14	−0.15	0.77	3	18.8 18.8	13.0 −13.0	141.4 141.4	329.8 −329.8	19.0 −19.0	27.3 27.3	7.6 −7.6
			114				0.08 3.8	−0.15 −0.06	0.72 0.83	17	36.7 36.7	7.0 −1.2	64.2 64.2	105.2 105.2	19.8 −6.6	20.9 −14.7	8.8 8.2
Ruan et al., (1993) [26]	Round	9.0	44.3	water	0.0045 0.0317	26 203	0.12 5.7	−0.07 −0.01	−0.05 1.08	20	11.9 4.1	19.6 −18.9	44.2 43.0	24.9 5.2	65.5 −65.5	22.6 −12.1	17.0 −15.0
DeBortoli et al., (1957) [27]	Rect. 25.4 W, 2.46 H	(4.49)	153	water	0.6251	205 978	4.2 250	−0.20 −0.04	0.10 0.97	9	10.0 −6.4	18.3 5.5	50.0 −50.0	390.7 390.7	13.8 −0.2	24.0 −24.0	14.3 11.0
			68.1			313 457	9 18	−0.13	0.15	2	18.6 18.6	20.8 −20.8	69.3 −69.3	251.9 251.9	73.8 −73.8	57.0 −57.0	19.8 −19.9
	Rect. 1.27 W, 2.4 H	(2.42)	126			457 768	12 31	−1.31 −1.22	0.21 0.34	4	12.6 12.6	15.1 −15.1	72.7 −72.7	379.8 379.8	62.3 −62.3	65.2 −65.2	18.8 −18.8
Gambill & Bundy (1961) [3]	Rect. 2.5 W × 2.5 H	(2.5)	186	water	0.0500 0.1719	7465 21,761	1100 7700	−0.59 −0.33	−0.10 −0.06	7	18.1 9.7	16.8 4.7	63.6 −63.6	76.7 76.7	37.7 −37.7	65.5 −65.5	15.6 10.1
Pappel et al., (1966) [4]	Round	12.5	24.4	Nitrogen	0.2032 0.4859	119 434	9.5 81	−0.51 −0.19	−0.11 0.01	5	47.7 45.9	26.2 18.2	35.1 27.3	298.9 298.9	47.5 −47.5	122.9 122.9	69.6 69.6
					0.1016 0.4046	484 2557	100 2400	−0.51 −0.08	−0.36 −0.01	52	18.1 18.1	24.8 12.7	66.8 65.8	145.0 145.0	32.5 −20.3	173.6 173.6	26.5 26.5
Pappel (1972) [5]	Round	12.5	24.4	Nitrogen	0.1060 0.3004	168 455	16 98	0.00	0.14 0.33	12	19.1 17.8	20.3 20.3	78.4 78.4	192.4 192.4	33.6 −33.6	174.0 174.0	73.0 73.0
						488 2544	100 1900	0.00	0.04 0.12	40	12.2 11.3	60.9 60.9	179.7 179.7	221.9 221.9	28.4 13.3	334.9 334.9	24.0 12.0
Cumo et al., (1977) [7]	Round	7.8	282	R-12	0.2587 0.4231	130 1000	3.6 160	−0.44 0.28	0.37 1.1	74	20.4 13.8	32.1 29.4	31.4 −29.9	877.8 877.8	64.9 −18.8	33.4 27.4	14.9 4.8
Remizov et al., (1983) [10]	Round	10.0	234	Water	0.6209	350 500	18 34	−0.11 −0.06	0.76 0.88	6	23.5 23.5	39.9 39.9	37.9 −37.9	1181.4 1181.4	89.0 −89.0	16.8 16.8	25.9 25.9
			511			700	61	−0.03 −0.02	0.43 0.46	3	55.1 55.1	93.1 93.1	26.7 −26.7	1139.9 1139.9	94.0 −94.0	40.1 40.1	70.9 70.9
All sources	Round, rectangular	2.4 15.9	44 186	Water, R-12, R-113, N₂	0.0045 0.6251	15 21,761	0.05 7700	−1.31 0.00	−0.10 1.1	304	18.9 15.7	28.8 20.0	66.4 32.4	350.2 341.8	266.6 199.4	103.9 80.3	21.9 13.7

Calculation of CHF with the local condition part of the Shah correlation requires the insertion of critical quality x_c. As x_c depends on the critical heat flux which itself has to be determined, iterative calculations were done with assumed values of x_c until the assumed and calculated values converged to within 0.01. During these iterations, x_c is calculated with the heat balance equation:

$$x_c = x_{in} + 4BoL_c/D_{HP} \qquad (13)$$

where,

$$Bo = \frac{q_c}{Gi_{LG}} \qquad (14)$$

For the data in which $x_{in} > 0$, calculations for all correlations were done using the boiling length L_B in place of L_c. It is defined as:

$$\frac{L_B}{D_{HP}} = \frac{L_c}{D_{HP}} + \frac{x_{in}}{4Bo} \qquad (15)$$

As q_c is to be determined, calculations were done with assumed values of q_c until adequate convergence was achieved.

Properties were obtained from REFPROP 9.1, Lemmon et al., (2013) [24].

The deviations listed in Table 1 are defined as below.

Mean absolute deviation (MAD) of a data set is defined as:

$$MAD = \tfrac{1}{N}\sum_1^N ABS\left(\left(q_{c,predicted} - q_{c,measured}\right)/q_{c,measured}\right) \qquad (16)$$

Average deviation of a data set AD is defined as:

$$AD = \tfrac{1}{N}\sum_1^N \left(\left(q_{c,predicted} - q_{c,measured}\right)/q_{c,measured}\right) \qquad (17)$$

The results in Table 1 show that the correlations of Shah, Katto and Ohno, and Zhang et al., are in fairly good agreement with most data while the other correlations, including those for downflow, have large deviations with most data.

4. Discussion

4.1. Accuracy of Correlations

In Table 1, it is seen that only the correlations of Shah, Katto and Ohno, and Zhang et al., show reasonable agreement with the downflow data. These correlations were developed and verified with upflow data. The correlations of Darges et al., Hirose et al., and Sudo et al., which were stated to be applicable to downflow, have large deviations with most data. The correlation of Hirose et al., has fairly good agreement with many data sets. Its overall MAD is very large because it has very large deviations with the data of Pappel et al., (1966) [4] and Pappel (1972) [5] for nitrogen. Those data are 36% of the total 304 data points. If the nitrogen data are left out, the MAD of the Hirose et al., correlation goes down to 33%, which is much more reasonable. The data analyzed by Hirose et al., did not include any for nitrogen or other cryogens.

Among the upflow correlations, Katto and Ohno have the least MAD of 18.9%. The next lowest is the Shah correlation with MAD of 21.9%. If the data of nitrogen at $G < 460$ kg/m^{2}s are left out, the MAD of the Shah correlation becomes about the same as that of the Katto–Ohno correlation.

The Shah correlation also has large deviations with the data of Remizov et al., (1985) [10] for $G = 700$ kg/m^{2}s. These data are also overpredicted by the Katto–Ohno and Zhang et al., correlations.

From the above discussions, it is clear that the correlations of Katto–Ohno and Shah give the best agreement with downflow CHF data.

Figures 1–3 show a comparison of some CHF data for downflow in tubes with various correlations.

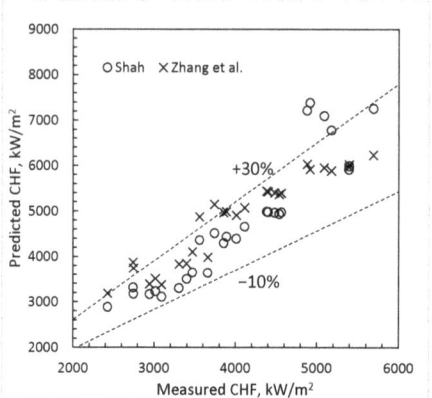

Figure 1. Data of Dougherty et al. [25] for downflow of water in a vertical tube compared to two correlations.

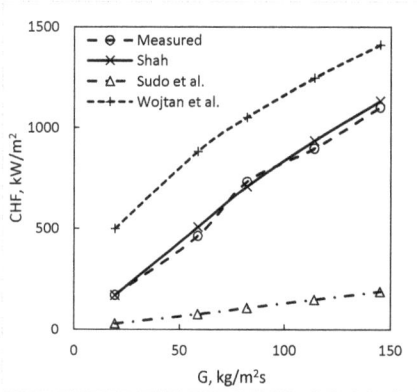

Figure 2. Data of Mishima et al., (1985) [9] for downflow of water in a round tube compared to some correlations. Pressure atmospheric, inlet quality −0.131.

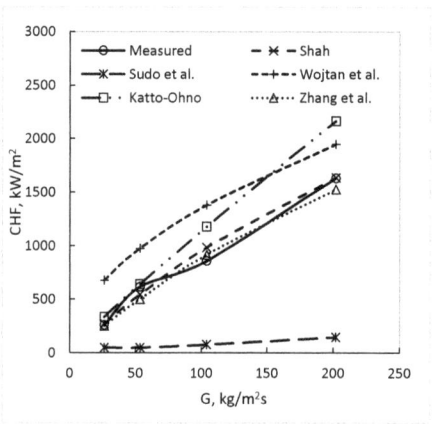

Figure 3. Data of Ruan et al., (1993) [26] for downward flow of water in a tube compared to various correlations. Pressure atmospheric, inlet quality −0.056.

4.2. Comparison of Upflow and Downflow Data

Some of the experimental studies on downflow CHF also included tests with upflow. Table 2 shows the deviations of Shah, Katto–Ohno, and Zhang et al., correlations with upflow and downflow data from those studies. The range of parameters during upflow was essentially the same as in the downflow listed in Table 1. The deviations of upflow and downflow data with the Shah correlation are seen to be comparable for all data except those of Pappel for nitrogen. If the data at low flow rate are left out, the MAD becomes about 25%, still significantly higher than about 16% for upflow. The results with the Zhang et al., correlation are similar. However, deviations of the Katto–Ohno correlations are about the same for upflow and downflow.

Table 2. Deviations of the best correlations with data from experimental studies in which both upflow and downflow CHF were measured.

Source	Channel Type	D_{HYD}	Fluid	Shah		Katto and Ohno		Zhang et al.	
				Downflow	Upflow	Downflow	Upflow	Downflow	Upflow
Pappel et al., (1966) [4]	Round tube	12.8	Nitrogen	30.3 30.3	17.0 12.4	20.7 20.5	13.2 5.6	25.0 13.2	21.2 −2.8
Pappel (1972) [5]	Round tube	12.8	Nitrogen	35.3 35.3	14.3 3.9	13.8 12.1	13.2 −9.4	51.5 51.5	26.9 21.9
Dougherty et al., (1994) [25]	Round tube	15.9	Water	15.8 15.8	14.5 12.7	17.3 17.3	14.4 13.7	21.1 21.1	21.4 21.4
Mishima et al., (1985) [9]	Round tube	6.0	Water	6.1 0.2	13.0 0.6	20.7 20.7	24.1 22.2	8.6 −5.1	15.0 −4.7
Lazarek & Black (1982) [8]	Round tube	3.1	R-113	26.1 −26.1	26.9 −26.9	4.3 −0.5	4.4 −1.9	12.3 −12.3	12.8 −12.8
Chang et al., (1991) [12]	Round tube	6.0	Water	8.8 8.2	10.6 8.9	36.7 36.7	37.9 37.9	7.0 −1.2	7.9 1.7
Remizov et al., (1983) [10]	Round tube	10.0	Water	40.9 40.9	42.2 42.2	34.1 34.1	35.7 35.7	57.6 57.6	59.5 59.5
DeBortoli et al., (1957) [27]	Rectangular channel	4.49	Water	15.3 −12.7	19.9 * −19.9	11.6 −8.6	9.0 * 9.0	18.7 0.71	17.5 * −17.5
		2.42	Water	18.8 −18.8	12.5 ** 0.2	12.6 12.6	19.8 ** 19.8	15.1 −15.1	2.8 ** −1.8
Cumo et al., (1977) [7]	Round tube	7.8	R-12	14.9 4.8	18.4 −7.6	20.4 13.8	17.7 −0.4	32.1 29.4	26.7 12.3
All sources				22.2 15.7	17.2 14.1	19.3 16.6	18.7 17.7	29.4 22.7	22.6 9.3

Note: * L/D = 58; ** L/D = 11, L/D for others same as in Table 1.

Figure 4 shows the data of Chang et al., (1991) [12] for up- and downflow together with predictions of some correlations. It is seen that there is really no difference in the CHF in the two directions, even at very low mass flux. While the Shah correlation predicts CHF a little higher at high flow rates, this cannot be attributed to flow direction as the measured CHF in both directions is about the same.

Figure 5 shows the data of Cumo et al., (1977) [7] for both upflow and downflow at the highest flow rate. It is seen that the downflow CHF at low inlet quality is a little lower than for upflow; meanwhile, at high inlet quality, they are about the same. The Shah correlation predictions are in-between the measured values in the two directions and, thus, in close agreement with both.

Figure 6 shows the data of Cumo et al., (1977) [7] at the lowest flow rate. CHF in downflow is about 15% lower than that in upflow; the two get close with increasing inlet quality. The correlations of Shah and Zhang et al., are within about −15% of data.

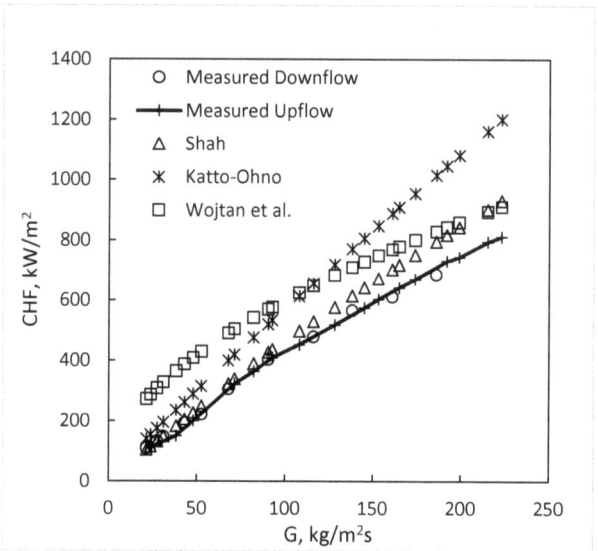

Figure 4. Data of Chang et al., (1991) [12] for up- and downflow of water in a vertical tube compared to various correlations. Pressure atmospheric, inlet quality −0.149, L/D = 114.

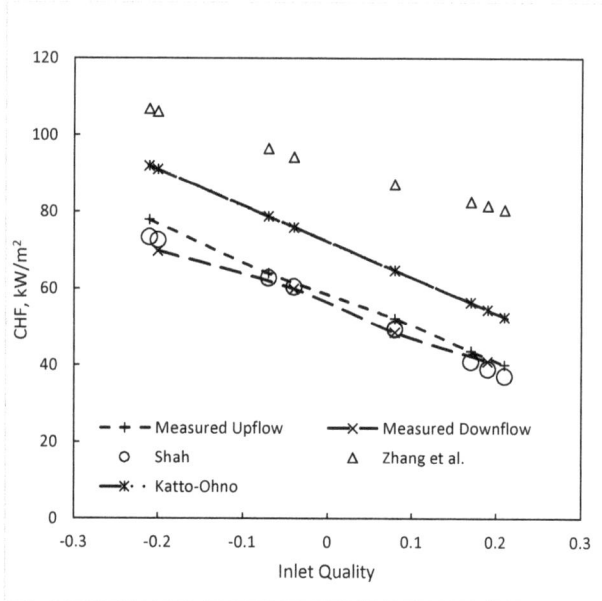

Figure 5. Data of Cumo et al., (1977) [7] at the highest mass flux compared to some correlations. $G = 1000$ kg/m^{2}s, pressure 10.5 bar.

Figure 7 shows the data of Pappel et al., (1966) [4] for nitrogen in both upflow and downflow. The data for downflow are considerably lower than upflow data at flow rates below about 500 kg/m^{2}s. Predictions of the Shah correlation are considerably higher than the downflow data for the lowest flow rates. On the other hand, the Katto–Ohno correlation gives good agreement throughout.

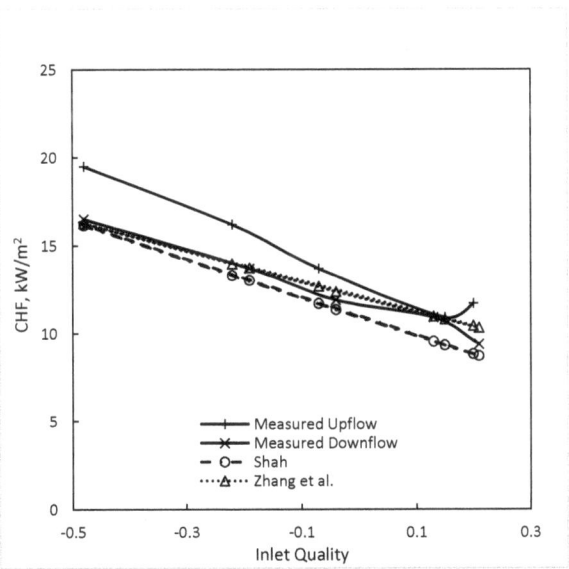

Figure 6. Data of Cumo et al., (1977) [7] for R-12 at the smallest flow rate compared to the Shah and Katto–Ohno correlations. Pressure 17.5 bar, G = 130 kg/m^{2}s.

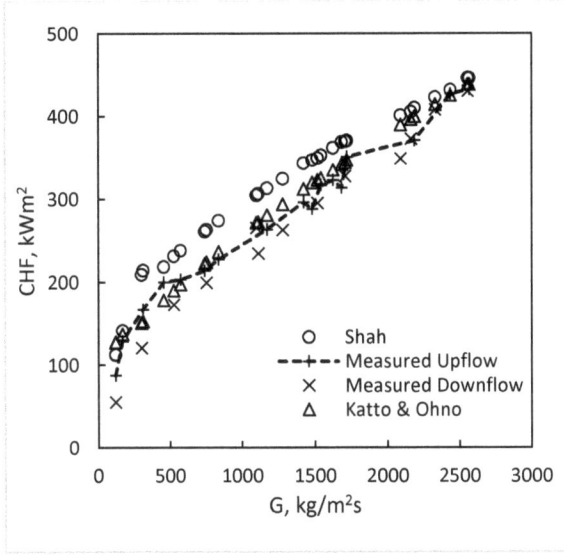

Figure 7. Data of Pappel et al., (1966) [4] for nitrogen compared with the Shah and Katto–Ohno correlations. T_{SAT} = 109 K, inlet subcooling 23.9 K.

Deviations of all three correlations are high for the downflow data of Remizov et al., but the deviations are also equally high for their upflow data. The data for flow in upward and downward directions cannot be directly compared as they provide critical quality at identical inlet quality and heat flux. Therefore, they were compared as the ratio of their deviations from the correlations of Shah and Katto and Ohno. This comparison is shown in Figure 8. It is seen that the downflow CHF is up to 12% higher than upflow CHF at the lowest mass flux, while it is up to 10% lower at the highest mass flux. Collier and Thome (1994) [28] have stated that the data of Remizov et al., show that downflow CHF is 10% to 30% lower than upflow CHF, the greatest difference being at the lowest flow rate. Remizov et al., did not make any such statement and the present analysis shows that CHF in downflow is up to 12% higher than in upflow at the lowest flow rate, and this is the maximum difference at any flow rate.

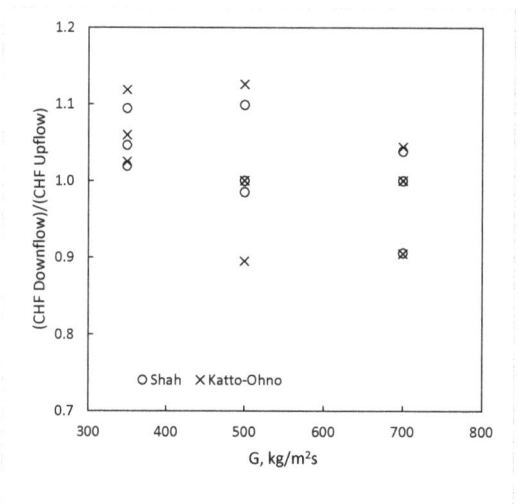

Figure 8. Ratio of CHF in downflow to that in upflow in the tests of Remizov et al., (1985) [10] estimated using the correlations of Shah and Katto–Ohno.

Figure 9 shows the ratio of CHF in downflow to that in upflow in the data of Lazarek and Black (1982) [8]. It is seen that the ratio is close to one over the entire range of mass flux. The inlet quality ranged from -0.25 to -0.02. Thus, inlet quality does not affect the ratio of upflow to downflow CHF, as indicated in the data of Cumo et al. Figure 6.

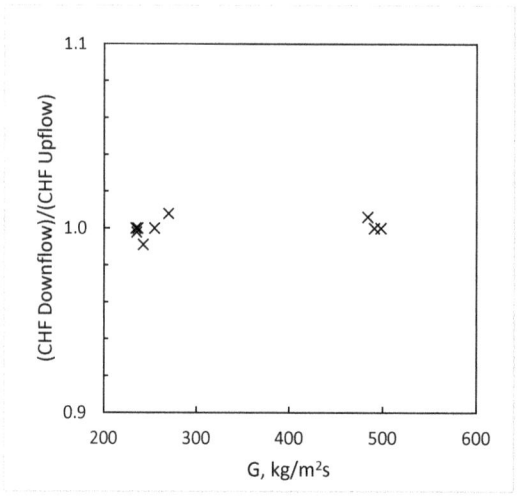

Figure 9. Ratio of CHF during downflow and upflow in the tests by Lazarek and Black (1982) [8].

Figure 10 shows the ratio of CHF in downflow to that in upflow in the tests by Ami et al., (2015) [14]. It is seen that the ratio increases with mass flux, with downflow CHF becoming larger than upflow CHF by up to 15%. The data for both upflow and downflow for higher flow rates are considerably lower than the correlations of Shah, Katto–Ohno, and Zhang et al. These three correlations are very well-verified with a vast amount of water data. This indicates that these data at a high flow rate are unusual and, hence, were not included in Tables 1 and 2.

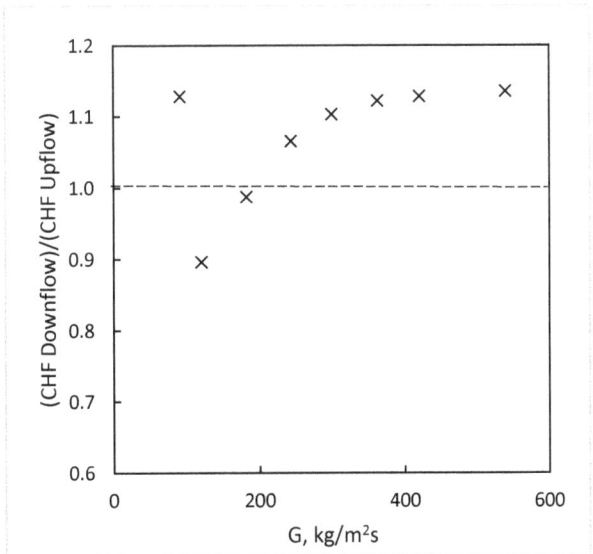

Figure 10. Ratio of measured CHF for water during downflow to that in upflow. D = 10 mm, L_c = 0.4 m, p = 3 bar, inlet temperature 60 °C. Data of Ami et al., (2015) [14].

In his tests with water, Barnett (1963) [1] found no effect of flow direction on CHF.

The previous discussions show that most of the experimental studies indicate that there is no or a small effect of flow direction on CHF. The only studies that show that CHF in downflow is much lower are Pappel et al., (1966) [4] and Pappel (1972) [5] for nitrogen. The two were done on the same test section and all parameters were the same except for inlet subcooling. Hence, it should be considered to be a single study.

4.3. Effect of Channel Shape

The data discussed earlier were all for round tubes. DeBortoli et al., (1957) [27] have listed data for CHF in rectangular channels in both directions. These are included in Tables 1 and 2. It is seen that the correlations of Shah, Katto–Ohno, and Zhang et al., are in good agreement with the data in both directions and deviations of each correlation are about the same in both directions. Figure 11 shows the comparison of some correlations with some downflow data from this source.

Gambill and Bundy (1961) [3] performed tests with water flowing downward in thin rectangular channels. As seen in Table 1, these are in good agreement with the correlations of Shah, Katto–Ohno, and Zhang et al. These data are shown in Figure 12.

It is seen that the correlations for downflow in tubes are in good agreement with the well-verified correlations for upflow CHF and there is no apparent effect of flow direction.

The effect of flow direction on CHF in shapes other than round and rectangular remains to be investigated.

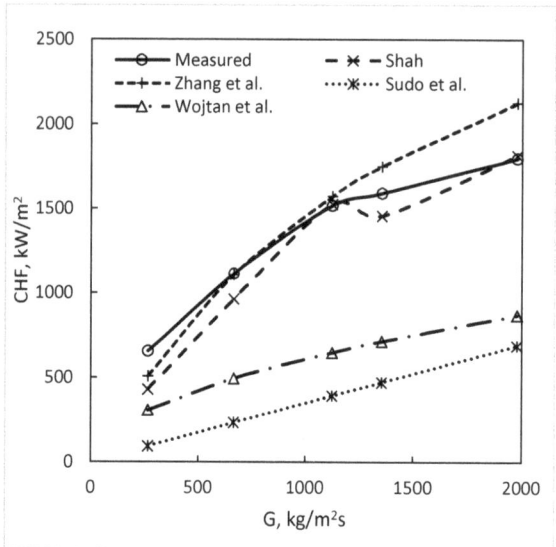

Figure 11. Data of DeBortoli et al., (1957) [27] for downflow of water in a rectangular channel 24.5 mm × 2.46 mm compared to some correlations. $L/D_{HYD} = 153$, $p = 13.79$ bar, $x_{in} = -0.2$.

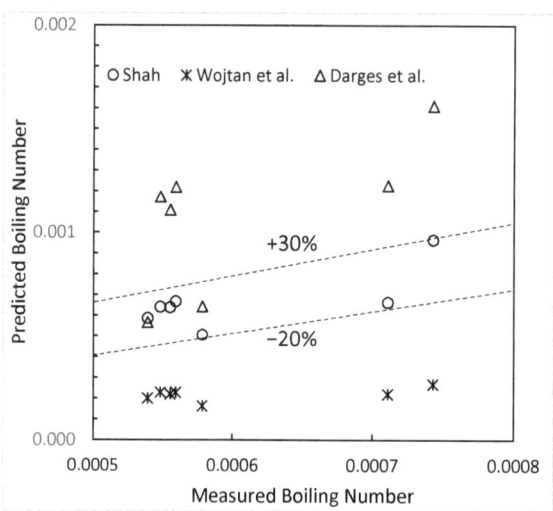

Figure 12. Data of Gambill and Bundy (1961) [3] for downflow in a rectangular channel compared to some correlations.

4.4. Recommendations for Design

The vast majority of the data analyzed show that there is no significant effect of flow direction on CHF and that CHF in downflow can be accurately calculated by reliable correlations for upflow CHF. While some data show decreases in CHF during downflow at low velocities, others (e.g., Remizov et al., and Ami et al.) show higher CHF in downflow. At near-zero mass flow rate, CHF will be due to flooding and then will be much lower than that predicted by the upflow correlations.

The recommendation for design is to use reliable upflow correlations to calculate CHF in downflow and apply a 15% safety factor. Also, calculate CHF due to flooding by a reliable correlation. Use the larger of the two calculated CHF values. The upflow CHF correlations recommended are Shah and Katto–Ohno.

5. Conclusions

1. Literature on CHF during downflow in vertical channels was studied. Some researchers reported up to 30% lower CHF in downflow compared to upflow at low flow rates. Many authors reported no effect of flow direction or even higher CHF during downflow.
2. Data were analyzed for CHF during downflow in fully heated channels from 11 sources. These included several diverse fluids (water, nitrogen, refrigerants) in round and rectangular channels, reduced pressure from 0.0045 to 0.625, mass flux from 15 to 21,761 kg/m^{2}s, inlet quality from -1.3 to 0, and exit quality from -0.2 to 1.09. These were compared to four correlations for upflow CHF and three applicable to downflow.
3. The correlations for CHF in downflow had large deviations with most data. The upflow correlations of Shah and Katto–Ohno gave good agreement with downflow data, their MAD being 21.9% and 18.9%, respectively for the 304 data points.
4. A comparison of data from studies in which CHF during both upflow and downflow was measured showed that most of them do not show any effect of orientation. Some show differences up to ±15%, with some having higher CHF in upflow and others having higher CHF in downflow. Such deviations are well within the accuracy of most correlations.
5. The correlations of Shah and Katto–Ohno are recommended for calculating CHF during downflow, subject to the minimum calculated with a flooding correlation.

Funding: This research received no external funding.

Data Availability Statement: All data used in this research were obtained from the publications cited in this paper.

Conflicts of Interest: The authors declare no conflicts of interest.

Nomenclature

A_F	Flow area, m^2
A_H	Heated area, m^2
Bd_θ	Bond number defined by Equation (12), dimensionless
Bo	Boiling number at CHF, $=q_c/(G\, i_{LG})$, dimensionless
C_{pL}	Specific heat of liquid at constant pressure, kJ/kg K
CHF	Critical heat flux
D	Diameter of channel, m
D_{HP}	Equivalent diameter based on heated perimeter, $=(4 \times$ flow area)/(heated perimeter), m
D_{HYD}	Hydraulic equivalent diameter, $=(4 \times$ flow area)/(wetted perimeter), m
Fr_θ	Froude number defined by Equation (11), dimensionless
g	Acceleration due to gravity, m/s^2
G	Mass flux, kg/m^{2}s
G^*	Dimensionless mass flux defined by Equation (5), dimensionless
i_{LG}	Latent heat of vaporization, kJ/kg
H	Height of channel, m
K	Constant in Kutateladze formula for pool boiling CHF, dimensionless
k_L	Thermal conductivity of liquid, W/(mK)
L, L_C	Heated length of channel from the entrance to the location of CHF, m
MAD	Mean absolute deviation, dimensionless
N	Number of data points, dimensionless
p	Pressure, Pa
p_c	Critical pressure, Pa
p_r	Reduced pressure $= p/p_c$, dimensionless

q^*	Dimensionless CHF defined by Equation (4), dimensionless
q_c	Critical heat flux, kW/m²
T	Temperature, K
ΔT_{SC}	$=(T_{SAT} - T_L)$, K
W	Width of channel, m
We_D	Weber number defined by Equation (10), dimensionless
x	Thermodynamic vapor quality, dimensionless
x_c	Critical quality, i.e., quality at CHF, dimensionless
x_{in}	Quality at inlet to heated section, dimensionless
Y	Parameter for correlating CHF in Shah correlation, dimensionless
Greek Symbols	
λ	Characteristic length defined by Equation (6), dimensionless
ρ	Density, kg/m³
μ	Dynamic viscosity, Pa·s
σ	Surface tension, N/m
θ	Inclination of flow direction from horizontal, degree (0° is horizontal, 90° is vertical up)
Subscripts	
G	vapor
L	liquid
SAT	at saturated condition
SC	at subcooled condition
wall	of wall

References

1. Barnett, P.G. *An Investigation into the Validity of Certain Hypotheses Implied by Various Burnout Correlations*; Rep. AEEW-R-214; United Kingdom Atomic Energy Authority: Abingdon, UK, 1963.
2. Chen, B. The upflow and downflow critical heat flux of water and freon in a vertical tube and its flow direction factor. *At. Energy Sci. Technol.* **1993**, *27*, 112–119. (In Chinese)
3. Gambill, W.R.; Bundy, R.D. *HFIR Heat-Transfer Studies of Turbulent Water Flow in Thin Rectangular Channels*; ORNL-3079, TID-4500; Oak Ridge National Laboratory for the US Atomic Energy Commission: Oak Ridge, TN, USA, 1961.
4. Papell, S.S.; Simoneau, R.J.; Brown, D.D. *Buoyancy Effects on Critical Heat Flux of Forced Convective Boiling in Vertical Flow*; NASA Technical Note D-3672; NASA: Washington, DC, USA, 1966.
5. Papell, S.S. *Combined Buoyancy and Flow Direction Effects on Saturated Boiling Critical-Heat Flux in Liquid Nitrogen*; NASA TM X-68086; NASA: Washington, DC, USA, 1977.
6. Kirby, G.J.; Stainforth, R.; Kinneir, J.H. *A Visual Study of Forced Convection Boiling. Part 2. Flow Patterns and Burnout for a Round Test Section*; AEEW-R 506, Quoted in Cumo et al., (1977); United Kingdom Atomic Energy Authority, Reactor Group: Winfrith, UK, 1967.
7. Cumo, M.; Bertoni, R.; Cipriani, R.; Palazzi, G. *Up-Flow and Down-Flow Burnout*; Mechanical Engineering Pub for the Institution of Mechanical Engineers: London, UK, 1977.
8. Lazarek, G.M.; Black, S.H. Evaporative heat transfer, pressure drop and critical heat flux in a small vertical tube with R-113. *Int. J. Heat Mass Transf.* **1982**, *25*, 945–960.
9. Mishima, K.; Nishihara, H.; Michiyoshi, I. Boiling burnout and flow instabilities for water flowing in a round tube under atmospheric pressure. *J. Heat Mass Transf.* **1985**, *28*, 1115–1129.
10. Remizov, O.; Sergeev, V.; Yurkov, Y. Experimental investigation of deterioration in heat-transfer with up-flow and down-flow of water in a tube. *Therm. Eng.* **1983**, *30*, 549–551.
11. Deqiang, S.; Hong, J.; Junkai, F.E.N.G. An experimental study of upward and downward flow critical heat flux in a vertical round tube. In Proceedings of the 4th Miami International Symposium on Multi-Phase Transport Particulate Phenomena (Condensed Papers), Miami Beach, FL, USA, 15–17 December 1987.
12. Chang, S.H.; Baek, W.P.; Bae, T.M. A study of critical heat flux for low flow of water in vertical round tubes under low pressure. *Nucl. Eng. Des.* **1991**, *132*, 225–237.
13. Kays, W.M. Boiling. In *Handbook of Heat Transfer*; Rohsenow, W.M., Hartnett, J.P., Eds.; McGraw-Hill: New York, NY, USA, 1973; pp. 13-50–13-75.
14. Ami, T.; Harada, T.; Umekawa, H.; Ozawa, M. Influence of tube diameter on critical heat flux in downward flow. *Multiph. Sci. Technol.* **2015**, *27*, 77–97. [CrossRef]
15. Shah, M.M. *Two-Phase Heat Transfer*; John Wiley & Sons: Hoboken, NJ, USA, 2021.
16. Sudo, Y.; Miyata, K.; Ikawa, H. Experimental study of differences in dnb heat flux between upflow and downflow in vertical rectangular channel. *J. Nucl. Sci. Technol.* **1985**, *22*, 604–618. [CrossRef]

17. Hirose, Y.; Sibamoto, Y.; Takashi Hibiki, T. Critical heat flux for downward flows in vertical round pipes. *Prog. Nucl. Energy* **2024**, *168*, 105027.
18. Darges, S.J.; Devahdhanush, V.S.; Mudawar, I. Assessment and development of flow boiling critical heat flux correlations for partially heated rectangular channels in different gravitational environments. *Int. J. Heat Mass Transf.* **2022**, *196*, 123291.
19. Shah, M.M. Improved general correlation for critical heat flux in uniformly heated vertical tubes. *Int. J. Heat Fluid Flow* **1987**, *8*, 326–335.
20. Katto, Y.; Ohno, H. An improved version of the generalized correlation of critical heat flux for the forced convection boiling in uniformly heated vertical tubes. *Int. J. Heat Mass Transf.* **1984**, *27*, 1641–1648.
21. Shah, M.M. Applicability of general correlations for CHF in conventional tubes to mini/macro channels. *Heat Transf. Eng.* **2017**, *38*, 1–10.
22. Wojtan, L.; Revellin, R.; Thome, J.R. Investigation of saturated critical heat flux in a single uniformly heated microchannel. *Exp. Therm. Fluid Sci.* **2006**, *30*, 765–774.
23. Sripada, R.; Mendu, S.S.; Tentu, D.; Varanasi, S.S.; Veeredhi, V.R. Development of correlation for critical heat flux for vertically downward two-phase flows in round tubes. *Exp. Heat Transf.* **2021**, *34*, 393–410. [CrossRef]
24. Lemmon, E.W.; Huber, M.L.; McLinden, M.O. *NIST Reference Fluid Thermodynamic and Transport Properties, REFPROP Version 9.1*; NIST: Gaithersburg, MD, USA, 2013.
25. Dougherty, T.; Fighetti, C.; Reddy, G.; Yang, B.W. Critical heat flux for vertical upflow and downflow in uniform tubes at low pressures. In Proceedings of the Third International Symposium on Multi-Phase Flow and Heat Transfer, Xi'an, China, September 1994.
26. Ruan, S.W.; Bartsch, G.; Yang, S.M. Characteristics of the critical heat flux for downward flow in a vertical tube at low flow rate and low pressure conditions. *Exp. Therm. Fluid Sci.* **1993**, *7*, 296–306. [CrossRef]
27. DeBartoli, R.A.; Green, S.J.; LeTourneau, B.W.; Troy, M.; Weiss, A. *Forced Convection Heat Transfer Burnout Studies for Water in Rectangular Channels and Round Tubes at Pressures above 500 Psia*; WAPD-188, TID-4500; US Department of Commerce: Washington, DC, USA, 1958.
28. Collier, J.G.; Thome, J.R. *Convective Boiling & Condensation*, 3rd ed.; Oxford University Press: Oxford, UK, 1994.

Disclaimer/Publisher's Note: The statements, opinions and data contained in all publications are solely those of the individual author(s) and contributor(s) and not of MDPI and/or the editor(s). MDPI and/or the editor(s) disclaim responsibility for any injury to people or property resulting from any ideas, methods, instructions or products referred to in the content.

Article

Hartmann Flow of Two-Layered Fluids in Horizontal and Inclined Channels

Arseniy Parfenov *, Alexander Gelfgat, Amos Ullmann and Neima Brauner

School of Mechanical Engineering, Faculty of Engineering, Tel-Aviv University, Tel-Aviv 6997801, Israel; gelfgat@tau.ac.il (A.G.); ullmann@tauex.tau.ac.il (A.U.); brauner@tauex.tau.ac.il (N.B.)
* Correspondence: ae2parfenov@gmail.com

Abstract: The effect of a transverse magnetic field on two-phase stratified flow in horizontal and inclined channels is studied. The lower heavier phase is assumed to be an electrical conductor (e.g., liquid metal), while the upper lighter phase is fully dielectric (e.g., gas). The flow is defined by prescribed flow rates in each phase, so the unknown frictional pressure gradient and location of the interface separating the phases (holdup) are found as part of the whole solution. It is shown that the solution of such a two-phase Hartmann flow is determined by four dimensionless parameters: the phases' viscosity and flow-rate ratios, the inclination parameter, and the Hartmann number. The changes in velocity profiles, holdups, and pressure gradients with variations in the magnetic field and the phases' flow-rate ratio are reported. The potential lubrication effect of the gas layer and pumping power reduction are found to be limited to low magnetic field strength. The effect of the magnetic field strength on the possibility of obtaining countercurrent flow and multiple flow states in concurrent upward and downward flows, and the associated flow characteristics, such as velocity profiles, back-flow phenomena, and pressure gradient, are explored. It is shown that increasing the magnetic field strength reduces the flow-rate range for which multiple solutions are obtained in concurrent flows and the flow-rate range where countercurrent flow is feasible.

Keywords: magnetohydrodynamics (MHD); Hartmann flow; two-phase; inclined stratified flow; gas–liquid; holdup; multiple solutions

Citation: Parfenov, A.; Gelfgat, A.; Ullmann, A.; Brauner, N. Hartmann Flow of Two-Layered Fluids in Horizontal and Inclined Channels. *Fluids* **2024**, *9*, 129. https://doi.org/10.3390/fluids9060129

Academic Editors: Dengwei Jing, Hemant J. Sagar and Nguyen Van-Tu

Received: 1 May 2024
Revised: 21 May 2024
Accepted: 27 May 2024
Published: 30 May 2024

Copyright: © 2024 by the authors. Licensee MDPI, Basel, Switzerland. This article is an open access article distributed under the terms and conditions of the Creative Commons Attribution (CC BY) license (https://creativecommons.org/licenses/by/4.0/).

1. Introduction

Flows of electrically conducting fluids in an electromagnetic field are found across a wide variety of phenomena, which are extensively investigated in the field of magnetohydrodynamics (MHD) and possess important technological applications (e.g., [1,2]). While much of the existing research on MHD focuses on single-phase flows, numerous practical challenges involve multiphase flow systems, particularly within the nuclear and petroleum industries, geophysics, and MHD power generation [3,4]. Among the applications are two-phase liquid metal magnetohydrodynamics (LMMHD) generators, which attracted attention in the industry due to their relatively simple structure, absence of moving parts, high-efficiency conversion, and reduced environmental pollution [5]. Alongside the liquid metal, another fluid phase (gas or liquid) is employed to convert thermal energy into kinematic energy. MHD flows can also transport weakly conducting fluids in microscale systems, e.g., in the microchannel networks, such as the microchannel networks of lab-on-a-chip devices [6,7], where the presence of a second non-conductive fluid enhances the mobility of the conducting fluid. Magnetic field-driven micropumps are in increasing demand due to their long-term reliability in generating flow, absence of moving parts, low power requirements, flow reversibility, and efficient mixing [8,9]. In all those applications, the system performance is dependent on the gas–liquid flow pattern in the channel, and the stability of the interface between the phases, which have been subject to experimental and numerical investigations (e.g., [10–14]).

Since the pioneering works of Hartmann and Lazarus in 1937 [15] the Poiseuille flow of an electrically conducting fluid in a transverse magnetic field (i.e., Hartmann flow) has been thoroughly studied in the literature (e.g., [16–19]). Hartmann [20] presented an analytical solution for the velocity profile of an isothermal laminar flow in an electrically insulated channel subjected to a wall-normal uniform magnetic field. It was demonstrated that the magnetic field flattens the velocity profile, a phenomenon often termed the 'Hartmann effect' in the literature. However, the presence of a second phase can drastically affect the velocity profile and the associated pressure gradient.

Due to the density difference between the phases, a stratified flow configuration is commonly encountered in both horizontal and inclined conduits. The part of the flow cross-section area occupied by the heavier fluid is referred to as the holdup, being an additional characteristic of the flow pattern. Shail [21] investigated the Hartmann flow of a conducting fluid between two horizontal insulating plates, with a layer of non-conducting fluid separating the top wall from the conducting fluid. It was discovered that the flow rate of the conducting fluid can be increased by approximately 30 percent for suitable ratios of the depths and viscosities of the two fluids used in an electromagnetic pump. Owen et al. [22] introduced two mechanistic models: a simple homogeneous model and a more sophisticated film flow model for computing the pressure drop in an MHD two-phase flow at high Hartmann numbers. Lohrasbi and Sahai [23] derived analytical solutions for the velocity and temperature profiles in a two-phase laminar steady MHD flow between two horizontal plates with one conducting phase. Malashetty and Umavathi [24], analytically investigated a similar system in an inclined channel, solving the nonlinear coupled momentum and energy equations in both phases. They found that, for a constant thickness of the layers, increasing the magnetic field has a dampening effect on the velocity of the conducting layer, akin to a single-phase flow. They also noted an increase in the velocity of the conducting fluid with a decrease in the thickness of the upper electrically non-conducting layer and/or with an increase in the channel inclination angle. A subsequent study by Umavathi et al. [25] addressed the magnetohydrodynamic Poiseuille–Couette flow and heat transfer of two immiscible fluids between inclined parallel plates. However, for isothermal flow, the effect of the different gravity driving force in the two layers is not considered in the model equations. Recently, Shah et al. [26] derived approximate analytical solutions for the velocity and temperature fields of unsteady MHD generalized Couette flows of two immiscible and electrically conducting fluids flowing between two horizontal electrically insulated plates subjected to an inclined magnetic field and an axial electric field. In all of these studies, the problem is solved for a pre-defined flow configuration (i.e., in situ holdup of the heavy layer and the pressure gradient).

An important issue in the context of two-phase flows is the occurrence of gravity-driven multiple solutions in inclined channels for a specified two-phase system and fixed operational conditions. It has been established in the two-phase flow literature [27–30] that there always exist two possible solutions for the holdup of steady countercurrent flow and up to three distinct steady-state solutions can be obtained within a limited range of the flow parameters in concurrent upward- and downward-inclined flows. The existence of multiple solutions for the holdup for some range of the superficial velocities in liquid–liquid flows was verified experimentally [28,29], where experimental verification of the existence of multiple solutions for the holdup within a certain range of superficial velocities in liquid–liquid flows has been conducted. However, the feasibility of achieving countercurrent flow in the presence of a magnetic field, and its impact on the potential for multiple stratified flow configurations in concurrent and countercurrent scenarios has not yet been explored.

In this study, we investigate the Hartmann flow of a conducting fluid between two parallel insulating plates, with a layer of non-conducting fluid separating the top wall from the conducting fluid. The presented analytical solution allows for the first time determination of the in situ holdup and the pressure gradient in horizontal or inclined channels for a specified two-phase system and fixed operational conditions. We identify the corresponding dimensionless input parameters that dictate the in situ stratified flow configuration. Furthermore, we utilize the solution to examine the impact of the presence of a second non-conductive layer and the intensity of the magnetic field on the flow characteristics in concurrent and countercurrent flows of the fluids. Special attention is given to the potential lubrication and pumping power saving achieved by introducing the gas phase, as well as to the operational conditions for which multiple solutions can be obtained for the stratified flow configuration in inclined flows.

2. Two-Phase Horizontal Flow

We consider isothermal, axial, steady and fully developed laminar stratified flow of two immiscible fluids between two infinite electrically insulating plates positioned at $z = 0$ and $z = H$ (see Figure 1). Under these conditions, the pressure gradient, $G = dp/dx$, is the same in both layers and is constant (e.g., [31]). For horizontal flow, the channel inclination to the horizontal is $\beta = 0$. A plane and smooth interface between the phases is located at $z = h$. The lower layer (1) is occupied by an electrically conducting fluid (e.g., mercury), which is affected by a constant transverse magnetic field $B = B_0 e_z$, while the upper lighter fluid is assumed to be non-conductive (e.g., air). Then, the x-component of the momentum equations in the two layers are as follows (e.g., Shail [21]):

$$\eta_1 \frac{d^2 u_1(z)}{dz^2} - \sigma_1 B_0^2 u_1(z) = G \qquad 0 \leq z \leq h \qquad (1)$$

$$\eta_2 \frac{d^2 u_2(z)}{dz^2} = G \qquad h \leq z \leq H \qquad (2)$$

where η_1, η_2 are the fluids' viscosities and σ_1 is the heavier fluid's electric conductivity. The solutions for the velocity profiles are as follows:

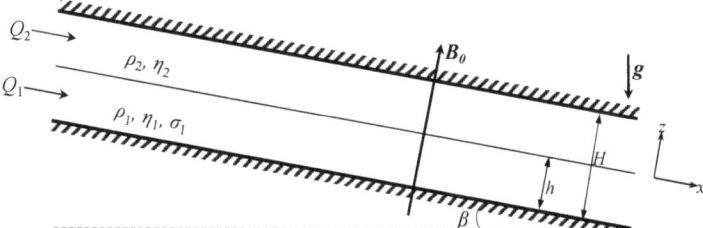

Figure 1. Sketch of the flow configuration and coordinates.

$$u_1(\tilde{z}) = A_1 \cosh(Ha\, \tilde{z}) + A_2 \sinh(Ha\, \tilde{z}) - \frac{G}{\sigma_1 B_0^2} \qquad 0 \leq \tilde{z} \leq \tilde{h} \qquad (3)$$

$$u_2(\tilde{z}) = \frac{GH^2}{2\eta_2} \tilde{z}^2 + B_1 H \tilde{z} + B_2 \qquad \tilde{h} \leq \tilde{z} \leq 1 \qquad (4)$$

where $\tilde{z} = z/H$, $\tilde{h} = h/H$ is the conductive (heavier) fluid holdup and Ha is the Hartmann number $Ha = B_0 H \sqrt{\frac{\sigma_1}{\eta_1}}$. Applying no-slip boundary conditions at $\tilde{z} = 0$ and $\tilde{z} = 1$, and the continuity of velocities and viscous shear stresses at the fluids' interface $\tilde{z} = \tilde{h}$ yields the following:

$$A_1 = \frac{G}{\sigma_1 B_0^2}, \quad B_2 = -\frac{GH^2}{2\eta_2} - B_1 H \qquad (5)$$

$$A_2 = \frac{GH^2}{\eta_2} \frac{-\frac{(1-\tilde{h})}{2} - \frac{1}{Ha^2\eta_{12}(1-\tilde{h})}\left[\cosh\left(Ha\tilde{h}\right) - 1\right] - \frac{1}{Ha}\sinh\left(Ha\tilde{h}\right)}{Ha\eta_{12}\cosh\left(Ha\tilde{h}\right) + \frac{1}{(1-\tilde{h})}\sinh\left(Ha\tilde{h}\right)} \quad (6)$$

$$B_1 = \frac{GH}{\eta_2}\left[-\tilde{h} + \frac{1}{Ha}\sinh\left(Ha\tilde{h}\right) + Ha\frac{\eta_1 A_2}{GH^2}\cosh\left(Ha\tilde{h}\right)\right] \quad (7)$$

The volumetric flow rates (per unit channel width) in the lower and upper layers are as follows:

$$Q_1 = H\int_0^{\tilde{h}} u_1(\tilde{z})d\tilde{z} = \frac{GH^3}{Ha^3\eta_1}\sinh\left(Ha\tilde{h}\right) + \frac{A_2 H}{Ha}\left[\cosh\left(Ha\tilde{h}\right) - 1\right] - \frac{GH^3\tilde{h}}{Ha^2\eta_1} \quad (8)$$

$$Q_2 = H\int_{\tilde{h}}^1 u_2(\tilde{z})d\tilde{z} = \frac{GH^3}{6\eta_2}\left(-2 - \tilde{h}^3 + 3\tilde{h}\right) + \frac{B_1 H^2}{2}\left(-1 - \tilde{h}^2 + 2\tilde{h}\right) \quad (9)$$

As the left hand side (l.h.s) of both Equations (8) and (9) are proportional to G, the ratio of Q_1/Q_2 yields an implicit equation of calculation of the holdup of the conductive layer in the channel, $\tilde{h} = h/H$:

$$Q_{12} = \frac{Q_1}{Q_2} = \frac{U_{1s}}{U_{2s}} = \frac{1}{Ha^3}\frac{\frac{1}{\eta_{12}}\sinh\left(Ha\tilde{h}\right) + Ha^2\frac{A_2\eta_2}{GH^2}\left[\cosh\left(Ha\tilde{h}\right) - 1\right] - Ha\frac{\tilde{h}}{\eta_{12}}}{\frac{1}{6}\left(-2 - \tilde{h}^3 + 3\tilde{h}\right) - \frac{B_1\eta_2}{2GH}\left(1 - \tilde{h}\right)^2} \quad (10)$$

where $U_{1,2s} = Q_{1,2}/H$ is the superficial velocity of the fluid in the channel cross-section. In view of Equations (6) and (7), the following expressions are introduced:

$$\frac{A_2\eta_2}{GH^2} = \frac{-\frac{(1-\tilde{h})}{2} - \frac{1}{Ha^2\eta_{12}(1-\tilde{h})}\left[\cosh\left(Ha\tilde{h}\right) - 1\right] - \frac{1}{Ha}\sinh\left(Ha\tilde{h}\right)}{Ha\eta_{12}\cosh\left(Ha\tilde{h}\right) + \frac{1}{(1-\tilde{h})}\sinh\left(Ha\tilde{h}\right)} \quad (11)$$

$$\frac{B_1\eta_2}{GH} = -\tilde{h} + \frac{1}{Ha}\sinh\left(Ha\tilde{h}\right) + Ha\frac{\eta_1 A_2}{GH^2}\cosh\left(Ha\tilde{h}\right) \quad (12)$$

and $\frac{A_2\eta_1}{GH^2} = \eta_{12}\frac{A_2\eta_2}{GH^2}$.

The calculation of the flow-rate ratio for a specified holdup via Equation (10) (with (11) and (12)) is straightforward. However, in practice, the input flow rates of the fluids are known, whereas the location of the interface is unknown. Equation (10) indicates that $\tilde{h} = \tilde{h}(Q_{12}, \eta_{12}, Ha)$. In the limit of $Ha \to 0$, Equation (10) converge to the expression obtained for $Q_{12}\left(\tilde{h}, \eta_{12}\right)$ for a two-layer Poiseuille flow in the absence of a magnetic field (e.g., [32,33]).

Once Equation (10) is solved for the holdup, the corresponding pressure gradient G can be determined. For example, using the solution for the holdup in Equation (9) yields the following:

$$G = \frac{dp}{dx} = \frac{dp_f}{dx} = \frac{U_{2s}}{\frac{H^2}{6\eta_2}\left(-2 - \tilde{h}^3 + 3\tilde{h}\right) - \frac{B_1 H}{2G}\left(1 - \tilde{h}\right)^2} \quad (13)$$

where B_1/G is independent of G (see Equation (7)). Obviously, in horizontal flow, the total pressure gradient is identical to the frictional pressure gradient. The corresponding dimensionless pressure gradient, $P_f^{2,0}\ P_f^{1,0}$, can be obtained by normalizing dp/dx with respect to the frictional pressure gradient of the single-phase flow of either the lighter or the heavier fluid obtained in the absence of a magnetic field, $\left(dp_f/dx\right)_{2s,1s}^0 = -12\eta_{2,1} U_{2s,1s}/H^2$, respectively, whereby

$$P_f^{2,0} = \frac{dp/dx}{\left(dp_f/dx\right)_{2s}^0} = \frac{1}{2} \frac{1}{\left(\tilde{h}^3 - 3\tilde{h} + 2\right) + \frac{3\eta_2 B_1}{GH}\left(1 - \tilde{h}\right)^2} \tag{14}$$

$$P_f^{1,0} = \frac{dp/dx}{\left(dp_f/dx\right)_{1s}^0} = \frac{P_f^{2,0}}{Q_{12}\,\eta_{12}} \tag{15}$$

where the first superscript (1 or 2) represents the phase selected to scale the two-phase pressure gradient, and the second superscript (0) indicates that the pressure gradient in the absence of a magnetic field is used for the scaling (see also the nomenclature list). Note that $\frac{B_1\,\eta_2}{GH}$ is also determined by (Q_{12}, η_{12}, Ha) (see Equation (7)).

The dimensionless pressure gradient that would be obtained in the case of a single-phase (SP) flow of the conductive (heavier) fluid under the same magnetic field strength (i.e., the same Ha) is as follows:

$$P_{fs}^{1,0} = \frac{(dp/dx)_{1s}}{\left(dp_f/dx\right)_{1s}^0} = \frac{Ha^2}{12\left[1 - \frac{2}{Ha} tgh\left(\frac{Ha}{2}\right)\right]} \tag{16}$$

Hence, when scaling the two-phase pressure gradient with respect to the SP flow of the heavy phase in the presence of the same magnetic field, the dimensionless pressure gradient is given by the following:

$$P_f^1 = \frac{P_f^{1,0}}{P_{fs}^1} = 12\frac{\tilde{G}^{1,0}}{Ha^2}\left[1 - \frac{2}{Ha} tgh\left(\frac{Ha}{2}\right)\right] \tag{17}$$

The corresponding dimensionless pumping power (power factor) required for the given flow rates is as follows:

$$Po = P_f^1 \frac{(U_{1s} + U_{2s})}{U_{1s}} = P_f^1(1 + Q_{21}) \tag{18}$$

Hence, the dimensionless pressure gradient (either $P_f^{2,0}$, $P_f^{1,0}$ or P_f^1) and the power factor are also determined in terms of (Q_{12}, η_{12}, Ha).

Finally, upon scaling the velocity with respect to U_{1s}, the dimensionless velocity profiles are also obtained in terms of those three dimensionless parameters:

$$\tilde{u}_1(\tilde{z}) = \frac{u_1(z)}{U_{1s}} = \tilde{A}_1\left[\cosh\left(Ha\tilde{z}\right) - 1\right] + \tilde{A}_2\sinh\left(Ha\tilde{z}\right) \tag{19}$$

$$\tilde{u}_2(\tilde{z}) = \frac{u_2(z)}{U_{1s}} = \tilde{B}_2\left(\tilde{z}^2 - 1\right) + \tilde{B}_1\left(\tilde{z} - 1\right) \tag{20}$$

with

$$\tilde{A}_1 = -\frac{12 P_f^{1,0}}{Ha^2};\ \tilde{B}_2 = -6\eta_{12} P_f^{1,0} \tag{21}$$

$$\widetilde{A}_2 = -12\eta_{12}P_f^{1,0} \frac{-\frac{(1-\widetilde{h})}{2} - \frac{1}{Ha^2\eta_{12}(1-\widetilde{h})}\left[\cosh\left(Ha\widetilde{h}\right) - 1\right] - \frac{1}{Ha}\sinh\left(Ha\widetilde{h}\right)}{Ha\eta_{12}\cosh\left(Ha\widetilde{h}\right) + \frac{1}{(1-\widetilde{h})}\sinh\left(Ha\widetilde{h}\right)} \qquad (22)$$

$$\widetilde{B}_1 = -12\eta_{12}P_f^{1,0}\left[-\widetilde{h} + \frac{1}{Ha}\sinh\left(Ha\widetilde{h}\right) + Ha\frac{\eta_1 A_2}{GH^2}\cosh\left(Ha\widetilde{h}\right)\right] \qquad (23)$$

3. Two-Phase Inclined Flow

In the case of inclined flow, the momentum equations in the two layers are

$$\eta_1\frac{d^2u_1(z)}{dz^2} - \sigma_1 B_0^2 u_1(z) = \frac{dp}{dx} - \rho_1 g\sin\beta = G_1 \qquad 0 \le z \le h \qquad (24)$$

$$\eta_2\frac{d^2u_2(z)}{dz^2} = \frac{dp}{dx} - \rho_2 g\sin\beta = G_2 \qquad h \le z \le H \qquad (25)$$

where β is the angle of the channel downward inclination to the horizontal (x is in the downward direction) and ρ_1, ρ_2 are the fluids' densities. The model assumptions are the same as those indicated in the case of the horizontal channel (see Section 2). The terms describing the gravity body force, which drives the flow along with the imposed pressure gradient, are added to the equations. The solution of (24) and (25) for the velocity profile that satisfies the no-slip boundary conditions at $\widetilde{z} = 0$ and $\widetilde{z} = 1$ is as follows:

$$u_1(\widetilde{z}) = \frac{G_1}{\sigma_1 B_0^2}\left[\cosh\left(Ha\,\widetilde{z}\right) - 1\right] + A_2\sinh\left(Ha\,\widetilde{z}\right) \qquad 0 \le \widetilde{z} \le \widetilde{h} \qquad (26)$$

$$u_2(\widetilde{z}) = \frac{G_2}{2\eta_2}\left(\widetilde{z}^2 - 1\right) + B_1\left(\widetilde{z} - 1\right) \qquad \widetilde{h} \le z \le 1 \qquad (27)$$

Imposing the boundary conditions of velocities and shear stress continuity at the interface yields the following:

$$B_1 H = -\frac{G_2 H^2 \widetilde{h}}{\eta_2} + \frac{G_1 H^2}{\eta_2}\frac{1}{Ha}\sinh\left(Ha\widetilde{h}\right) + A_2 Ha\eta_{12}\cosh\left(Ha\widetilde{h}\right) \qquad (28)$$

$$A_2 = \frac{-\frac{G_2 H^2}{2\eta_2}\left(1 - \widetilde{h}\right) - \frac{G_1 H^2}{Ha^2\eta_1(1-\widetilde{h})}\left[\cosh\left(Ha\widetilde{h}\right) - 1\right] - \frac{G_1 H^2}{Ha\eta_2}\sinh\left(Ha\widetilde{h}\right)}{\eta_{12}Ha\cosh\left(Ha\widetilde{h}\right) + \frac{1}{(1-\widetilde{h})}\sinh\left(Ha\widetilde{h}\right)} \qquad (29)$$

The volumetric fluxes in the lower and upper layers are as follows:

$$Q_1 = U_{1s}H = H\int_0^{\widetilde{h}}u_1(\widetilde{z})d\widetilde{z} = \frac{G_1 H^3}{Ha^3\eta_1}\sinh\left(Ha\widetilde{h}\right) + \frac{A_2 H}{Ha}\left[\cosh\left(Ha\widetilde{h}\right) - 1\right] - \frac{G_1 H^3 \widetilde{h}}{Ha^2\eta_1} \qquad (30)$$

$$Q_2 = U_{2s}H = H\int_{\widetilde{h}}^1 u_2(\widetilde{z})d\widetilde{z} = -\left(1 - \widetilde{h}\right)^2\left[\frac{G_2 H^3}{6\eta_2}\left(\widetilde{h} + 2\right) + \frac{B_1 H^2}{2}\right] \qquad (31)$$

Substituting $B_1 H$ (Equation (28)) in Equation (31) and dividing Equations (30) and (31) by U_{1s} yields the following:

$$1 = -12\frac{\widetilde{G}_1^{1,0}}{Ha^3}\sinh\left(Ha\widetilde{h}\right) + \frac{\widetilde{A}_2}{Ha}\left[\cosh\left(Ha\widetilde{h}\right) - 1\right] + 12\frac{\widetilde{G}_1^{1,0}\widetilde{h}}{Ha^2}, \qquad (32)$$

$$\frac{U_{2s}}{U_{1s}} = 2\left(1-\widetilde{h}\right)^2\left\{\widetilde{G}_2^{1,0}\eta_{12}\left(\widetilde{h}+2\right) + \left[-3\widetilde{G}_2^{1,0}\eta_{12}\widetilde{h} + \frac{3\widetilde{G}_1^{1,0}\eta_{12}}{Ha}\sinh\left(Ha\widetilde{h}\right) - \frac{\widetilde{A}_2 Ha\eta_{12}}{4}\cosh\left(Ha\widetilde{h}\right)\right]\right\} \qquad (33)$$

where the following relations were introduced:

$$\frac{G_2}{U_{1s}\eta_2/H^2} = -\frac{12\eta_{12}G_2}{\left(dp_f/dx\right)^0_{1s}} = -12\eta_{12}\widetilde{G}_2^{1,0} \qquad (34)$$

$$\frac{G_1}{U_{1s}\eta_1/H^2} = -\frac{12G_1}{\left(dp_f/dx\right)^0_{1s}} = -12\widetilde{G}_1^{1,0} \qquad (35)$$

and $\widetilde{A}_2 = A_2/U_{1s}$, which by Equation (29) reads as follows:

$$\widetilde{A}_2 = 12\frac{\frac{\widetilde{G}_2^{1,0}\eta_{12}\left(1-\widetilde{h}\right)}{2} + \frac{\widetilde{G}_1^{1,0}}{Ha^2\left(1-\widetilde{h}\right)}\left[\cosh\left(Ha\widetilde{h}\right) - 1\right] + \frac{\widetilde{G}_1^{1,0}\eta_{12}}{Ha}\sinh\left(Ha\widetilde{h}\right)}{Ha\eta_{12}\cosh\left(Ha\widetilde{h}\right) + \frac{1}{\left(1-\widetilde{h}\right)}\sinh\left(Ha\widetilde{h}\right)} \qquad (36)$$

In principle, once U_{1s}, U_{2s} are known, Equations (32) and (33) can be solved for the unknown holdup, $\widetilde{h}$, and pressure gradient, dp/dx, embedded both in $\widetilde{G}_1^{1,0}$ and $\widetilde{G}_2^{1,0}$. However, to reveal the dimensionless parameters of the solution in inclined flows, the solution form is further manipulated.

The total pressure gradient is a sum of the frictional and gravitational (hydrostatic) pressure gradients; hence:

$$\frac{dp}{dx} = \left(\frac{dp}{dx}\right)_f + \left(\frac{dp}{dx}\right)_g = \left(\frac{dp}{dx}\right)_f + \left[\rho_1\widetilde{h} + \rho_2\left(1-\widetilde{h}\right)\right]g\sin\beta \qquad (37)$$

Therefore:

$$G_2 = \frac{dp}{dx} - \rho_2 g\sin\beta = \left(\frac{dp}{dx}\right)_f + \widetilde{h}(\rho_1 - \rho_2)g\sin\beta \qquad (38)$$

and

$$\widetilde{G}_2^{1,0} = \frac{G_2}{\left(dp_f/dx\right)^0_{1s}} = \left[\widetilde{P}_f^{1,0} + \widetilde{h}Y^{1,0}\right]; \quad Y^{1,0} = \frac{(\rho_1-\rho_2)g\sin\beta}{\left(dp_f/dx\right)^0_{1s}} \qquad (39)$$

where $P_f^{1,0}$ is the dimensionless frictional pressure gradient in the inclined two-phase flow and $Y^{1,0}$ is the (a priori known) inclination parameter. Similarly,

$$G_1 = \frac{dp}{dx} - \rho_1 g\sin\beta = \left(\frac{dp}{dz}\right)_f - \left(1-\widetilde{h}\right)(\rho_1-\rho_2)g\sin\beta \qquad (40)$$

and
$$\tilde{G}_1^{1,0} = \frac{G_1}{\left(dp_f/dx\right)_{1s}^0} = \left[\tilde{P}_f^{1,0} - \left(1-\tilde{h}\right)Y^{1,0}\right] \quad (41)$$

Upon substituting (39) and (41) in Equations (32) and (33) (and in (36)), a solution for the holdup $\tilde{h}$ and the dimensionless frictional gradient $P_f^{1,0}$ can be obtained in terms of the four (a priori known) dimensionless parameters ($\eta_{12}, Q_{12}, Ha, Y^{1,0}$). The values of those parameters are a priori known for a specified two-flow system (i.e., channel geometry and inclination, fluid properties, and magnetic field intensity).

In fact, Equation (32) can be used to obtain an explicit expression for $P_f^{1,0}$ in terms of $\left(\eta_{12}, Ha, Y^{1,0}, \tilde{h}\right)$, which can then be substituted into Equation (33) to obtain an implicit equation for the holdup. This step is important to explore the possibility of multiple solutions for prescribed values of the dimensionless parameters, since this becomes rather straightforward once a single algebraic equation for the holdup is obtained.

In view of Equation (37), the total pressure gradient in inclined flow (normalized with respect to the frictional pressure gradient of mercury flow in the absence of a magnetic field) is given by the following:

$$P^{1,0} = \frac{dp/dx}{\left(dp_f/dx\right)_{1s}^0} = P_f^{1,0} + P_g^{1,0} = P_f^{1,0} + \tilde{h}Y^{1,0} + \frac{\rho_2}{\Delta\rho}Y^{1,0} \quad (42)$$

where the sum of the last two terms on the right hand side (r.h.s) represents the dimensionless hydrostatic pressure gradient, $P_g^{1,0}$. When normalized with respect to the frictional pressure gradient obtained for SP flow of the conductive layer under the same magnetic field, $\left(dp_f/dx\right)_{1s}$ (see Equation (16)), the dimensionless total pressure gradient is then $P^1 = P^{1,0}/P_{fs}^{1,0}$, ($P_f^1$ and P_g^1 are the frictional and hydrostatic components). Note that the density ratio, $\rho_{12} = \rho_1/\rho_2$, is an additional dimensionless parameter that has to be prescribed in order to obtain the total pressure and power factors. However, in the case the lighter (non-conductive) fluid is a gas, the last term on the r.h.s of Equation (42) is negligible and practically $P^{1,0} \approx \tilde{G}_2^{1,0}$.

To examine the effect of the gas layer on the pressure gradient, its value compared to the total pressure gradient in SP flow of the conductive fluid which develops under the same Ha is of interest. It is given by the following:

$$P^{1T} = \frac{P^1}{1 + \frac{\rho_2}{\Delta\rho}\frac{Y^{1,0}}{P_{fs}^{1,0}}} \quad (43)$$

and the corresponding power factor is $Po = P^{1T}(1 + Q_{21})$. Once the holdup and pressure gradient have been obtained, the corresponding dimensionless velocity profile in the two layers are given by the following:

$$\tilde{u}_1(\tilde{z}) = \frac{u_1(\tilde{z})}{U_{1s}} = \tilde{A}_1\left[\cosh\left(Ha\tilde{z}\right) - 1\right] + \tilde{A}_2\sinh\left(Ha\tilde{z}\right) \quad (44)$$

$$\tilde{u}_2(\tilde{z}) = \frac{u_2(\tilde{z})}{U_{1s}} = \tilde{B}_2\left(\tilde{z}^2 - 1\right) + \tilde{B}_1\left(\tilde{z} - 1\right) \quad (45)$$

The dimensionless coefficients are given by the following:

$$\widetilde{A}_1 = -\frac{12\widetilde{G}_1^{1,0}}{Ha^2}; \quad \widetilde{B}_2 = \frac{G_2 H^2}{2\eta_2 U_{1s}} = -6\eta_{12}\widetilde{G}_2^{1,0} \tag{46}$$

$\widetilde{A}_2$ is given in Equation (36) and by using Equation (28):

$$\widetilde{B}_1 = 12\left[\widetilde{G}_2^{1,0}\eta_{12}\widetilde{h} - \frac{\widetilde{G}_1^{1,0}\eta_{12}}{Ha}\sinh\left(Ha\widetilde{h}\right)\right] + \widetilde{A}_2 Ha\eta_{12}\cosh\left(Ha\widetilde{h}\right) \tag{47}$$

4. Results and Discussion

The analytical solution presented in Sections 2 and 3 enables determining the conductive liquid holdup and the dimensionless pressure gradient in MHD two-phase flow in terms of the set of dimensionless ($\eta_{12}, Q_{21}, Ha, Y^{1,0}$) parameters. The values of these parameters are determined by the channel geometry, fluid properties, and flow rates, and are independent of the in situ flow characteristics. Worth noting is that using the above analytical solution for the calculation of the holdups and velocity profiles appears to be non-trivial. Due to the hyperbolic functions involved, it is necessary to work with very large and very small numbers, which poses an accuracy problem for the usual calculations with 16, and even 32, decimal places. To obtain correct results we had to require 128 or even 256 decimal places, which was possible when the calculations were performed using Maple (https://www.maplesoft.com).

The analytical solution obtained is used to explore the effect of the presence of a second non-conductive (gas) layer and the intensity of the magnetic field on the flow characteristics in horizontal flows and in concurrent and countercurrent inclined flows of the fluids. With this aim, we selected two representative two-phase systems—mercury–air and sodium–argon. Their physical properties are provided in Tables 1 and 2, respectively. The viscosity ratio of the sodium–argon and the mercury–air systems are $\eta_{12} = 11.83$ and $\eta_{12} = 81.87$, respectively. The higher value of the latter results from the lower viscosity of sodium compared to mercury (and somewhat higher viscosity of argon compared to air). Worth noting is that, for the same Ha number, the magnetic field intensity applied in the case of mercury flow is about five times larger than that in argon (e.g., $B_0 = 0.01$ T corresponds to Ha = 5.18 in mercury and to Ha = 25.23 in sodium).

Table 1. Properties of mercury (Hg) and air at room temperature.

Name	Notation	Mercury (Hg)	Air
Density	ρ	13,534 kg/m^3	1.2 kg/m^3
Dynamic viscosity	η	0.00149 kg/(m·s)	1.8×10^{-5} kg/(m·s)
Kinematic viscosity	$\nu = \eta/\rho$	1.1×10^{-7} m^2/s	1.8×10^{-5} m^2/s
Electric conductivity	σ	10^6 1/(Ω·m)	
Surface tension	γ	0.4589 N/m	

Table 2. Properties of liquid sodium (Na) and argon (Ar) at 400 °C.

Name	Notation	Sodium (Na)	Argon (Ar)
Density	ρ	856 kg/m^3	0.713
Dynamic viscosity	η	0.000284 kg/(m·s)	2.4×10^{-5}
Kinematic viscosity	$\nu = \eta/\rho$	3.32×10^{-7} m^2/s	3.37×10^{-5}
Electric conductivity	σ	4.52×10^6 1/(Ω·m)	
Surface tension	γ	0.161 N/m	

4.1. Horizontal Channel

In a horizontal channel, $Y^{1,0} = 0$, whereby, for specified fluids (i.e., specified η_{12}), the solutions for the holdup $\tilde{h}$ and the dimensionless pressure gradient $P_f^{1,0}$ are a function of the phases' flow-rate ratio, Q_{12}, and the Hartmann number, Ha.

Figure 2 shows the effect of the magnetic field intensity variation of the holdup with the gas-to-liquid flow-rate ratio $Q_{21} = U_{2s}/U_{1s}$ for mercury–air and sodium–argon two-phase flows. Obviously, the liquid metal holdup decreases with the increase in the gas flow rate, and is higher in the mercury–air system of the larger viscosity ratio. The effect of the magnetic field on the holdup becomes significant for Ha > ~1. Due to the slowing down of the flow by the magnetic damping force, increasing the magnetic field intensity results in a higher holdup of the liquid metal and attenuates its decline with the increase in the gas flow rate. These trends are more pronounced for the same Ha in the more viscous mercury–air system.

Figure 3 shows the effect of the magnetic field intensity on the dimensionless frictional pressure gradient $P_f^{1,0}$ (normalized with respect to the frictional pressure gradient of a single-phase liquid metal flow with the same U_{1s}, but without a magnetic field, see Equation (15)). As expected, increasing Ha results in a higher frictional pressure gradient, whereby the $P_f^{1,0}$ factor reaches a value of ~1000 for Ha = 103.7 (corresponding to $B_0 = 0.2$ T and 0.041 T for mercury and sodium, respectively). It is well known that, in the absence of a magnetic field, the introduction of a gas flow to the flow of a viscous liquid can result in a lubrication effect (e.g., [31]), where $P_f^{1,0} < 1$ values are obtained for some range of low Q_{21}. The lubrication effect increases with increasing η_{12} (approaching a value of 0.25 in the TP geometry for large η_{12}). For example, in mercury–air flow and Ha = 0, $P_f^{1,0}$ can reach a value of ~0.35 (65% reduction of the pressure gradient) for $Q_{21} = 0.05$, and $P_f^{1,0} < 1$ up to $Q_{21} < ~10$ (see also Figure 4b below for low Ha, which shows the power reduction). Indeed, a reduction of the pressure gradient factor $P_f^{1,0}$ by the gas flow at low Q_{21} values is also obtained for low values of Ha < ~1. However, the figure shows that, in the examined range of Ha > 5, the presence of a gas layer further augments the pressure gradient factor, $P_f^{1,0}$, in particular, when $Q_{21} > 1$ and Ha is relatively low. The sensitivity of the $P_f^{1,0}$ factor to the gas flow is higher for the less viscous sodium–argon system. In both systems and high Ha, the gas flow rate has a rather small effect on the $P_f^{1,0}$ factor over a wide range of gas flow rates.

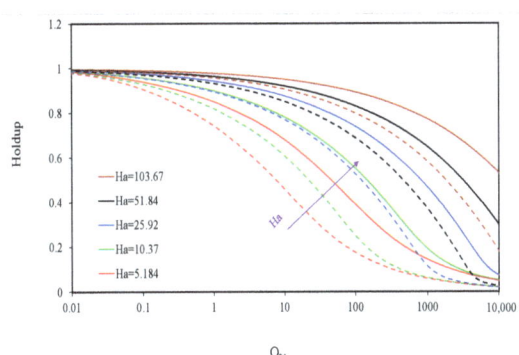

Figure 2. The effect of the magnetic field intensity on the liquid metal holdup, $\tilde{h}$. Comparison of holdup vs. the gas-to-liquid flow-rate ratio in the mercury–air (solid line) and sodium–argon (dashed line) systems for the same Ha. The range of Ha considered corresponds to $B_0 = 0.01$ T to 0.2 T and $B_0 = 0.0020544$ T to 0.041088 T in mercury and sodium, respectively.

The lubrication effect of the gas flow is noticed over a wider range of Ha when examining the P_f^1 factor (Equation (17)), which is the dimensionless frictional pressure gradient normalized with respect to a single-phase liquid metal flow under the same magnetic field intensity. As shown in Figure 4a, the lubrication effect is more pronounced in the case of the more viscous mercury, where values of $P_f^1 < 1$ are obtained up to Ha~25 for $Q_{21} < 1$. The adverse effect of the gas flow on the pressure gradient at $Q_{21} > ~10$ is also reflected in the P_f^1 factor, which, for the same Ha, is higher for the sodium–argon system. The lubrication region is more visible in Figure 4b, where the pumping power factor, Po, is also shown for the mercury–air system for lower Ha ($B_0 = 0.001$ T, Ha = 0.5183). Values of Po = $P_f^1(1 + Q_{21})$ <1 indicate that pumping power saving can be achieved by introducing air flow to the conductive liquid flow. Note that for low Q_{21} Po $\sim P_f^1$ (both approach 1 at $Q_{21} = 0$). As shown in Figure 4b, significant power reduction can be achieved for low Ha. For example, for Ha = 0.5183, the minimal value Po = 0.3586 is obtained by adding 3.5% air to the mercury flow. For such low magnetic field strength, the $P_f^1 \sim P_f^{1,0}$ and the value of both is almost the same as that obtained for Ha = 0. The potential for power saving by the air flow is obviously reduced with increasing the magnetic field strength (negligible for Ha > 25), and the minimum Po is shifted to lower Q_{21} values.

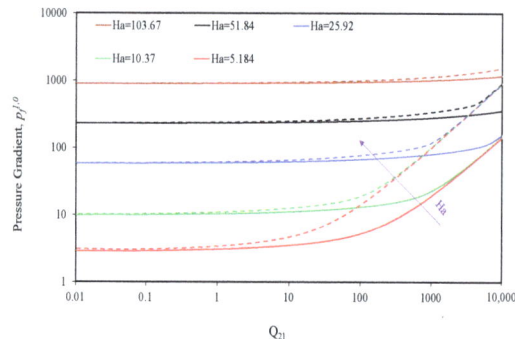

Figure 3. The effect of the magnetic field intensity on the $P_f^{1,0}$ factor (with respect to the pressure gradient in single-phase liquid metal flow without a magnetic field)—comparison of $P_f^{1,0}$ vs. the gas-to-liquid flow-rate ratio in the mercury–air (solid line) and sodium–argon (dashed line) systems for the same Ha.

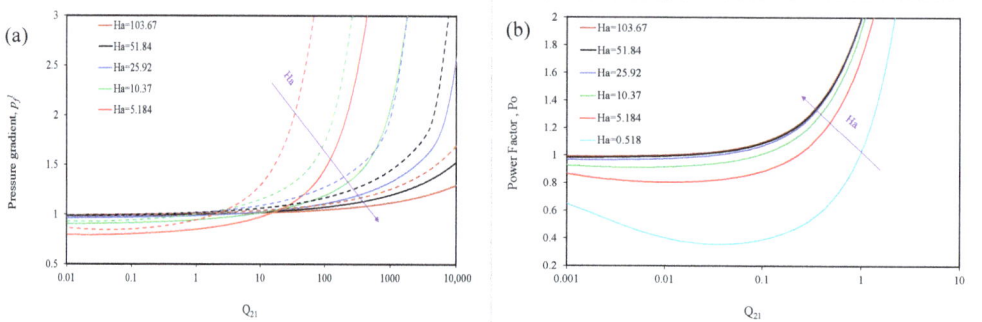

Figure 4. The effect of the magnetic field strength on (**a**) the P_f^1 factor (with respect to the pressure gradient in single-phase liquid metal flow under the same magnetic field, see Equation (17))—comparison of P_f^1 vs. the gas-to-liquid flow-rate ratio in the mercury–air (solid line) and sodium–argon (dashed line) systems for the same Ha; (**b**) the power factor, Po (Equation (18)), vs. the gas-to-liquid flow-rate ratio in the mercury–air system.

Further insight into the effect of the gas flow on the holdup and pressure gradient can be obtained by examining the dimensionless velocity profiles (normalized by the conductive liquid superficial velocity, U_{1s}). Figure 5a demonstrates the change in the velocity profile in the mercury–air system when the gas flow rate is increased from $Q_{21} = 1$ to 10 for a constant Ha (=25.92, B_0 = 0.05 T). As shown, in both cases, the air moves much faster than the mercury and its velocity profile is practically parabolic. The velocity gradients at the upper wall (hence the shear stress) are sharp and obviously increase with the gas flow rate. The velocity profiles in the mercury are depicted in Figure 5b for a wider range of Q_{21}. The magnetic field results in a flat velocity profile in the bulk of the mercury layer, with sharp velocity gradients at the wall and at the interface. As shown, except for very low Q_{21}, the maximum velocity is in the air layer, whereby, near the interface, the mercury is dragged by the air flow. The sharp velocity gradients at the interface indicate that high interfacial shear stresses are involved. Increasing the gas flow rate also results in a higher near-wall velocity gradient in the mercury. The increase in the wall shear stresses on both channel walls with the gas flow is reflected in the increased pressure drop factors, $P_f^{1,0}$ and P_f^1 (in particular for $Q_{21} > \sim 10$, Figures 3 and 4).

Figure 6 shows that, with increasing the intensity of the magnetic field (for a fixed value of $Q_{21} = 1$), even though the mercury holdup increases, the mercury velocity gradient near the wall and at the interface becomes steeper, resulting in an increased pressure gradient at higher Ha. The mercury velocity rapidly grows from the bottom wall and quickly attains an almost flat profile, like in the single-phase Hartmann flow. However, contrary to the latter, in the upper part of the mercury layer the velocity does not decay, but steeply grows towards the air velocity at the interface. The thickness of the gas layer decreases at increased applied magnetic fields, while its maximum velocity and interfacial shear stress increase. In the experimental study by Lu et al. [10], such a flow configuration was characterized as a gas jet. This configuration is likely to enhance mixing between the two phases, potentially causing the flow regime to gradually transition to a mixed flow.

The effect of Q_{21} and Ha on the shape of the velocity profile can be deduced by examining the variation in the maximal velocity and the interfacial velocity. Figure 7a shows the variation in the maximal velocity in the velocity profile vs. Q_{21} for various Ha (B_0 ranging from 0.001 to 0.1). The range of Q_{21} where the maximal velocity is in the mercury layer is indicated by a dashed line, corresponding to the range of flow rates where the mercury flow drags the air at the interface. In this range of flow rates, the interfacial velocity is obviously lower than the maximal mercury velocity. However, as shown, this flow range is limited to low Q_{21} and diminishes with increasing Ha. Examining the interfacial velocity in mercury–air flow vs. Q_{21} for various values of Ha (see Figure 7b) shows that it increases with increasing Q_{21}, with a rather low sensitivity to the magnetic field strength.

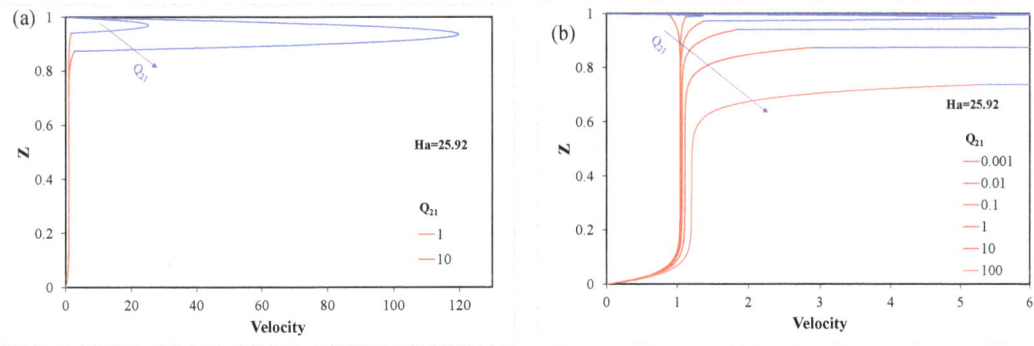

Figure 5. (**a**) The effect of the air-to-mercury flow-rate ratio on the dimensionless velocity profile (scaled with respect to mercury superficial velocity). (**b**) Enlargement of the mercury domain. Ha = 25.92. Mercury—red curves, air—blue curves.

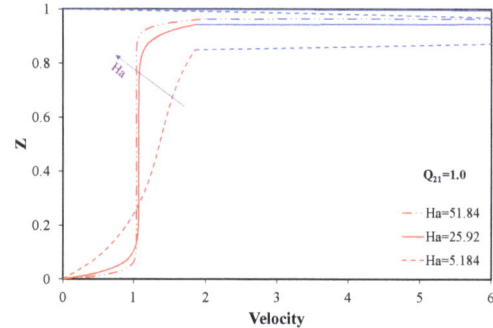

Figure 6. The effect of the magnetic field intensity, Ha, on the mercury dimensionless velocity profile (scaled with respect to mercury superficial velocity), $Q_{21} = 1$. Mercury—red curves, air—blue curves.

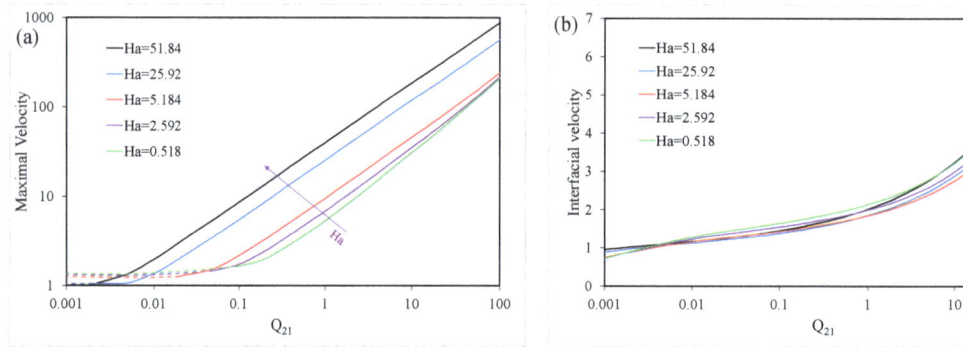

Figure 7. (a) The dimensionless maximal velocity vs. Q_{21} for various Ha. The range of Q_{21} where the maximal velocity is in the mercury layer is indicated by the dashed line. For large Q_{21} values, the maximal velocity is always in the air and increases proportionally to $Q_{21}^{0.6737}$. (b) The interfacial velocity in mercury–air flow vs. Q_{21} for various values of Ha.

4.2. Inclined Channel

In inclined channels, the holdup and (dimensionless) frictional pressure gradient $\left(P_f^{1,0}, P_f^1\right)$ are determined by four dimensionless parameters $\left(\eta_{12}, Q_{21}, Ha, Y^{1,0}\right)$. To obtain the total (dimensionless) pressure gradient factors ($P^{1,0}$, P^1, or P^{1T}; see Equations (42) and (43)) the hydrostatic pressure gradient contribution should be accounted for. It can be calculated based on the holdup; however, the density ratio, $\rho_{12} = \rho_1/\rho_2$, is then an additional required parameter. Depending on the values set for Q_{21} and $Y^{1,0}$, concurrent up-flow, concurrent down-flow or countercurrent flow of the liquid metal and gas are considered. In the following, the effects of the system parameters on the flow characteristics are demonstrated by referring to the mercury–air system, thus fixing the values of $\eta_{12} = (81.87)$ and ρ_{12} (=13,534).

4.2.1. Concurrent Upward Flow

According to the selected flow configuration and coordinates (Figure 1), concurrent upward flow corresponds to negative values of the superficial velocities, $U_{2s}, U_{1s} < 0$; hence, $Q_{21} > 0$ and $Y^{1,0} > 0$.

Figure 8a shows the effect of the magnetic field intensity on the variation in the holdup with the air-to-mercury flow-rate ratio for a fixed inclination parameter value of $Y^{1,0} = 2075.25$. The latter corresponds to a constant flow rate of the mercury ($U_{1s} = -0.005$ m/s) and a slight upward inclination of the channel, $\beta = 0.2°$ (the channel height is 0.02 m). The variation in

Q_{21}, in this case, actually corresponds to the change in the air flow rate. A comparison of Figure 8 with Figure 2 shows the drastic effect of the slight inclination on the holdup. Due to the retarding gravity force, the metal flow is slowed down. Consequently, even for very shallow channel inclination and low mercury flow rate, the mercury occupies most of the flow cross-section over a wide range of air flow rates ($\tilde{h} > 0.9$ up to $Q_{21} = \sim 200$ with low sensitivity to Ha). The holdup (and the hydrostatic pressure drop) steeply declines with the air flow rate for $Q_{21} > 2000$. Then, for some range of high air flow rates, three different solutions for the holdup are obtained for the same air (and mercury) flow rates: the high holdup solution and two additional solutions of lower holdup values. An example of the triple solution is indicated by the three circles in Figure 8b. With increasing the magnetic field strength (i.e., higher Ha), the range of air flow rates where a triple solution is obtained diminishes and is shifted to higher air flow rates (see Figure 8a). For the tested parameter set, the triple solution is feasible up to Ha$\sim$52 ($B_o \sim 0.1$ T). For even higher air flow rates, beyond the triple-solution region, only a single (low) holdup solution is obtained, and the sensitivity of its value to the magnetic field intensity is rather low. Figure 8b shows that, with increasing the mercury flow rate (i.e., reducing $Y^{1,0}$), the range of gas flow rates where triple solutions are obtained is shifted to a lower Q_{21}, but becomes narrower. For $|U_{1s}| \geq 0.1$ m/s, only a single solution for the holdup is obtained even for low Ha (=0.518). For the same flow-rate ratio (Q_{21}) the holdup of the mercury layer increases with reducing the mercury flow rate (i.e., increasing $Y^{1,0}$).

The possibility of obtaining multiple holdup solutions is typical to gravity-dominated two-phase systems, where the body force is of the order of the frictional pressure gradient (or higher), and the forces acting on the fluids can be balanced in more than one flow configuration. A demonstration of the velocity profiles associated with the three different solutions for the holdup is shown in Figure 9. The velocity profile in the air flow cross-section is almost parabolic, whereby the air flow practically determines the values of the pressure gradient and the interfacial shear stress. In the low holdup solution, the gravity force acting on the thin mercury layer is low, and the pressure gradient and air shear at the interface are capable of carrying upward the flow over the entire mercury layer. In the high holdup solution, the mercury experiences the highest counter-flow body force, whereby the pressure gradient and the air interfacial shear are insufficient to carry the entire mercury layer upward. Consequently, a back-flow of the mercury is observed near the lower channel wall, and the wall shear direction is reversed. As shown in the figure, the back-flow is reduced in the middle holdup solution, which results in a lower reversed wall shear.

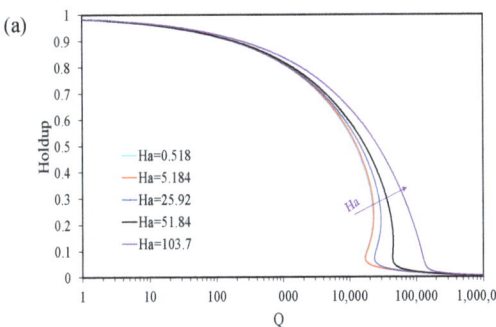
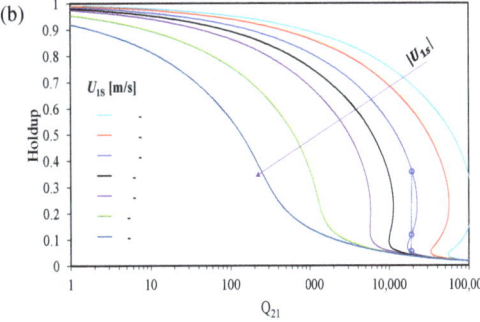

Figure 8. Variation in the mercury holdup in concurrent up-flow of mercury and air: (**a**) Effect of the magnetic field strength, Ha ($Y^{1,0} = 2075.25$, corresponding to $U_{1s} = -0.005$ m/s $\beta = 0.2°$ and H = 0.02 m). (**b**) Effect of the mercury superficial velocity for Ha = 0.518 ($Y^{1,0}$ values in the range of 20.75 to 10376). An example of triple solutions is shown by the 3 circles on the holdup curve corresponding to $U_{1s} = -0.005$ m/s.

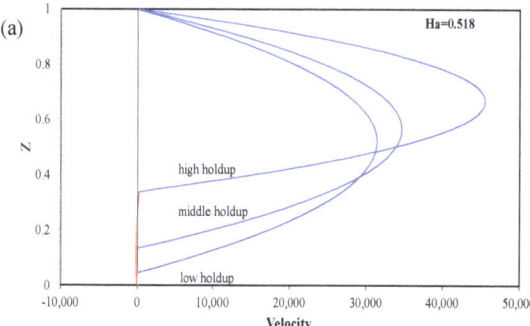 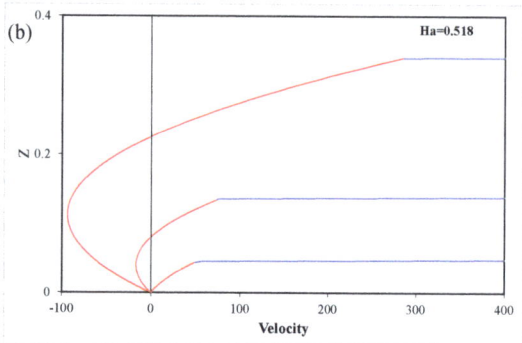

Figure 9. Velocity profiles (dimensionless) in the 3-holdup solutions of mercury–air up-flow. (**a**) (Ha = 0.518, $Y^{1,0} = 2075.25$, $Q_{21} = 2 \times 10^4$ $U_{1s} = -0.005$ m/s, $U_{2s} = -100$ m/s. (**b**) Enlargement of the velocity profiles in the mercury layer. Mercury—red curves, air—blue curves.

Back-flow of the heavier fluid is a source of instability of the stratified flow configuration, as it introduces a large disturbance at the inlet, where the fluids are introduced into the channel. The effect of the magnetic field strength on the velocity profiles of the high and middle holdup solutions is demonstrated in Figure 10 (for $Y^{1,0} = 2075.25$). As shown, due to the magnetic field damping force, increasing Ha reduces the back-flow in both flow configurations, although the mercury holdup corresponding to the high holdup solution increases with Ha, and, therefore, is associated with a higher backward gravity force. For the examined parameters, triple-holdup solutions are obtained up to Ha~18 (B_0 = 0.035 T). For higher Ha, a single (relatively high) holdup solution is obtained, yet with a significant region of mercury back-flow (e.g., Ha = 51.83 in Figure 10b). The effect of the air flow rate on the velocity profile in the mercury layer for Ha = 51.84 is demonstrated in Figure 10c,d. For this magnetic field intensity, a single solution for the holdup is obtained for any Q_{21}. As shown, although the mercury holdup is reduced with increasing the air flow rate, the back-flow intensity increases. Note that the same dimensionless velocity profiles would be obtained for lower superficial velocities of the conductive layer and the gas, provided the value of $Y^{1,0}$ is the same (e.g., by referring to a smaller channel size and/or inclination, and/or density difference), and Q_{21} and η_{12} are unchanged.

The effect of the magnetic field strength and the air flow rate on the pressure gradient is examined in light of Figure 11. As expected, the frictional pressure gradient factor $P_f^{1,0}$ (scaled with the frictional pressure gradient of single-phase mercury flow for Ha = 0) increases with Ha (see Figure 11a). Introducing the air flow further increases the frictional pressure gradient. In fact, at high Q_{21} (beyond the triple-solution region, corresponding to the loop in the pressure gradient curves) the increase in the frictional pressure gradient is dominated by the gas flow and is insensitive to the magnetic field intensity. Examining the frictional pressure gradient factor P_f^1 (i.e., scaled with the frictional pressure gradient of mercury flow under the same Ha, Figure 11b) reveals that the lubrication effect, obtained in the horizontal channel at low air flow rates (Figure 7b), is lost at even such a slight upward inclination of the channel. The variation in the hydrostatic pressure gradient factor, P_g^1 (see Equation (42)), is shown in Figure 11c. Obviously, for a constant Ha, it follows the trend of the holdup variation with Q_{21} (shown in Figure 7). With increasing Ha, the hydrostatic pressure gradient factor decreases, and values < 1 are reached at high Ha and Q_{21}.

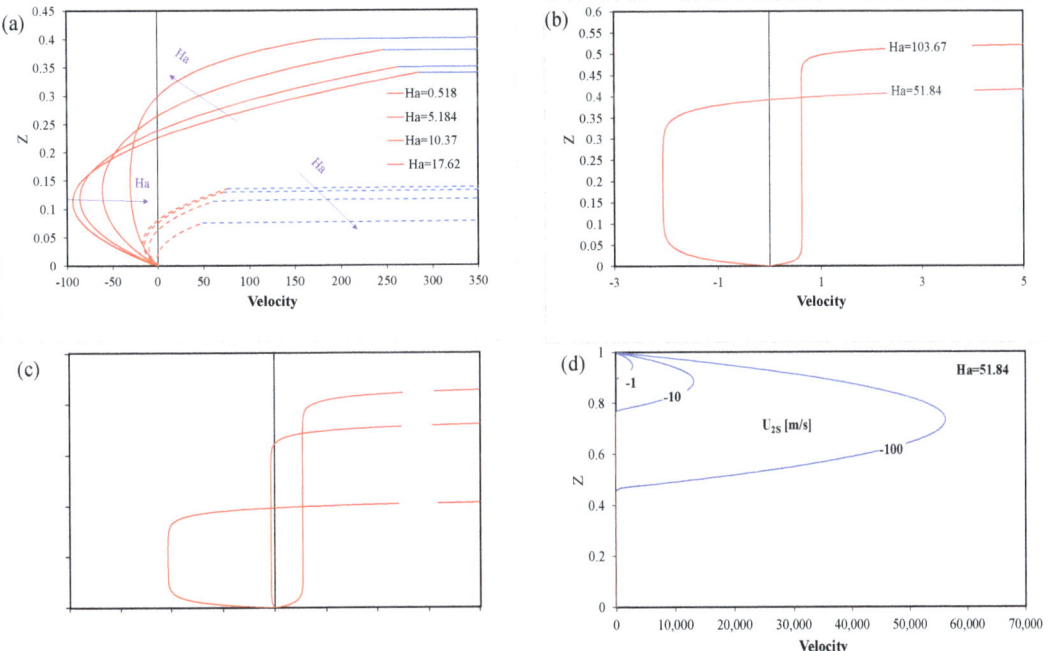

Figure 10. Velocity profiles (dimensionless) in mercury–air up-flow ($Y^{1,0} = 2075.25$, $U_{1s} = -0.005$ m/s). (**a**) Effect of Ha on the mercury velocity profiles in the high and middle holdup solutions (for $U_{2s} = -100$ m/s). Triple solutions are obtained up to Ha~18 (**b**) Velocity profiles in the mercury layer for higher Ha for $U_{2s} = -100$ m/s, where a single holdup solution is obtained. (**c**) Effect of the air superficial velocity on the mercury velocity profiles at high magnetic field strength, Ha = 51.84 ($B_0 = 0.1$T). (**d**) Velocity profile in the air layer. Mercury—red curves, air—blue curves.

From the perspective of reducing the pressure gradient by the air flow the values of the total pressure gradient factor, P^{1T} (i.e., dp/dx normalized with respect to the total pressure gradient of single-phase mercury flow with the same magnetic field, Equation (43)), should be examined. Figure 11d shows that, at low Q_{21}, introducing the air flow practically does not affect the total pressure gradient. However, in the triple-solution region and in its vicinity, the steep reduction in the mercury holdup (and thus the hydrostatic pressure gradient), results in a pronounced reduction in the total pressure gradient compared to the mercury single-phase up-flow (e.g., by a factor of 0.122 at low Ha, and by a factor of 0.285 for $B_0 = 0.1$ T, Ha = 51.83). The potential for pressure drop reduction in this region is reduced with the increase in the magnetic field strength.

The trends of the variation in the holdup with the channel upward inclination are similar to those obtained upon changing the mercury flow rate (Figure 8b). Obviously, an increase in $Y^{1,0}$ can be affected by reducing the flow rate of the conductive phase (mercury), by increasing the channel upward inclination, or by increasing the channel size. Figure 12a shows that, for the same Ha and maintaining the same flow rates of both phases (i.e., same Q_{21}), the holdup of the conductive layer is higher at a steeper upward channel inclination. Also, with increasing β, the triple-solution region is shifted to higher air flow rates. In this region, owing to the lower values of the low and middle holdup solutions at steeper inclinations, smaller values of the total pressure-gradient factor, P^{1T}, can be obtained (e.g., 0.00425 for β = 5° in Figure 12b). Obviously, the same range of $Y^{1,0}$ examined in Figure 12a can be obtained for a lower superficial velocity of the conductive layer in a smaller channel size, in which case the Q_{21} range of the triple-solution region corresponds to lower gas dimensional velocity.

The concurrent down-flow corresponds to $Y^{1,0} < 0$ $U_{1s}, U_{2s} > 0$ ($Q_{21} > 0$). In countercurrent flow, a net downward flow of conductive heavier fluid is considered, namely $U_{1s} > 0$ and $Y^{1,0} < 0$, while the light fluid flows upward ($U_{2s} < 0$); hence, $Q_{21} < 0$. Countercurrent flow is a basic configuration in many heat and mass transfer systems.

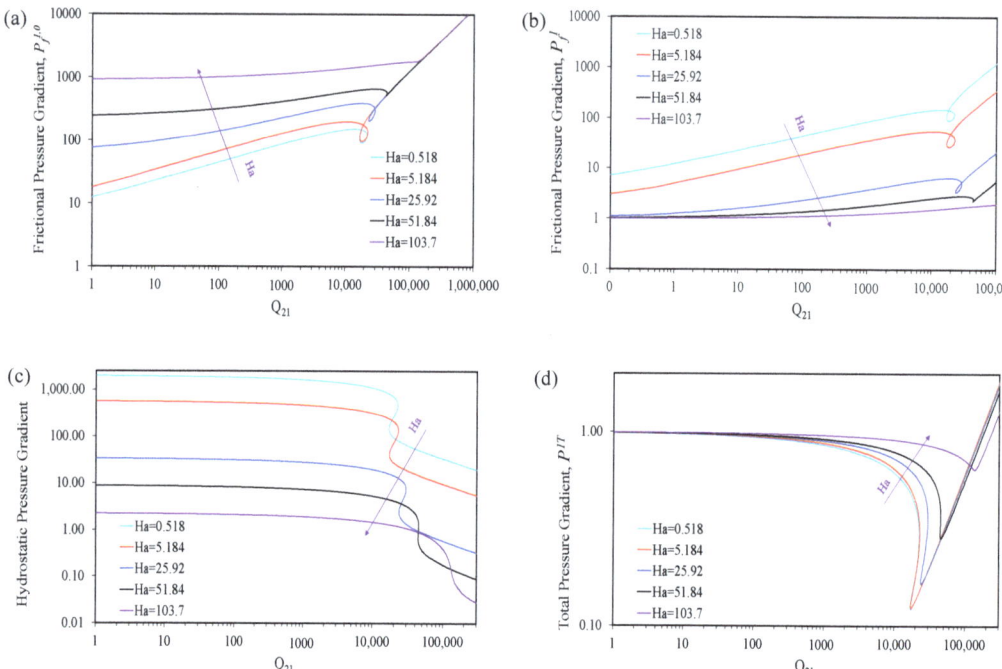

Figure 11. Effect of the magnetic field strength, Ha, on the pressure gradient in concurrent up-flow of mercury and air ($Y^{1,0} = 2075.25$, corresponding to $U_{1s} = -0.005$ m/s, $\beta = 0.2°$ and H = 0.02 m). (**a**) Frictional pressure gradient factor $P_f^{1,0}$ (with respect to the frictional pressure gradient of SP mercury flow and Ha = 0). (**b**,**c**) Frictional and hydrostatic pressure gradient factor, P_f^1 and P_g^1, respectively (with respect to the frictional pressure gradient of SP mercury flow with the same Ha). (**d**) Total pressure gradient normalized with respect to the total pressure gradient of SP mercury flow with the same Ha, $P^{1T} = P^1 / \left(1 + \frac{\rho_2}{\Delta\rho} \frac{Y^{1,0}}{P_{fs}^{1,0}}\right)$.

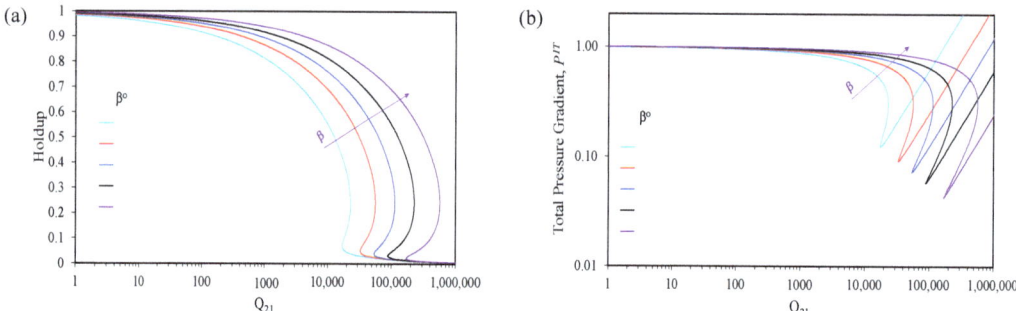

Figure 12. Effect of the channel inclination on the holdup (**a**) and on the total pressure gradient factor, P^{1T} (**b**), for mercury–air flow, Ha = 5.18, $U_{1s} = -0.005$ m/s, and H = 0.02 m ($Y^{1,0}$ values in the range of 2075.25 to 51815.6, corresponding to β ranging from 0.2° to 5°).

4.2.2. Countercurrent and Concurrent Downward Flows

Figure 13 demonstrates the effect of the magnetic field intensity on the countercurrent flow characteristics of mercury and air for a given inclination parameter value of $Y^{1,0} = -259.1$. The latter corresponds to a constant flow rate of the mercury ($U_{1s} = 1 \text{ m/s}$) and a channel inclination of $\beta = 5°$ (the channel height is 0.02 m). The variation in Q_{21} can then be attributed to variation in U_{2s}. As shown in the figure, countercurrent flow can be established for a limited range of sufficiently low (upward) air flow ($Q_{21} < 0$) and magnetic field strength. In this range, two distinct solutions for the holdup (and the corresponding flow characteristics) are obtained for fixed flow rates of the air and mercury, which merge to a single solution at the flooding point, beyond which countercurrent flow is not feasible. The countercurrent flow region diminishes with increasing Ha. For Ha > 53.7, countercurrent flow of mercury and air is not feasible (see also Figure 15b below). It is worth noting that the possibility of establishing the two high and low holdup configurations in the countercurrent region was demonstrated experimentally by Ullmann et al. [28] for Ha = 0. The flow configuration realized actually depends on the resistance at the heavy phase outlet. The high holdup configuration is obtained by increasing the resistance at the outlet while maintaining the same flow rates.

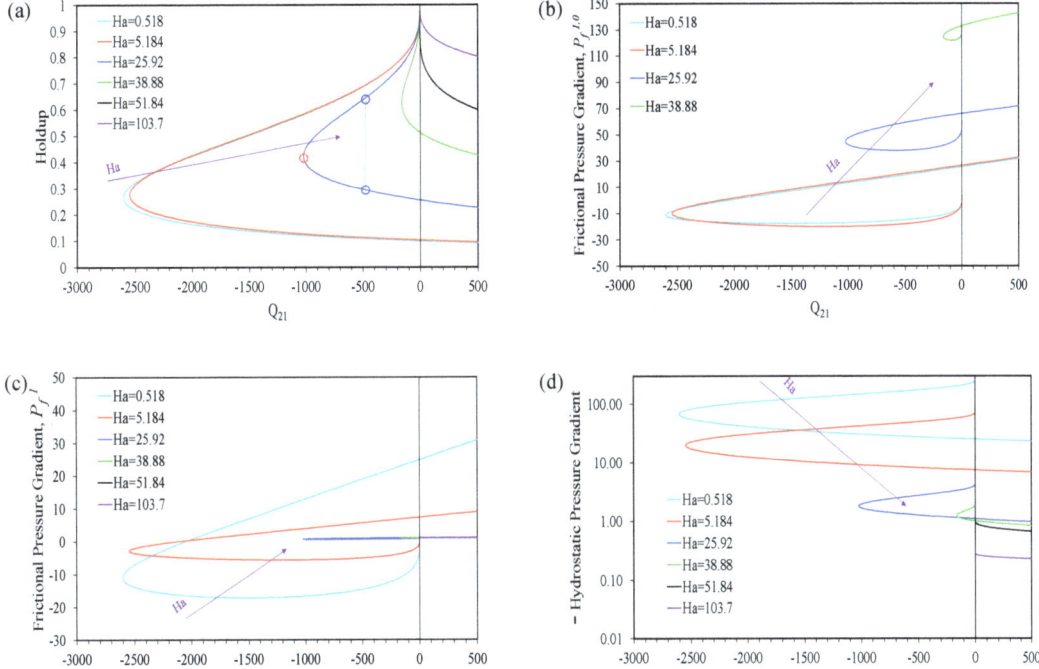

Figure 13. Countercurrent mercury–air flow—effect of the magnetic field intensity (Ha) and air flow rate on the countercurrent flow characteristics: (**a**) Mercury holdup (where the two solutions and the flooding point for Ha = 25.92 are indicated by circles). (**b**–**d**) Frictional and hydrostatic pressure gradient factors, $P_f^{1,0}$, P_f^1, and $-P_g^1$, respectively. ($Y^{1,0} = -259.1$, corresponding to $U_{1s} = 1 \text{ m/s}$, $\beta = 5°$, H = 0.02 m).

Figure 13b,c show the variation in frictional pressure gradient factors $P_f^{1,0}$, P_f^1 with Q_{21} and Ha. In the countercurrent flow region, the positive values of the frictional pressure gradient factors are associated with the low-holdup configuration, indicating that the frictional pressure gradient is dominated by the downward mercury flow. Negative values of these factors are obtained (usually, but not exclusively, in the high holdup configuration) for low Ha (e.g., Ha = 0.518 and 5.18 in Figure 13b), which indicates that the frictional pressure gradient is dominated by the upward gas flow. With increasing Ha, the difference between the $P_f^{1,0}$ and P_f^1 values associated with the low and high holdup configurations diminishes (both $P_f^{1,0}$ values are positive). With high Ha values, the frictional pressure gradients become close to that obtained in single-phase mercury flow with the same Ha (i.e., the P_f^1 values are close to 1). The corresponding $P_f^{1,0}$ values at high Ha are in the concurrent flow outside the range of Figure 13b. The downward flow of the mercury is obviously assisted by the hydrostatic pressure, $(dp/dx)_g > 0$ ($P_g^1 < 0$, Figure 13c, and, when $-P_g^1 > P_f^1$, no pump is needed to drive the mercury flow.

The difference between the flow characteristics associated with the two-holdup solutions in countercurrent flow can be elucidated by examining the velocity profiles. Figure 14 shows the effect of Ha on the velocity profiles of the two holdup solutions obtained for a downward flow of mercury (U_{1s} = 1 m/s) and a low upward gas flow ($U_{2s} = -0.05$ m/s, $Q_{21} = -0.05$). The velocity profiles in the lower holdup solution (Figure 14a) show that the downward flow of the mercury drags the air downward near the interface (i.e., air back-flow region). With increasing Ha, the mercury holdup increases and its (downward) velocity is reduced, resulting in lower air back-flow near the interface. Figure 14b shows the velocity profiles in the corresponding upper holdup solution. For the low Q_{21} considered, the mercury occupies most of the channel, and the air flows in a very thin layer. Here, the air flow drags the mercury upward near the interface (i.e., mercury back-flow region). Such velocity profiles are associated with negative values of $P_f^{1,0}$, P_f^1 (see Figure 13b,c). As shown, the mercury back-flow in the upper holdup solution diminishes with increasing Ha (for Ha = 25.92, the mercury flows downward over the entire layer). As shown in Figure 13a, for higher upward air flow rates, the thickness of the air layer in the high holdup solution is much larger. Yet, for low Ha, the mercury back-flow phenomenon near the interface is sustained over a wide range of negative Q_{21} corresponding to $P_f^{1,0}$, $P_f^1 < 0$ (Figure 13b,c).

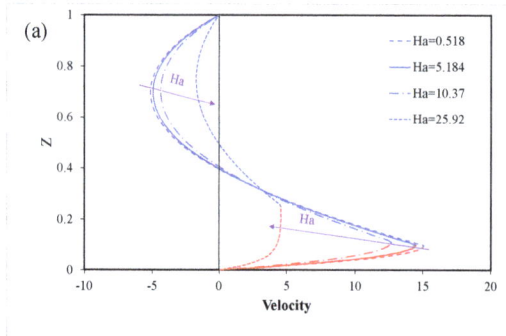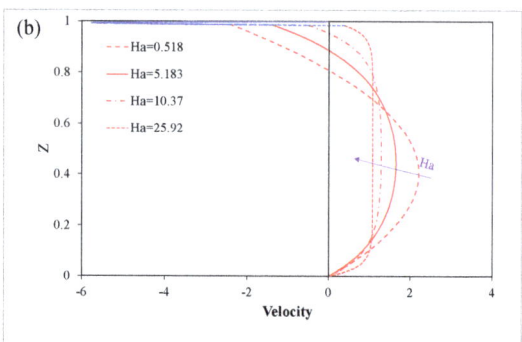

Figure 14. Effect of the magnetic field intensity on the (dimensionless) velocity profiles in the 2-holdup solutions of mercury–air countercurrent flow: (**a**) lower holdup solution; (**b**) upper holdup solution ($Y^{1,0} = -259.1$, corresponding to $U_{1s} = 1$ m/s, $U_{2s} = -0.05$ m/s β = 5° H = 0.02 m). Mercury- red curves, air-blue curves.

The variation in the holdup with the magnetic field intensity in concurrent down-flow of mercury and air is elucidated in view of Figure 15 ($Y^{1,0} = -259.1$, corresponding to $U_{1s} = 1$ m/s, β=5°, H = 0.02 m). Figure 15a implies that, apparently, a single solution for the holdup is obtained for specified Ha and Q_{21}. As expected, the holdup increases with Ha and is reduced by increasing the air flow rate. However, the enlargement of the region of high holdups and low Q_{21} (Figure 15b) shows that two additional solutions of higher holdups can be obtained in the range of Ha < 53.92. This range of Ha is slightly higher than the maximal Ha value for obtaining countercurrent flow (Ha~53.7). However, independently of the magnetic field intensity, for the tested inclination parameter value ($Y^{1,0} = -259.1$), already, only the low holdup solution branch persists in the concurrent flow for $Q_{21} > 0.006$. Note that the same inclination parameter can be obtained for lower mercury superficial velocity, lower channel inclination, or smaller channel height.

The frictional pressure gradient factor P_f^1 is shown in Figure 15c. For low Ha, and (even low) Q_{21} values, which are already outside the triple-solution region (see Figure 15a), the low mercury holdup solution is associated with higher frictional pressure gradients compared to single-phase mercury flow under the same Ha. The effect of the airflow on the frictional pressure gradient diminishes with increasing Ha. Enlargement of the low Q_{21} (triple solution) region is shown in Figure 15c, where P_f^1 values associated with the two additional high holdup solutions at low Q_{21} are shown (e.g., for Ha = 25.92 and 51.84). The two high holdup solutions in the triple-solution regions correspond to $P_f^1 \sim 1$ (only slightly lower than the frictional pressure gradients of single-phase mercury under the same Ha) and are lower than that obtained for the low holdup solution. For sufficiently high Ha, only a single holdup solution is obtained in the entire concurrent down-flow region, for which the dimensionless frictional pressure gradient values are only slightly lower than 1 at low Q_{21}, due to the air lubrication effect. Practically, for large Ha, the frictional pressure gradient factor, P_f^1, in the concurrent region, is of the order of 1 in a wide range of Q_{21}. For example, for $Q_{21} = 200$, the frictional pressure gradient is only about 5% larger than the single-phase mercury flow. Obviously, it then keeps increasing with further increase in the air flow rate.

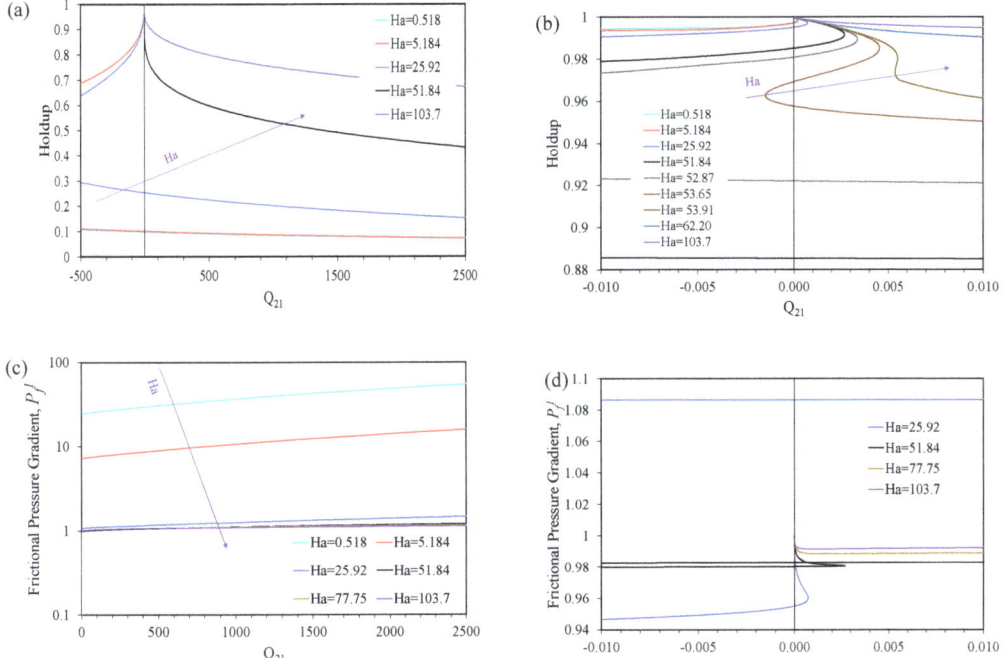

Figure 15. Concurrent mercury–air down-flow—effect of the magnetic field intensity (Ha) and air flow rate on the concurrent flow characteristics: (**a**,**b**) Mercury holdup. (**c**,**d**) Frictional pressure gradient factor, P_f^1. The r.h.s. figures show enlargement of the low Q_{21} region. ($Y^{1,0} = -259.1$, corresponding to $U_{1s} = 1$ m/s, $\beta = 5°$, H = 0.02 m.)

The difference between the velocity profiles of the three holdup solutions in concurrent down-flow is demonstrated in Figure 16. As shown, the velocity profile of the lower holdup solution (Figure 16a) is similar to that of the lower solution in countercurrent flow (Figure 14a). The difference is that here the net flow of air is downward, while there is a region of upward (back-flow) of the air near the upper channel surface. The steep mercury velocity gradient at the wall at low Ha is responsible for the large frictional pressure gradient associated with the lower holdup solution discussed above (with reference to Figure 15c). With increasing Ha, the holdup of the mercury increases, its velocity is reduced, and the velocity profile is flatter and becomes more similar to that observed in single-phase mercury flow under the same Ha. The back-flow in the thinner air layer diminishes, and eventually, for sufficiently large Ha, the flow in the entire gas layer is downward, and only one solution exists. A similar effect of Ha on the mercury velocity profile is observed in the middle and high holdup solutions (the latter is not shown), as both correspond to high mercury holdup and the velocity profiles in the mercury layer are similar. The main difference in the velocity profile of the middle and high holdup configurations is in the air layer (shown in Figure 16c,d). While in the upper solution, the flow in the entire air layer is downward for all Ha; in the middle holdup solution, back-flow of the air is still obtained. The increase in the mercury holdup with Ha results in a higher (positive) hydrostatic pressure gradient, which facilitates the downward flow of the heavier conductive liquid. Due to the opposite sign of the hydrostatic and frictional pressure gradient in concurrent down-flow, and also in countercurrent flow, the interpretation of the total pressure gradient value is more complicated than in concurrent up-flow. For example, for a specified single-phase flow rate of mercury in the particular channel size considered, the hydrostatic (dimensional) pressure gradient can be larger than the (negative) frictional pressure gradient. Hence, the total pressure gradient can attain positive values, indicating that the specified mercury flowrate is entirely driven by gravity. In fact, a restriction (valve) at the channel outlet is required to maintain the specified flow rate.

To analyze the effect of the air flow and the magnetic field intensity on the total pressure gradient, in concurrent down-flow and countercurrent flow, the values of dp/dx are normalized with respect to the hydrostatic pressure gradient in single-phase mercury flow, i.e., $\pi_T = \left(\frac{dp}{dx}\right)/(\rho_1 g \sin\beta)$. The obtained π_T values are shown in Figure 17. The effect of the air flow rate on the total pressure gradient is demonstrated in Figure 17a, where the π_T values are depicted vs. Q_{21}. Positive values are obtained in the countercurrent flow region (as long as the magnitude of the magnetic field is low enough to enable countercurrent flow), indicating that the specified mercury downward flowrate is also entirely driven by gravity in the presence of air upward gas flow. Figure 17b shows the π_T values when presented vs. the mercury holdup. The range of holdups that correspond to countercurrent flow extends from a holdup of almost 1 (marked by a square) to the holdup for which $\pi_T = 0$ (marked by a diamond). For this holdup, the frictional pressure gradient for the air–mercury system with the specified mercury flow rate (i.e., specified $Y^{1,0}$) is exactly balanced by the hydrostatic pressure gradient. This point corresponds to the lowest holdup solution obtained for $Q_{21} = 0$ (see Figure 13a), where the air is circulating in the channel with a zero net flow. The high holdup configuration (indicated by the squares) at $Q_{21} = 0$ also corresponds to gas circulation in the channel, however, with $\pi_T > 0$ (see Figure 17a). As shown in Figure 17b, the holdup range of countercurrent flow diminishes with increasing Ha. Countercurrent flow is feasible for magnetic field intensities below a critical value, for which the π_T value for single-phase mercury flow (i.e., holdup = 1) is 0 (i.e., the frictional pressure gradient for the single-phase mercury flow is balanced by the hydrostatic pressure gradient). For a higher Ha, the single-phase dimensionless pressure gradient is <1, and a pump is needed to drive the mercury flow, whereby countercurrent flow of air becomes unfeasible.

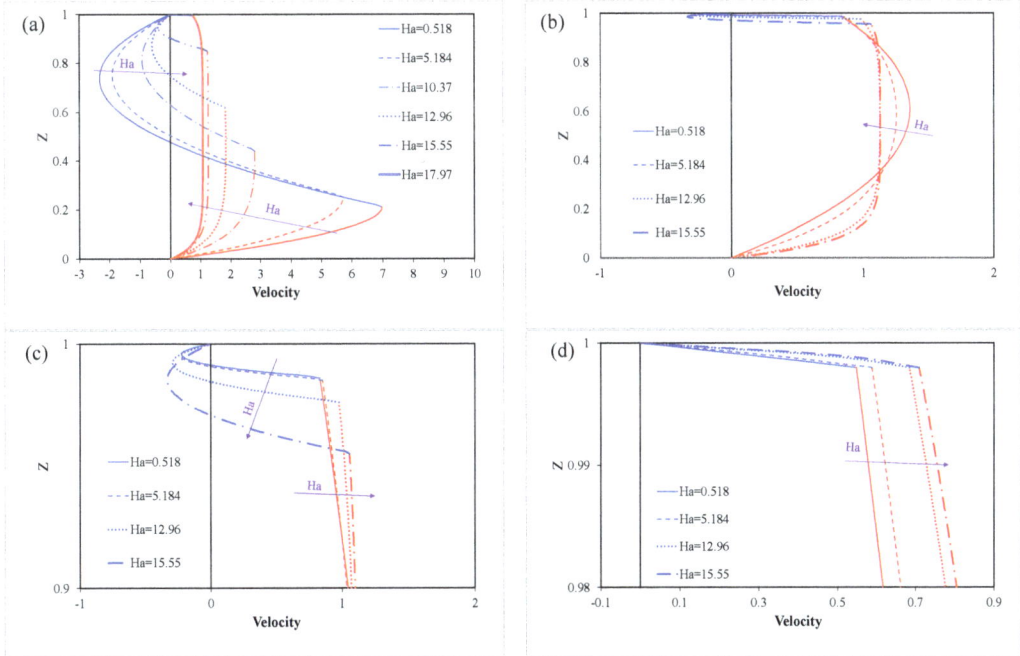

Figure 16. Effect of the magnetic field intensity on the (dimensionless) velocity profiles in the 3-holdup solutions of mercury–air concurrent down-flow: (**a**) lower holdup solution (for Ha = 17.97 only one solution exists); (**b**) middle holdup solution; (**c**,**d**) enlargement of the air velocity profile in the middle and upper holdup solutions. ($U_{1s} = 0.005$ m/s and $Y^{1,0} = -25.91$, corresponding to $U_{1s} = 10$ m/s, β = 5°, H = 0.02 m.) Mercury—red curves, air—blue curves.

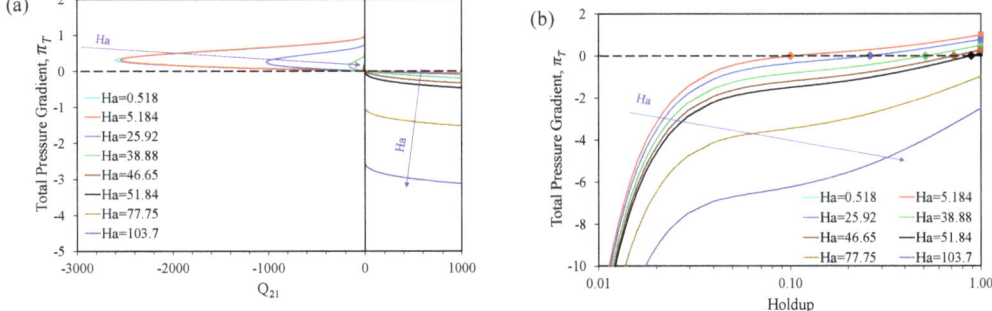

Figure 17. The effect of the magnetic field intensity (Ha) and the air flow rate (Q_{21}) on the dimensionless total pressure gradient factor π_T (normalized by the hydrostatic pressure gradient in single-phase mercury flow) in concurrent down-flow and countercurrent flow ($Y^{1,0} = -259.1$, corresponding to $U_{1s} = 1$ m/s, β = 5°, H = 0.02 m). (**a**) π_T vs. Q_{21} for various Ha. (**b**) π_T vs. the holdup. For each Ha the countercurrent flow extends from a holdup of almost 1 (marked by squares) to the holdup for which $\pi_T = 0$ (marked by diamonds).

The negative values of π_T in Figure 17 correspond to concurrent down-flow, and in the triple-solution region they correspond to the lower holdup solution. The figure shows that these π_T values become more negative as the magnitude of the magnetic field strength and/or the downward air flow are increased, indicating that higher pumping power is needed to drive the concurrent downward mercury–air flow. For Ha values for which triple solutions in concurrent down-flow are feasible, the π_T values of the two high holdup solutions (which are close to 1, and cannot be distinguished in Figure 17b) are positive, and are only slightly lower than the π_T value of single-phase mercury flow for the same Ha. This indicates that in these configurations the mercury–air down-flow is driven by gravity. It is worth noting that the correspondence of $\pi_T = 0$ and the low holdup solution of countercurrent flow at $Q_{21} = 0$ is valid in cases of $\rho_2/\Delta\rho \ll 1$. Otherwise, values $\pi_T > 0$ can also be obtained in the low holdup solution at $Q_{21} = 0$ as well as in concurrent flow ($Q_{21} > 0$).

The effect of channel inclination on the holdup and the total pressure gradient factor, π_T, in both countercurrent and concurrent downward flow is illustrated in Figure 18. The range of β = 0.5° to 10° corresponds to $Y^{1,0} = -25.94$ to -516.2. Decreasing the channel inclination reduces the range of airflow rates for countercurrent flow (Figure 18a). With decreasing channel inclination, the thickness of the mercury layer associated with the high holdup solution decreases, while thicker mercury layers are associated with the low holdup solution. Conversely, at shallower inclinations, the range where a triple-holdup solution is obtained in concurrent downward flow extends over a wider range of downward airflow rates (refer to Figure 18b, which depicts an enlargement of the high holdup region at low Q_{21}; the corresponding low holdup solution is shown in Figure 18a). While the total pressure gradient factor (π_T) of the high and middle solutions is positive (values close to 1 for all the examined inclinations, Figure 18d), the π_T value of the low holdup solution becomes more negative at shallower channel inclinations (see Figure 18c). This indicates that a pump is required to propel the concurrent downward mercury–air flow over most of the range of $Q_{21} > 0$, where the low holdup is the only solution of the flow equations.

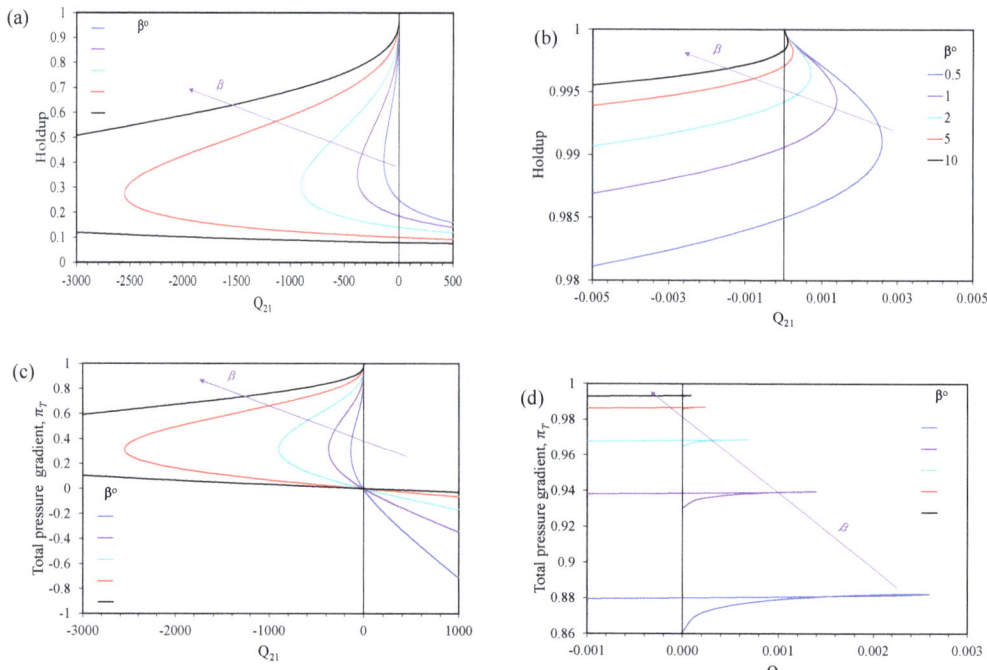

Figure 18. The effect of the channel inclination and the air flow rate (Q_{21}) on the holdup (**a**,**b**) and dimensionless total pressure gradient factor π_T (**c**,**d**) in countercurrent and concurrent down-flow. The r.h.s. figures are enlargement of the low Q_{21} to show the holdup and π_T values of the high and middle holdup solutions in the triple-solution region of concurrent down-flow (Ha = 5.184, $U_{1s} = 0.005$ m/s; the values of $Y^{1,0}$ correspond to β = 0.5°, 1°, 2°, 5°, 10°).

The influence of channel inclination on the flow characteristics illustrated in Figure 18 can be used to anticipate the effect of the mercury flow rate or the density of the conductive fluid. This is evident from the fact that the same variation in $Y^{1,0}$ can be achieved by adjusting the mercury flow rate while keeping β constant, or by altering the density of the conductive layer (ρ_2 is significantly less than ρ_1 and has minimal impact on the density difference). It is noteworthy that increasing the viscosity of the lighter layer (i.e., reducing η_{12}) has a similar effect to channel inclination (though not depicted), as it diminishes the countercurrent flow region and marginally expands the range of Q_{21} for triple solutions in concurrent down-flow. However, for higher η_2, the value of the single (low) holdup solution in concurrent down-flow (at a specified Q_{21}) is reduced. On the other hand, the impact of altering η_1 is more complex, as it also affects the values of Ha and $Y^{1,0}$, and thus necessitates specific examination of the particular two-phase system of interest.

4.3. Numerical Calculations

As already mentioned above, the hyperbolic functions involved in the analytical solutions pose an accuracy problem for usual calculations with 16, and even 32, decimal places. To obtain correct results, the calculations were carried out in Maple with 128 or 256 decimal places. For example, consider the mercury–air system with $H = 0.2$ m, $B_0 = 0.1$ T. The holdup noticeably decreases below 1 for $Q_{21} = 100$ and larger. The correct holdup values are calculated using 256 floating point digits, which yields $h = 0.962, 0.919, 0.829$, and 0.645 for $Q_{121} = 10^2, 10^3, 10^4$, and 10^5, respectively. The inaccurate results calculated by Maple using 128 digits are 0.594, 0.590, 0.585, and 0.579, respectively.

An alternative method of calculation of the holdups and the associated velocity profiles and pressure gradients is a numerical solution of the flow, Equations (1), (2), (24), and (25). The numerical schemes do not involve evaluation of the hyperbolic functions, and therefore can be carried out with a usual floating-point precision. We obtained numerical solutions using the finite difference and Chebyshev collocation methods. The Chebyshev collocation method is the same as that presented in [33], and the finite difference method applies central differences on an arbitrary stretched grid. The holdups and pressure drops are calculated using the superposition principle and the secant method as described in [34,35].

For the above example, using the finite difference method for the above parameters and at all values of Q_{21} considered, the holdup converges to the value obtained by the analytical solution within the fourth decimal place with 200 uniformly distributed grid points, and within the fifth place with 700 points. Trying to improve the convergence, we applied the *tanh* stretching,

$$x \leftarrow 0.5 + 0.5 \frac{tanh[s(x-0.5)]}{tanh(0.5s)}, \tag{48}$$

near the boundaries and the interface. However, with the stretching parameter gradually increased up to $s = 3$, we did not observe any noticeable improvement in the convergence.

The Chebyshev collocation method exhibits much faster, spectral convergence. Thus, for the same test cases, convergence within four decimal places is reached with the truncated series of 50 polynomials, while beyond 90 polynomials eight decimal places are readily converged. Obviously, the analytical solution is important for the validation of the numerical results and for assuring that all possible solutions are considered in inclined flows, where multiple steady-state configurations are feasible. An example of the holdup convergence study and comparison with the analytic solution is presented in Appendix A.

It is emphasized that, for example, the numerical study of stability of these flows is preferably conducted by using the Chebyshev collocation method, as demonstrated in [32,33]. At the same time, consideration of two-phase flows in bounded rectangular ducts or circular pipes (see [34,35]) will require lower-order methods. In these cases, the present results on convergence of the finite difference method will help estimate the computational resources required.

5. Concluding Remarks

We analyzed the characteristics of two-phase laminar stratified flows in horizontal and inclined channels under the effect of a transverse magnetic field, when the lower, heavier liquid is electrically conducting while the upper fluid is an electric insulator. The simplified channel geometry of two infinite parallel plates is considered, allowing for analytical solutions, and thus facilitating the analysis of system parameters' effects on flow characteristics.

The flow is defined by prescribed flow rates of the fluids so that the pressure gradient needed to drive the flow, as well as the conducting layer holdup, is obtained together with the solution of the whole problem. We focused on two main issues:

(a) The effect of the magnetic field on the liquid metal holdup and the pressure gradient driving the flow for prescribed flow rates of the fluids.

(b) The effect of the magnetic field on the multiple states (i.e., different holdups, pressure gradients, etc., for the same operational conditions), which, under certain conditions, may exist in inclined stratified two-phase flows.

The main conclusions are as follows:

(1) In horizontal flow, there exists a single state whose holdup increases with the increase in magnetic field strength. This increase is explained by the slowing down of the flow by the magnetic damping force. This effect is also valid in concurrent upward and downward flows for operational conditions that correspond to a single holdup solution of the flow governing equations.

(2) In inclined concurrent flows, the high-density difference between the liquid metal and the gas results in high sensitivity of the flow configuration to the channel inclination, which allows for multiple states (three) for fixed flow rates under low-to-moderate magnetic field intensities.

(3) In the triple-solution region, growth of the magnetic field strength leads to an increase in the lower and upper holdup values, but to a decrease in the holdup of the middle solution.

(4) The triple-solution regions are associated with the possibility of a back-flow region (downward) of the heavy phase near the lower wall in upward flow, and a back-flow region (upward) of the gas near the upper wall. However, back-flow is not exclusively associated with multiple solutions and can also be obtained under conditions where a single solution exists.

(5) Back-flow is a source of instability of the stratified flow configuration, as it introduces a large disturbance at the inlet, where the fluids are introduced into the channel. It can result in the formation of liquid metal slugs in an upward flow or entrainment of the gas into the liquid in a downward flow.

(6) The magnetic field weakens the back-flow, and the velocity profile of the liquid metal flattens. Consequently, the ranges of flow rates where multiple solutions exist narrow, and then disappear in strong magnetic fields. In this respect, a magnetic field stabilizes the single holdup flow configuration.

(7) In countercurrent flow, two distinct states exist up to a certain value of the phase's flow-rate ratio, beyond which countercurrent flow is impossible. In most cases, in the state of the smaller holdup, the heavier liquid metal drags the lighter one near the interface, while in the other state of a larger holdup, the thinner gas layer drags the liquid metal near the interface.

(8) With the increase in magnetic field strength, the countercurrent region diminishes, and beyond a certain Hartmann number, countercurrent flow is not feasible.

(9) Compared to the classical single-phase Poiseuille flow, the magnetic field damping force increases the pressure gradient needed to reach the same flow rate. In liquid metal–gas horizontal stratified flows, the addition of a gas layer may yield a lubrication effect so that, compared to the single-phase Hartmann flow, the same flow rate of the metal liquid can be reached with a smaller pressure gradient.

(10) In horizontal flows, the lubrication effect increases with increasing the liquid–metal/gas viscosity ratio and can reach about 75% pressure gradient reduction with high viscosity liquid metal under low magnetic field strength. The maximal lubrication effect is reached by adding a few percent gas flow rate to that of the liquid metal flow rate, and therefore can also be associated with pumping power reduction. However, the lubrication effect diminishes with increasing the Hartmann number (e.g., for mercury–air flow, it becomes insignificant for Ha > 25).

(11) In upward concurrent flow, the introduction of gas flow does not result in a lubrication effect and, in fact, increases the frictional pressure gradient. However, as the hydrostatic pressure gradient is reduced in the presence of a gas layer, the total pressure gradient can be significantly reduced. Indeed, a reduction in the total pressure gradient compared to the single-phase flow of the liquid metal is obtained in the triple-solution region and its vicinity. As this region is associated with high gas/liquid flow-rate ratios, power reduction may not be obtained. In any case, the potential for a pressure drop reduction in this region is reduced with the increase in the magnetic field strength.

(12) In concurrent downward flow, the addition of the gas layer also has a small effect on the frictional pressure gradient in the triple-solution region. However, as the driving force of the hydrostatic pressure is reduced in the presence of a gas layer, there is no benefit to adding gas to the flow of the liquid metal.

(13) In countercurrent flows, the frictional pressure gradient is higher in the state of the smaller holdup. In both states, it is larger than in single-phase mercury flow under the same magnetic field strength.

The analytical solution has been used to validate the results obtained by the numerical solution of the flow equations, which is intended to be used for studying the effect of the magnetic field on the stability of the laminar stratified flow pattern in horizontal and inclined flows.

Author Contributions: Conceptualization, A.P., A.G., N.B. and A.U.; Software, A.P. and A.G.; Validation, A.P. and A.G.; Formal analysis, A.P., A.G. and N.B.; Resources, A.G. and N.B.; Data curation, A.P. and A.U.; Writing—original draft, N.B. and A.U.; Writing—review & editing, A.P.; Visualization, A.U.; Supervision, A.G.; Funding acquisition, N.B. All authors have read and agreed to the published version of the manuscript.

Funding: This research was supported by Israel Science Foundation (ISF) grant No 1363/23 and by the Israel Ministry of Aliyah and Integration (for A. Parfenov).

Data Availability Statement: The original contributions presented in the study are included in the article, further inquiries can be directed to the corresponding author.

Conflicts of Interest: The authors declare no conflict of interest.

Nomenclature

Latin Symbols

B_0	magnetic field flux	kg/(s^2–A)
g	gravitational acceleration	m/s^2
G	pressure gradient, $= dp/dx$	Pa/m
h	height of interface plane	m/s
$\tilde{h}$	Holdup, $= h/H$	-
H	channel height	-
Ha	Hartmann number, $= B_0 H \sqrt{\sigma_1/\eta_1}$	-
p	pressure	Pa
P	dimensionless pressure gradient	-
Po	dimensionless pumping power	-
Q	volumetric flow rate	m^3/(m–s)
u	velocity	m/s
$\tilde{u}$	dimensionless velocity, $= u/U_{1s}$	-

$U_{1s,2s}$	superficial velocity, $= Q_{1,2}/H$	m/s
x	horizontal coordinate	m
Y	inclination parameter, defined in Equation (39)	-
z	vertical coordinate	m
$\tilde{z}$	dimensionless vertical coordinate, $= z/H$	-
Greek letters		
β	channel inclination angle to the horizontal	
η	dynamic viscosity	kg/(m-s)
π	dimensionless pressure gradient normalized by the hydrostatic pressure gradient of single-phase flow	-
ρ	density	kg/m^3
σ	electric conductivity	1/(Ω-m)
Subscripts		
1	lower phase (conductive fluid)	
2	upper phase (gas)	
12, 21	ratio (e.g., $\eta_{12} = \eta_1/\eta_2$)	
f	frictional pressure	
g	gravitational (hydrostatic) pressure	
s	single phase	
Superscripts		
0	value without magnetic field is used for normalization	
1,2	value in single-phase flow of fluid 1,2 is used for normalization	
T	normalized by the total pressure gradient in single-phase flow	

Appendix A

Comparison between analytical and numerical solution for $B_0 = 0.1$T, $s = 3$.

	Finite Differences		Chebyshev Collocation		Analytical
Q_{12}	Number of Grid Nodes	Holdup	Number of Collocation Points	Holdup	Holdup
10^{-6}	20	0.2953	10	0.2067	
	50	0.2944	20	0.2885	
	100	0.2942	50	0.2940	
	200	0.2941	100	0.2940	
	500	0.2941	150	0.2940	
	1000	0.2940	200	0.2940	0.2940
10^{-4}	20	0.8303	10	0.7694	
	50	0.8296	20	0.8253	
	100	0.8294	50	0.8293	
	200	0.8293	100	0.8293	
	500	0.8293	150	0.8293	
	1000	0.8293	200	0.8293	0.8293
10^{-2}	20	0.9626	10	0.9582	
	50	0.9624	20	0.9621	
	100	0.9624	50	0.9624	
	200	0.9624	100	0.9624	
	500	0.9624	150	0.9624	
	1000	0.9624	200	0.9624	0.9624
1	20	0.9919	10	0.9918	
	50	0.9919	20	0.9919	
	100	0.9919	50	0.9919	
	200	0.9919	100	0.9919	
	500	0.9919	150	0.9919	
	1000	0.9919	200	0.9919	0.9919

References

1. Branover, H. *Magnetohydrodynamic Flow in Ducts*; John Wiley: New York, NY, USA, 1978.
2. Kadid, F.Z.; Drid, S.; Abdessemed, R. Simulation of magnetohydrodynamic and thermal coupling in the linear induction MHD pump. *J. Appl. Fluid Mech.* **2011**, *4*, 51–57.
3. Nikodijević, D.; Stamenković, Ž.; Milenković, D.; Blagojević, B.; Nikodijevic, J. Flow and heat transfer of two immiscible fluids in the presence of uniform inclined magnetic field. *Math. Probl. Eng.* **2011**, *2011*, 132302. [CrossRef]
4. Kim, S.-H.; Park, J.-H.; Choi, H.-S.; Lee, S.-H. Power Generation Properties of Flow Nanogenerator with Mixture of Magnetic Nanofluid and Bubbles in Circulating System. *IEEE Trans. Magn.* **2017**, *53*, 4600904. [CrossRef]
5. Wang, Y.; Cheng, K.; Xu, J.; Jing, W.; Huang, H.; Qin, J. A rapid performance prediction method for Two-Phase liquid metal MHD generators based on Quasi-One-Dimensional model. *Therm. Sci. Eng. Prog.* **2024**, *47*, 102258. [CrossRef]
6. Haim, H.B.; Jianzhong, Z.; Shizhi, Q.; Yu, X. A magneto-hydrodynamically controlled fluidic network. *Sens. Actuators B Chem.* **2003**, *88*, 205–216.
7. Hussameddine, S.K.; Martin, J.M.; Sang, W.J. Analytical prediction of flow field in magnetohydrodynamic-based microfluidic devices. *J. Fluids Eng.* **2008**, *130*, 091204.
8. Yi, M.; Qian, S.; Bau, H. A magnetohydrodynamic chaotic stirrer. *J. Fluid Mech.* **2002**, *468*, 153–177. [CrossRef]
9. Weston, M.C.; Gerner, M.D.; Fritsch, I. Magnetic fields for fluid motion. *Anal. Chem.* **2010**, *82*, 3411–3418. [CrossRef] [PubMed]
10. Lu, P.; Ye, Q.; Yan, X.; Wei, J.; Huang, H. Study on the gas-liquid two-phase flow regime in the power generation channel of a liquid metal MHD system. *Int. Commun. Heat Mass Transf.* **2022**, *137*, 106217. [CrossRef]
11. Wang, Y.; Huang, H.; Lu, P.; Liu, Z. Numerical Investigation of Gas–Liquid Metal Two-Phase Flow in a Multiple-Entrance Magnetohydrodynamic Generator. *Ind. Eng. Chem. Res.* **2022**, *61*, 4980–4995. [CrossRef]
12. Lee, C.; Ho Kim, H. Velocity measurement of magnetic particles simultaneously affected by two-phase flow and an external magnetic field using dual-sided SPIM-mPIV. *Chem. Eng. Sci.* **2022**, *252*, 11727. [CrossRef]
13. Khan, H.A.; Goharzadeh, A.; Jarrar, F. Effect of a uniform magnetic field on a two-phase air-ferrofluid slug flow. *J. Magn. Magn. Mater.* **2023**, *580*, 170944. [CrossRef]
14. He, Y.; Wen, G.; Li, Q.; Jiao, F. Field-induced interfacial instabilities in a two-phase ferrofluid flow. *Chem. Eng. J.* **2024**, *485*, 14995. [CrossRef]
15. Hartmann, J.; Lazarus, F. Experimental investigations on the flow of mercury in a homogeneous magnetic field. *K. Dan. Vidensk. Selsk. Math. Fys. Medd.* **1937**, *15*, 1–45.
16. Davidson, P.A. *An Introduction to Magnetohydrodynamics*; Cambridge University Press: Cambridge, UK, 2001.
17. Molokov, S.; Moreau, R.; Moffatt, K. *Magnetohydrodynamics Historical Evolution and Trends Fluid Mechanics and Its Applications*; Springer: Berlin/Heidelberg, Germany, 2007; FMIA; Volume 80.
18. Qian, S.; Bau, H.H. Magneto-hydrodynamics based microfluidics. *Mech. Res. Commun.* **2009**, *36*, 10–21. [CrossRef] [PubMed]
19. Kundu, B.; Saha, S. Review and Analysis of Electro-Magnetohydrodynamic Flow and Heat Transport in Microchannels. *Energies* **2022**, *15*, 7017. [CrossRef]
20. Hartmann, J. Theory of the laminar flow of an electrically conductive liquid in a homogeneous magnetic field. *K. Dan. Vidensk. Selsk. Mat. Fys. Medd.* **1937**, *15*, 1–28.
21. Shail, R. On laminar two-phase flow in magnetohydrodynamics. *Int. J. Eng. Sci.* **1973**, *11*, 1103–1108. [CrossRef]
22. Owen, R.G.; Hunt, J.C.R.; Collier, J.G. Magnetohydrodynamic pressure drop in ducted two-phase flows. *Int. J. Multiph. Flow* **1976**, *3*, 23–33. [CrossRef]
23. Lohrasbi, J.; Sahai, V. Magnetohydrodynamic heat transfer in two-phase flow between parallel plates. *Appl. Sci. Res.* **1988**, *45*, 53–66. [CrossRef]
24. Malashetty, M.S.; Umavathi, J.C. Two-phase magnetohydrodynamic flow and heat transfer in an inclined channel. *Int. J. Multiph. Flow* **1997**, *23*, 545–560. [CrossRef]
25. Umavathi, J.C.; Liu, I.-C.; Prathap Kumar, J. Magnetohydrodynamic Poiseuille-Couette flow and heat transfer in an inclined channel. *J. Mech.* **2010**, *26*, 525–532. [CrossRef]
26. Shah, N.A.; Alrabaiah, H.; Vieru, D.; Yook, S.-J. Induced magnetic field and viscous dissipation on flows of two immiscible fluids in a rectangular channel. *Sci. Rep.* **2022**, *12*, 39. [CrossRef] [PubMed]
27. Landman, M.J. Non-unique holdup and pressure drop in two-phase stratified inclined pipe flow. *Int. J. Multiph. Flow* **1991**, *17*, 377–394. [CrossRef]
28. Ullmann, A.; Zamir, M.; Ludmer, Z.; Brauner, N. Stratified laminar countercurrent flow of two liquid phases in inclined tubes. *Int. J. Multiph. Flow* **2003**, *29*, 1583–1604. [CrossRef]
29. Ullmann, A.; Zamir, M.; Gat, S.; Brauner, N. Multi-holdups in co-current stratified flow in inclined tubes. *Int. J. Multiph. Flow* **2003**, *29*, 1565–1581. [CrossRef]
30. Ullmann, A.; Goldstien, A.; Zamir, M.; Brauner, N. Closure relations for the shear stresses in two-fluid models for stratified laminar flows. *Int. J. Multiph. Flow* **2004**, *30*, 877–900. [CrossRef]
31. Goldstein, A.; Ullmann, A.; Brauner, N. Characteristics of stratified laminar flows in inclined pipes. *Int. J. Multiph. Flow* **2015**, *75*, 267–287. [CrossRef]
32. Barmak, I.; Gelfgat, A.; Vitoshkin, H.; Ullmann, A.; Brauner, N. Stability of stratified two-phase flows in horizontal channels. *Phys. Fluids* **2016**, *28*, 044101. [CrossRef]

33. Barmak, I.; Gelfgat, A.Y.; Ullmann, A.; Brauner, N. Stability of stratified two-phase flows in inclined channels. *Phys. Fluids* **2016**, *28*, 084101. [CrossRef]
34. Gelfgat, A.; Brauner, N. Instability of stratified two-phase flows in rectangular ducts. *Int. J. Multiph. Flow* **2020**, *131*, 103395. [CrossRef]
35. Gelfgat, A.; Barmak, I.; Brauner, N. Instability of stratified two-phase flows in inclined rectangular ducts. *Int. J. Multiph. Flow* **2021**, *138*, 103586. [CrossRef]

Disclaimer/Publisher's Note: The statements, opinions and data contained in all publications are solely those of the individual author(s) and contributor(s) and not of MDPI and/or the editor(s). MDPI and/or the editor(s) disclaim responsibility for any injury to people or property resulting from any ideas, methods, instructions or products referred to in the content.

Article

Air Flow Monitoring in a Bubble Column Using Ultrasonic Spectrometry

Ediguer Enrique Franco, Sebastián Henao Santa, John Jairo Cabrera and Santiago Laín *

Engineering Faculty, Universidad Autónoma de Occidente, Cll 25 # 115-85, Cali 760030, Colombia; eefranco@uao.edu.co (E.E.F.); sebastian.henao_san@uao.edu.co (S.H.S.); jjcabrera@uao.edu.co (J.J.C.)
* Correspondence: slain@uao.edu.co

Abstract: This work demonstrates the use of an ultrasonic methodology to monitor bubble density in a water column. A flow regime with droplet size distribution between 0.2 and 2 mm was studied. This range is of particular interest because it frequently appears in industrial flows. Ultrasound is typically used when the size of the bubbles is much larger than the wavelength (low frequency limit). In this study, the radius of the bubbles ranges between 0.6 and 6.8 times the wavelength, where wave propagation becomes a complex phenomenon, making existing analytical methods difficult to apply. Measurements in transmission–reception mode with ultrasonic transducers operating at frequencies of 2.25 and 5.0 MHz were carried out for different superficial velocities. The results showed that a time-averaging scheme is necessary and that wave parameters such as propagation velocity and the slope of the phase spectrum are related to the number of bubbles in the column. The proposed methodology has the potential for application in industrial environments.

Keywords: bubble column; ultrasonic spectrometry; digital image processing; heterogeneous flow monitoring

Citation: Franco, E.E.; Henao Santa, S.; Cabrera, J.J.; Laín, S. Air Flow Monitoring in a Bubble Column Using Ultrasonic Spectrometry. *Fluids* 2024, 9, 163. https://doi.org/10.3390/fluids9070163

Academic Editors: Hemant J. Sagar and Nguyen Van-Tu

Received: 20 June 2024
Revised: 11 July 2024
Accepted: 12 July 2024
Published: 18 July 2024

Copyright: © 2024 by the authors. Licensee MDPI, Basel, Switzerland. This article is an open access article distributed under the terms and conditions of the Creative Commons Attribution (CC BY) license (https://creativecommons.org/licenses/by/4.0/).

1. Introduction

Bubbly flows are integral to a variety of industrial operations, including alloy production, two-phase heat exchangers, reactor aeration and agitation, flotation equipment, and bubble column reactors. Bubble columns, where numerous gas bubbles travel upward through a liquid, are commonly utilized in the chemical, petrochemical, and biotechnological industries. These reactors play a crucial role in chemical processes like Fischer–Tropsch synthesis, fine chemical manufacturing, oxidation reactions, coal liquefaction, and fermentation [1]. Bubble column reactors are favored due to their simple construction and the absence of mechanically moving parts, which facilitates easy maintenance and lowers operating costs. Additionally, these reactors offer large interfacial areas and high transport rates, resulting in superior heat and mass transfer efficiency. This makes them highly effective for processes requiring significant interaction between gas and liquid phases [2].

In a bubble column, there are two different flow regimes depending on the superficial gas velocity U (the volumetric flow of air divided by the cross-sectional area of the column). By increasing U, an increase in the gas holdup ε (the number of bubbles per unit of volume) is observed. At the beginning, this increase is almost proportional and the homogeneous bubbly flow regime occurs, where the distribution of bubble sizes is narrow (1–7 mm) and the gas rise velocity is low, although trajectories of individual bubbles experience non-linear instabilities [3]. Above a transition superficial gas velocity, the coalescence phenomenon becomes important and large bubbles form and rise at a higher velocity. In this case, a heterogeneous or churn-turbulent flow regime occurs, with small bubbles that coexist with much larger ones (20–70 mm), and important horizontal velocity components are present, generating the mixing of the liquid phase [4,5]. When the gas in a vertical pipe occupies almost the entire cross section, this bullet-shaped bubble is called a Taylor bubble [6].

In a homogeneous bubbly flow, the individual bubbles move through the continuous liquid phase at low liquid superficial velocities and the interaction between them is negligible [7]. These kind of flows are found in boiling water nuclear reactors, steam generators, and refrigeration and air conditioning equipment [8]. On the other hand, the addition of bubbles in some industrial processes has been shown to be beneficial, for example, by increasing the efficiency of mixing or heat transfer between fluids, and the reaction rate in chemical reactors [9]. Therefore, the characterization or monitoring of these flows is important for the chemical, pharmaceutical, nuclear, and petrochemical industries. Another important area is environmental sciences, where the measurement of greenhouse gases migrating from the seafloor is an important topic [10].

For the characterization or monitoring of bubbly flows optical, electrical, and acoustical techniques have been used. In the case of optics, laser scattering [11] and laser-induced fluorescence [12] allow the density and relative size of the bubbles to be inferred. The passage of bubbles at a position in the column can be determined using an optical fiber immersed in the liquid [13], and this frequency can be related to the density of bubbles in the column. The pulse-light velocimetry (PLV) technique allows more precise measurement of bubble size and velocity [14], but the implementation of this technique requires expensive equipment and laboratory conditions. On the other hand, a relatively cheap and easy-to-implement technique is based on the digital processing of images captured with high-speed cameras. Edge detection algorithms are used to calculate the bubble density [15]. Overlapping, grouping, and irregular shapes of the bubbles are problems that are not easy to solve. To obtain accurate values, more than one camera and elaborate processing algorithms are needed [16,17]. However, the main disadvantage of optical methods is the opaqueness of many flows of interest. In this case, the use of x-rays has allowed the characterization of multi-phase flows [18]. But the measurement process can be complicated, requiring the capture and analysis of several planes or prior knowledge of some flow parameters. Furthermore, X-rays are a form of ionizing radiation that is harmful to life.

In the electrical case, measurements of the electric impedance in pairs of electrodes and the conductance in wire meshes are the main sensing approaches. The measurement of electrical impedance through a set of electrodes in contact with the medium under study is a cheap and relatively easy-to-implement technique that has gained attention in recent years [19]. Works related to the characterization of the stratified bubble flows and the study of the cavitation phenomenon are interesting examples [20]. These works use a single pair of electrodes, or a small number of them, to determine the electrical impedance at a certain frequency or range of working frequencies. The electrical impedance data can be related to the physical properties of the medium, and the flow dynamics can be analyzed using the temporal signal obtained. When a large number of electrodes are used, an image can be generated by solving an inverse problem. This technique is called electrical impedance tomography and it has been used to characterize multiphase flows [21–23]. In the case of wire-mesh tomography, the electrodes are wires arranged in a mesh pattern. Each crossing point of the wires serves as a sensing point [24]. In this case, the measurement is direct, and therefore, no reconstruction algorithms are needed. The resolution depends on the number of wires, and the data processing is fast, allowing for the measurement of hundreds of frames per second [25]. Although the electrical technique has much potential, its main disadvantage is that small chemical changes in the medium and material deposits on the electrodes or wires can affect the measurement.

Acoustical techniques use ultrasonic waves to infer the physical properties of the medium. Their main advantages are their capability to penetrate opaque media where optical techniques are not useful, the absence of ionizing radiation, and the fact that the required equipment is relatively simple and cheap [26]. In this respect, the ultrasonic characterization of heterogeneous media such as gas–liquid mixtures and immiscible liquids (emulsions) has been a topic of interest in recent years. Most of the works reported in the literature have been carried out with the limits of low frequency ($r/\lambda \ll 1$) and a low concentration of the dispersed phase [27–29]. Under these conditions, the propagation of ultrasonic waves is well behaved and there are analytical models that allow us to determine the droplet size spectrum from the ultrasonic signals. For example, these methods have been fundamental in the development and study of the contrast agents used in medical ultrasonography. These agents are fluids containing gas-filled microbubbles [30]. However, when the dispersed-phase droplets are of comparable ($r/\lambda \sim 1$) or larger ($r/\lambda \gg 1$) size than the wavelength, which is called the high-frequency limit, the propagation of ultrasonic waves becomes complicated. In these cases, there is an important interaction between the waves and the bubbles, generating large variations in the amplitude, and to a lesser extent in its phase, of the receiving waves. This behavior can be almost chaotic and the reception signal could even disappear.

Some works dealing with relatively large droplet sizes have been published in the literature. For instance, ultrasonic devices were developed to detect bubbles in the bloodstream [31]. The possibility of using common ultrasonic flow meters to determine the size of gas bubbles flowing through the pipe was evaluated, but it was only possible to obtain qualitative results [32]. The use of ultrasound and neural networks for the interpretation of data in the characterization of bubble flows in a water column was also reported. The bubble size was large ($r/\lambda \sim 65$) and a normalization scheme for the amplitude spectra [33] was used. Another work compared the analysis of ultrasound images obtained with a phased array and optical images for the characterization of a bubble flow, achieving similar results with both methods [10].

In this work, an ultrasonic methodology for bubble density monitoring in a water column is proposed. Measurements in transmission–reception mode and working frequencies of 2.25 and 5.0 MHz were carried out for different values of superficial gas velocity. Digital image processing allowed the characterization of the bubble flow, showing a droplet size distribution between 0.4 and 2.0 mm ($0.6 \leq r/\lambda \leq 6.8$), almost independent of bubble density. By modifying the power supply voltage of the peristaltic pump it was possible to vary the amount of bubbles in the column. A signal-averaging scheme allowed us to circumvent the problem of large amplitude variations at reception. It was found that wave parameters such as the slope of the phase spectrum and the propagation velocity are closely related to the number of bubbles in the column, allowing the real-time monitoring of the bubbly flow. The proposed methodology is relatively simple and reliable, with potential for industrial application.

2. Materials and Methods

2.1. Experimental Setup

Figure 1 shows a scheme of the experimental setup, including the water column. The ultrasonic transducers were installed at half the height of the column, with metal brackets screwed to the acrylic wall to maintain alignment. The distance between the radiating surface of the transducers was approximately 151 mm. The column was filled with water to a level of approximately 160 mm above the transducers. A porous stone (diffuser) of the type used in decorative aquariums was installed at the bottom. The air was injected using a positive displacement pump (peristaltic pump) driven by a direct current (DC) motor powered by a laboratory power supply. The amount of bubbles in the column depended on the excitation voltage of the DC motor.

The transducers were driven by an ultrasonic pulse/receiver (Olympus 5077PR, Olympus NDT, Waltham, MA, USA), which excited the emitter with a high-voltage and short-duration pulse, and at the same time, amplified the signals that reached the receiver with gains of up to 40 dB. A digital oscilloscope with a bandwidth of 200 MHz (Keysight DSOX2022A, Keysight, Santa Rosa, CA, USA) synchronized with the pulser/receiver allowed the signals to be visualized and digitized. The ultrasonic signals were transferred to desktop computers through the LAN network and stored for later processing in Matlab (R2018b). All tests were carried out in a laboratory at room temperature, which was maintained at 23 ± 1.3 °C by the air conditioning system. Temperature was measured using a digital thermometer with accuracy of 0.1 °C. Figure 2 shows an image of the experimental setup, where all the components can be seen, except the desktop computer.

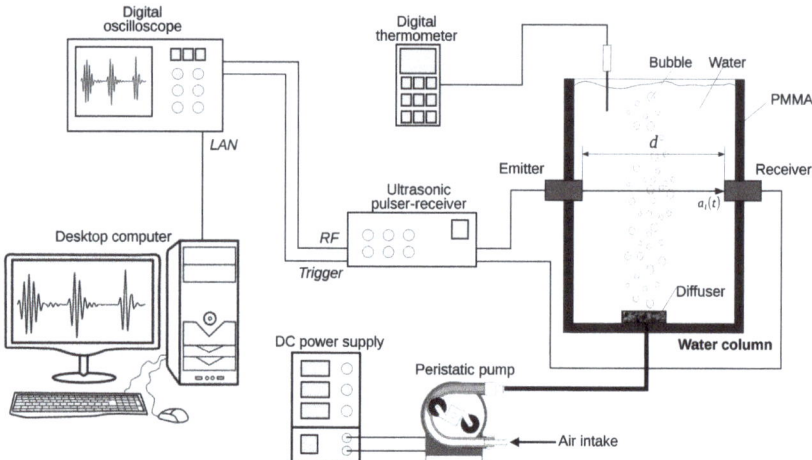

Figure 1. Scheme of the experimental setup.

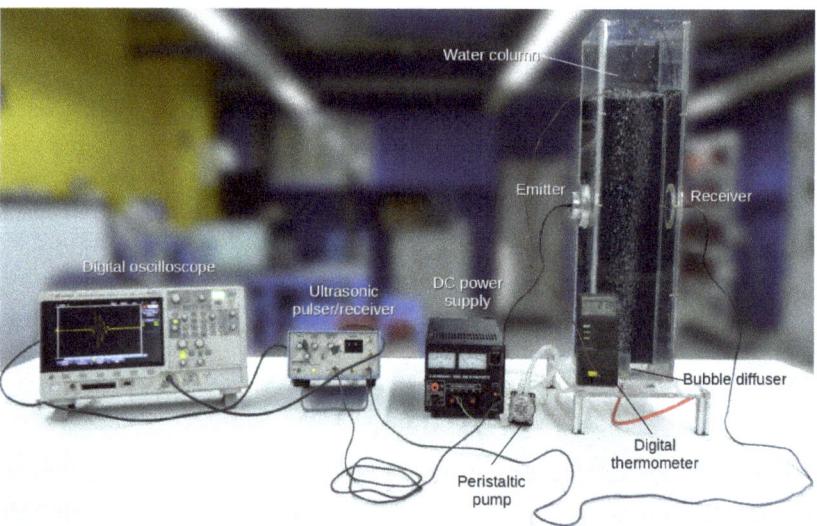

Figure 2. Image of the experimental setup, including the water column including the two ultrasonic transducers.

In this work, two pairs of ultrasonic transducers with a working frequency of 2.25 and 5.0 MHz from a well-known manufacturer (Krautkramer, Lewiston, PA, USA) were used. Table 1 reports the most relevant technical data, including the center frequency (f_c), the bandwidth (BW) of the signal acquired in water (without bubbles), calculated for a −6 dB amplitude drop, and the acoustic field parameters. These parameters are the near-field length, $Z_m = (\phi/2)^2/\lambda$, and the beam divergence angle, $\sin(\theta/2) = 1.22\lambda/\phi$, where ϕ is the diameter of the transducer radiating surface, $\lambda = c_w/f_c$ is the theoretical wavelength, and $c_w = 1480$ m/s is the propagation velocity in water at 20 °C. Figure 3 shows the waveform and the respective Fourier spectra of the ultrasonic pulses obtained in reception with water.

Table 1. Main characteristics of the ultrasonic transducers used is this work.

Transducer	f_c (MHz)	ϕ (mm)	BW (−6 dB)	Z_m (mm)	θ (degree)
Krautkrame 242–280	2.25	12.7	32%	61.3	3.62
Krautkrame 254–360	5.0	24	17%	486	0.86

The excitation signal, as seen on the oscilloscope without the transducer connected, is a square pulse with an amplitude and a width that can be varied by certain set values. The width allows the pulse to be tuned to the transducer's working frequency to achieve a better response. In the emission, we used an excitation pulse with an amplitude of 200 V, and gains between 0 and 10 dB were used in the receiver. Despite the high excitation voltage, the acoustic waves generated in pulse-echo mode are of low intensity. These are the waves used in ultrasonic non-destructive testing (UT-NDT), where other physical phenomena, such as cavitation or streaming, do not occur.

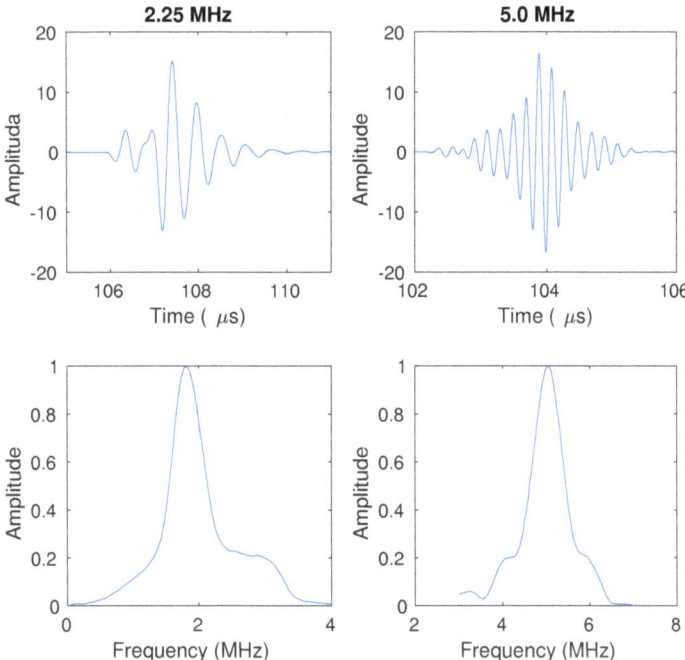

Figure 3. Waveforms and spectra of the ultrasonic transducers used in this work.

The transducer spacing is large compared to the wavelength, at 228λ and 508λ for the 2.25 and 5.0 MHz transducers, respectively. This allows clear reception of the ultrasonic pulses, without the problems of reverberation or spurious reflections. On the other hand, the diameter and frequency of the transducers cause the receiver to be located in the far field and the near field for the 2.25 and 5 MHz cases, respectively. This difference is not relevant due to the frequency domain normalization performed using the signal in the bubble-free case.

2.2. Signal Processing

Let $a_2(t)$ and $a_1(t)$ be the ultrasonic signals received in the cases with and without bubbles, respectively, where t is the time. In the case with bubbles, there is a drop in amplitude and a difference in the arrival time of the wave, which are related to diffraction, attenuation, and changes in the propagation velocity. The comparison between the cases with bubbles and the reference (without bubbles) is performed in the frequency domain using the loss coefficient (P):

$$P(f) = pe^{j\phi}, \tag{1}$$

where p and ϕ are the magnitude and phase of the loss coefficient, which are calculated from the ultrasonic signals as follows:

$$p(f) = \frac{|A_2(f)|}{|A_1(f)|} \tag{2}$$

and

$$\phi = \arg[A_2(f)] - \arg[A_1(f)] \tag{3}$$

where $A_2(f)$ and $A_1(f)$ are the Fourier transforms of the signals $a_2(t)$ and $a_1(t)$, respectively, and f is the frequency.

The effect of the presence of bubbles on ultrasonic waves is analyzed by means of attenuation and phase spectra. The attenuation spectrum is given by [34]:

$$\alpha(f) = \frac{1}{d} 20 \log[p(f)] \tag{4}$$

where d is the distance between the face of the transducers (see Figure 1). The attenuation spectrum quantifies the amplitude reduction of each spectral component in a suitable frequency range around the center frequency of the transducer.

The velocity spectrum is obtained by calculating the additional time (δ) that the wave takes, due to the presence of the bubbles, by means of the phase of the Fourier transform:

$$\delta = \frac{\phi}{2\pi} T = \frac{\phi}{2\pi f} \tag{5}$$

where $T = 1/f$ is the period. The velocity spectrum is calculated by dividing the distance traveled by the total time in the case with bubbles [34]:

$$v(f) = \frac{d}{\delta_0 + \delta} \tag{6}$$

where δ_0 is the arrival time in the case without bubbles. Replacing δ_0 and δ in (6), we obtain the expression for the velocity spectrum:

$$v(f) = \frac{dv_0(2\pi f)}{d(2\pi f) + v_0 \phi}, \tag{7}$$

where $v_0 = d/\delta_0$ is the velocity in the reference case.

2.3. Characterization of the Bubble Column

Figure 4 shows the image processing methodology used to estimate the bubble density. The images were captured with a reflex camera (Nikon D3200, Nikon, Tokyo, Japan) using a white background with a matte finish and high-intensity LED lighting. Due to the rise speed of the bubbles, a high shutter speed (1.25 ms) was necessary. First, a circular element was positioned on the axis that joins the center of the two ultrasonic transducers and a picture was taken. This circle of known diameter was a size reference for estimating the observed void area. Keeping the camera in the same position, pictures of the bubble column were taken. The air flow that generated the column of bubbles was controlled by the electrical voltage applied to the DC motor of the peristaltic pump.

To estimate the void fraction, a portion of the area observed by the camera was established. The transverse area illuminated by the surface transducer, including the divergence of the beam in the far field, was taken (see Figure 4 and Table 1). This area was defined with the intention of covering the area with the highest interaction between ultrasonic waves and bubbles. However, the results for both transducers were similar, and finally, only the area established with the larger-diameter transducer (5.0 MHz) was used.

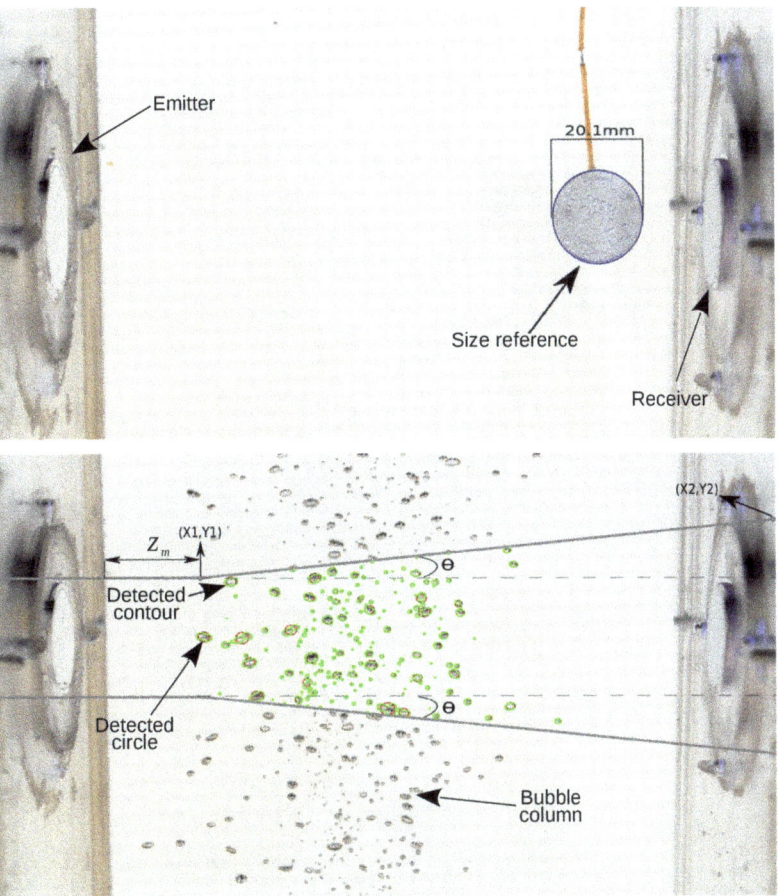

Figure 4. Image processing methodology used for the bubble density estimation.

The images with bubbles were analyzed using the $OpenCV$ Python library (https://opencv.org/ (accessed on 11 July 2024)). Processing began by converting each image to gray scale; then, segmentation and capture algorithms were applied. The processing was based on applying the Hough transform to detect circles [35]. The algorithm returned the coordinates of the centers and radii of the detected circles. Another algorithm was used to detect contours. The detected circles and contours are shown in red and green, respectively, in Figure 4 (bottom). The results of both algorithms presented similar values. However, contour detection presented more unexpected results, such as contours with areas of water inside, which required reprocessing or changing the image. The circle detection algorithm was more stable and probably more suitable for a possible practical application.

Figure 5 shows the histograms of the bubble size spectra for six excitation voltages of the peristaltic pump. The x-axis is the size range and the y-axis is the number of bubbles detected by the algorithm in each range. The results show that regardless of the excitation voltage, the highest count is within the 0.2 to 0.4 mm range. As the voltage is increased, the number of bubbles detected also increases, but it is the larger bubbles that show a more significant increase due to coalescence. The entire spectrum range remains almost the same (0–2 mm).

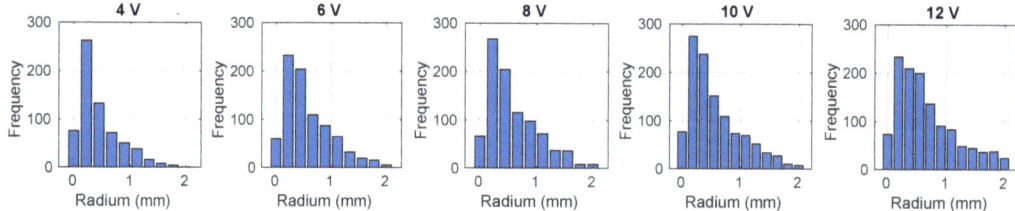

Figure 5. Bar graph showing the bubble radius distribution obtained by image processing for five excitation voltages of the peristaltic pump.

The area void fraction calculated by the image processing procedure is shown in Figure 6 (left). This value was calculated as the quotient between the sum of all the areas detected by the algorithm (A_b) and the area of influence of the acoustic beam (A_0) defined in Figure 4. These results show that the void fraction increases with the pump excitation voltage, as expected. This increase does not appear to be linear.

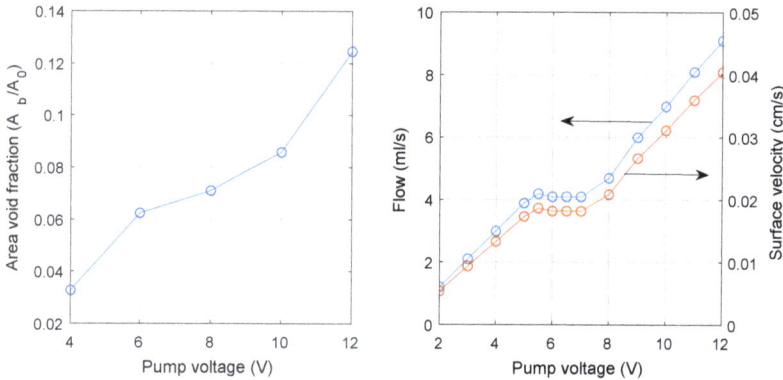

Figure 6. Area void fraction obtained with the image processing technique (**left**) and superficial gas velocity obtained in the characterization of the peristaltic pump (**right**).

The peristaltic pump was also characterized to determine the volumetric flow (Q) as a function of the excitation voltage. In a test, the time required to pump a certain volume of liquid was measured. Figure 6 (right) shows the injected air flow and the superficial velocity as a function of the pump voltage. The shape of the curve is similar to that of the void fraction obtained by digital image processing. Clearly, there is a linear behavior in the 2–5 and 8–12 V ranges. Between 5.2 and 7 V, approximately, the flow remains constant. These results are important because they relate the wave parameters to the actual air flow. By dividing the air flow by the cross-sectional area of the column (225 cm^2), the superficial velocity was obtained.

3. Results

The behavior of acoustic waves in reception is chaotic due to the relative size of the bubbles. For the analysis shown in this work, a diffuser (porous stone) that provides bubbles with an average radius of 400 μm and maximum radii close to 2 mm (see Figure 5) was used. These bubble sizes lead to values of $0.6 < r/\lambda < 3.0$ and $1.4 < r/\lambda < 6.8$ for the working frequencies for 2.25 and 5.0 MHz, respectively. Therefore, most bubbles are of similar size, and some others several times larger than the wavelength. In this measurement range, there is a high interaction of the ultrasonic waves with the bubbles, causing large variations in the amplitude observed in reception.

Figure 7 shows the mean and standard deviation of the loss coefficient as a function of the signal averages, calculated at the central frequency of each transducer. The results show how the mean value stabilizes and the standard deviation reduces dramatically as the averages increase. The mean value is very close to 0.6 for both frequencies. The repetition rate used in the ultrasonic pulser/receiver is 5 kHz. Therefore, the acquisition times are relatively short in spite of the high number of averages. For example, the acquisition times for the 2^{12} and 2^{16} averages are 0.82 and 13 s, respectively. These times are short enough to perform several measurements per minute. However, the transfer, storage, and processing time on the computer must be added, which, depending on the hardware, may be relevant.

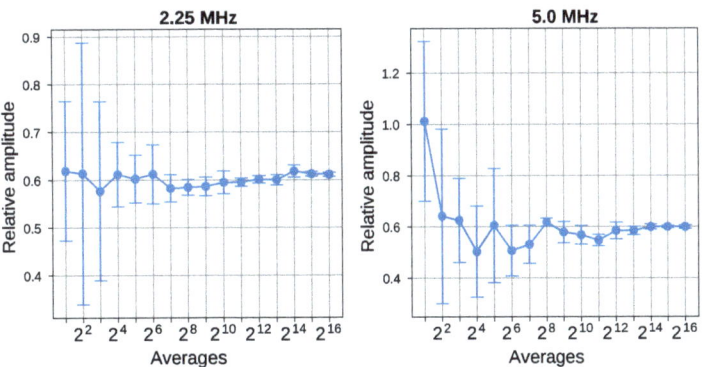

Figure 7. Mean and standard deviation of the loss coefficient as a function of the signal averages. $p(f)$ were calculated at the central frequency of each transducer.

Figure 8 shows the magnitude and phase of the loss coefficient as a function of frequency for the two transducers in a frequency band of -12 dB. The signals were acquired with 10^{15} averages and the temperature in the water column was 23.1 °C. In the case of the magnitude of the loss coefficient, the results show a complicated spectrum, with oscillations and increasing and decreasing trends. For 2.25 MHz, the size of the bubbles is closer to the wavelength and there is greater interaction, with increasing magnitude values as the frequency increases. For 5.0 MHz, all magnitudes decrease with frequency. However, for both working frequencies, decreasing magnitude with a decreasing amount of bubbles can be observed. When the average value or the area under the curve was calculated, the results were erratic. In the case of the phase of the loss coefficient, a more stable and almost linear behavior was observed. This result is the expected in the case of a receiving ultrasonic pulse, and clearly, the slope of the phase is related to the amount of bubbles.

Figure 9 shows the attenuation and propagation velocity spectra in the -12 dB band for the two working frequencies. Since the attenuation spectrum depends on the magnitude of the loss coefficient, its behavior is very similar to that shown in Figure 8. On the other hand, the propagation velocity can be calculated at any frequency. In this case, the velocity was calculated at the central frequency of the transducer (see Figure 3). These results show an approximately constant propagation velocity as a function of frequency, which increases with the number of bubbles. The propagation velocity measured without bubbles was 1489.1 m/s at 23.1 °C, and the value reported in the literature is 1491.5 m/s [36,37]. Even though the velocity increase due to the presence of bubbles is only 2 m/s for for an air flow variation of 6.0 mL/s, which is equivalent to 0.13%, perfectly separated curves are observed for the air flow valuer. This shows that the system has good resolution and stability for measuring propagation velocity.

These results allow us to conclude that in this heterogeneous medium with bubbles, the phase is more stable and useful for bubble density monitoring than the magnitude. This becomes clearer when it is recalled that the phase is related to the arrival time of the waves, and the magnitude to the measured acoustic pressure. This result is in agreement with that reported by other authors who worked with homogeneous media and emulsions [38].

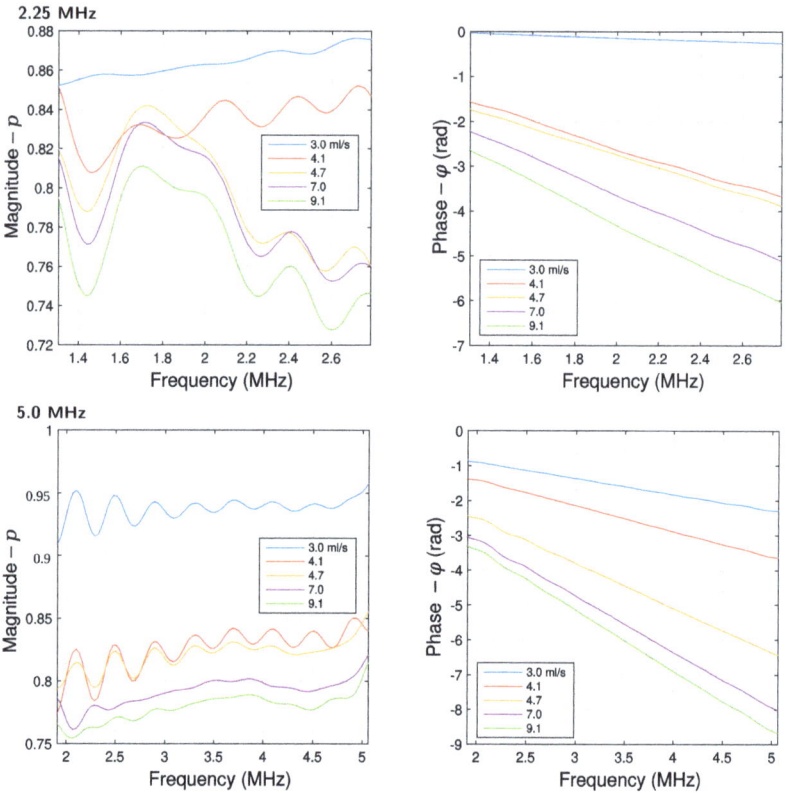

Figure 8. Magnitude (p) and phase (ϕ) of the loss coefficient as a function of frequency for all the values of the air flow (frequency band of −12 dB, 10^{15} averages, and test temperature of 23.1 °C).

Figure 10 (left) shows the slope of the phase spectrum and the propagation velocity as a function of the air flow in the column for the two working frequencies. In the case of the phase slope, both curves show a monotonically rising tendency with the air flow, with less variation at 5 MHz. Considering that the slope of the phase spectrum is zero for the bubble-free case due to normalization, the range of variation is 0.5 and 0.6 rad/MHz for 2.25 and 5.0 MHz, respectively. The slope of the phase spectrum was already used for monitoring of the water content in water-in-crude oil emulsions [39]. In this case, the phase slope variation was higher, up to 12 rad/MHz. This difference must be a consequence of the concentration in both heterogeneous media. However, such a value is also affected by some measurement parameters, for example, the distance at which the waves interact with the bubbles of the dispersed phase.

In the case of propagation velocity shown in Figure 10 (right), a similar behavior with a clearly increasing trend is observed. The behavior seems less stable, with points further away from this trend. It can be seen that the total variation in propagation velocity is 2.5 m/s; such small variations (0.17%) must have a considerable error component due to random noise. In measurements carried out in water-in-crude oil emulsions with a volumetric concentration of up to 40%, variations in the propagation velocity of up to 30 m/s were observed [38]. In that case, it can be stated that the difference in propagation velocity is exclusively a consequence of the concentration of the dispersed phase in the media.

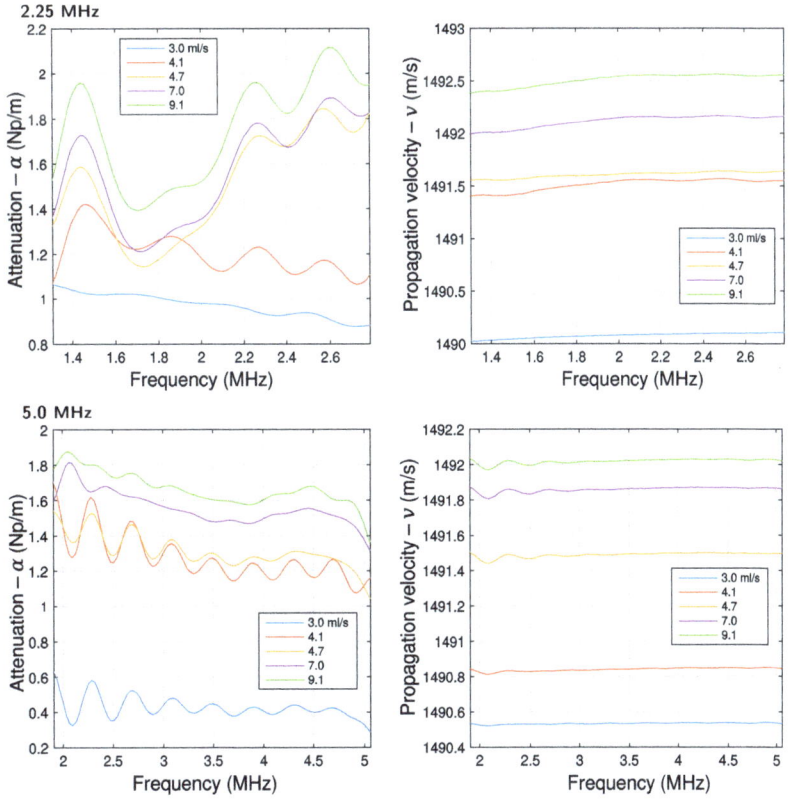

Figure 9. Attenuation (α) and velocity (v) spectra for all the values of the air flow (frequency band of -12 dB, 10^{15} averages, and test temperature of 23.1 °C).

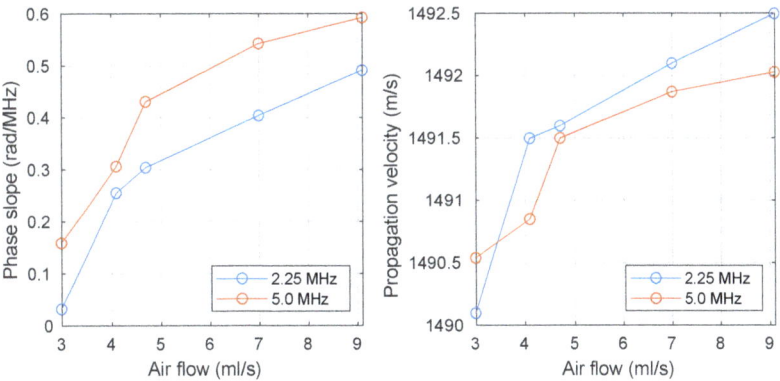

Figure 10. Phase slope (**left**) and propagation velocity (**right**) as a function of the air flow in the column.

These results show the possibility of monitoring the amount of bubbles in the water column using both the slope of the phase spectrum and the propagation velocity. Both properties can be used to obtain calibration curves that directly provide the air flow or superficial velocity. However, the analysis of the influence of temperature and the applicability of the technique with higher concentrations requires further research.

4. Conclusions

This work proposes a simple and inexpensive methodology based on ultrasonic spectrometry for monitoring the bubble density in a water column. Tests were carried out at two working frequencies in transmission–reception mode for different values of superficial velocity. The amount of bubbles was determined by characterizing the positive displacement pump used for air injection. On the other hand, digital image analysis allowed us to establish the droplet size spectrum, showing bubble radii comparable to or greater than the wavelength. Under these conditions, wave propagation is complicated, with large variations in amplitude, even leading to signal disappearance at reception. To overcome this problem, a signal averaging scheme proved to be an appropriate strategy.

Signal analysis was performed in the frequency domain using a loss coefficient, and attenuation and velocity spectra. The results were normalized using a reference case without bubbles. The amplitude of the loss coefficient and the attenuation spectrum showed an intricate behavior that could not be related to the number of bubbles. On the other hand, the phase of the loss coefficient and the velocity spectrum showed a more stable behavior, dependent on the number of bubbles. The best case was provided by the slope of the phase spectrum, which displays monotonic growth with the superficial velocity.

Our results showed the possibility of monitoring the density of bubbles in the water column using the phase spectrum of the loss coefficient. The proposed methodology is relatively simple and inexpensive, and the signal processing requires little computational power, making it possible to use low-cost microcontrollers. The studied regime, with the presence of large droplets compared to the wavelength, is interesting because it occurs in important industrial processes, for instance, in chemical reactors, with little information in the literature about non-destructive testing by ultrasound.

Finally, the main limitations of the proposed technique are its applicability at high temperatures and concentrations. This technique will probably require a calibration process for each working temperature. Additionally, at temperatures above 100 °C, conventional ultrasonic transducers may present problems, such as large temperature gradients that affect their operation or even permanent depolarization of the piezoelectric material. On the other hand, with concentrations higher than those used in this work, the signals must have a more complex behavior, and the proposed methodology will probably not be applicable. For these cases, additional research is required.

Author Contributions: Conceptualization, E.E.F.; methodology, E.E.F., S.H.S. and J.J.C.; measurements, S.H.S. and J.J.C.; results analysis, S.H.S., E.E.F., J.J.C. and S.L.; data visualization, S.H.S. and E.E.F.; writing, E.E.F. and S.L.; project administration, E.E.F. and S.L.; funding acquisition, E.E.F. and S.L. All authors have read and agreed to the published version of the manuscript.

Funding: This work was partially supported by the Vice-Rectorate for Research, Innovation and Entrepreneurship (VIIE) of the Universidad Autónoma de Occidente through grants 19INTER-306 and 23INTER-444.

Data Availability Statement: The original contributions presented in the study are included in the article, further inquiries can be directed to the corresponding author.

Conflicts of Interest: The authors declare no conflicts of interest.

References

1. Sokolichin, A.; Eigenberger, G.; Lapin, A. Simulation of buoyancy driven bubbly flow: Established simplifications and open questions. *AIChE J.* **2004**, *50*, 24–45. [CrossRef]
2. Lain, S.; Bröder, D.; Sommerfeld, M. Experimental and numerical studies of the hydrodynamics in a bubble column. *Chem. Eng. Sci.* **1999**, *54*, 4913–4920. [CrossRef]
3. Göz, M.; Laín, S.; Sommerfeld, M. Study of the numerical instabilities in Lagrangian tracking of bubbles and particles in two-phase flow. *Comput. Chem. Eng.* **2004**, *28*, 2727–2733. [CrossRef]
4. Krishna, R.; van Baten, J. Mass transfer in bubble columns. *Catal. Today* **2003**, *79–80*, 67–75. [CrossRef]
5. Laín, S.; Bröder, D.; Sommerfeld, M.; Göz, M. Modelling hydrodynamics and turbulence in a bubble column using the Euler–Lagrange procedure. *Int. J. Multiph. Flow* **2002**, *28*, 1381–1407. [CrossRef]

6. Ambrose, S.; Hargreaves, D.M.; Lowndes, I.S. Numerical modeling of oscillating Taylor bubbles. *Eng. Appl. Comput. Fluid Mech.* **2016**, *10*, 578–598. [CrossRef]
7. Etminan, A.; Muzychka, Y.S.; Pope, K. A Review on the Hydrodynamics of Taylor Flow in Microchannels: Experimental and Computational Studies. *Processes* **2021**, *9*, 870. [CrossRef]
8. Asiagbe, K.S.; Fairweather, M.; Njobuenwu, D.O.; Colombo, M. Large Eddy Simulation of Microbubble Transport in Vertical Channel Flows. In *Computer Aided Chemical Engineering*; Espuña, A., Graells, M., Puigjaner, L., Eds.; Elsevier: Amsterdam, The Netherlands, 2017; Volume 40, pp. 73–78. [CrossRef]
9. Dabiri, S.; Tryggvason, G. Heat transfer in turbulent bubbly flow in vertical channels. *Chem. Eng. Sci.* **2015**, *122*, 106–113. [CrossRef]
10. Takimoto, R.Y.; Matuda, M.Y.; Oliveira, T.F.; Adamowski, J.C.; Sato, A.K.; Martins, T.C.; Tsuzuki, M.S. Comparison of Optical and Ultrasonic Methods for Quantification of Underwater Gas Leaks. *IFAC-PapersOnLine* **2020**, *53*, 16721–16726. [CrossRef]
11. Abbaszadeh, M.; Alishahi, M.M.; Emdad, H. A new bubbly flow detection and quantification procedure based on optical laser-beam scattering behavior. *Meas. Sci. Technol.* **2020**, *32*, 025202. [CrossRef]
12. Alméras, E.; Cazin, S.; Roig, V.; Risso, F.; Augier, F.; Plais, C. Time-resolved measurement of concentration fluctuations in a confined bubbly flow by LIF. *Int. J. Multiph. Flow* **2016**, *83*, 153–161. [CrossRef]
13. Ma, Y.; Muilwijk, C.; Yan, Y.; Zhang, X.; Li, H.; Xie, T.; Qin, Z.; Sun, W.; Lewis, E. Measurement of Bubble Flow Frequency in Chemical Processes Using an Optical Fiber Sensor. In Proceedings of the 2018 IEEE SENSORS, New Delhi, India, 28–31 October 2018; pp. 1–4. [CrossRef]
14. Bröder, D.; Sommerfeld, M. Planar shadow image velocimetry for the analysis of the hydrodynamics in bubbly flows. *Meas. Sci. Technol.* **2007**, *18*, 2513. [CrossRef]
15. Shamoun, B.; Beshbeeshy, M.E.; Bonazza, R. Light extinction technique for void fraction measurements in bubbly flow. *Exp. Fluids* **1999**, *26*, 16–26. [CrossRef]
16. Fu, Y.; Liu, Y. Development of a robust image processing technique for bubbly flow measurement in a narrow rectangular channel. *Int. J. Multiph. Flow* **2016**, *84*, 217–228. [CrossRef]
17. Karn, A.; Ellis, C.; Arndt, R.; Hong, J. An integrative image measurement technique for dense bubbly flows with a wide size distribution. *Chem. Eng. Sci.* **2015**, *122*, 240–249. [CrossRef]
18. Lau, Y.; Möller, F.; Hampel, U.; Schubert, M. Ultrafast X-ray tomographic imaging of multiphase flow in bubble columns—Part 2: Characterisation of bubbles in the dense regime. *Int. J. Multiph. Flow* **2018**, *104*, 272–285. [CrossRef]
19. Cabrera-López, J.J.; Velasco-Medina, J. Structured Approach and Impedance Spectroscopy Microsystem for Fractional-Order Electrical Characterization of Vegetable Tissues. *IEEE Trans. Instrum. Meas.* **2020**, *69*, 469–478. [CrossRef]
20. George, D.L.; Iyer, C.O.; Ceccio, S.L. Measurement of the Bubbly Flow Beneath Partial Attached Cavities Using Electrical Impedance Probes. *J. Fluids Eng.* **1999**, *122*, 151–155. [CrossRef]
21. Huang, C.; Lee, J.; Schultz, W.W.; Ceccio, S.L. Singularity image method for electrical impedance tomography of bubbly flows. *Inverse Probl.* **2003**, *19*, 919. [CrossRef]
22. de Moura, B.F.; Martins, M.F.; Palma, F.H.S.; da Silva, W.B.; Cabello, J.A.; Ramos, R. Nonstationary bubble shape determination in Electrical Impedance Tomography combining Gauss–Newton Optimization with particle filter. *Measurement* **2021**, *186*, 110216. [CrossRef]
23. Zhu, Z.; Li, G.; Luo, M.; Zhang, P.; Gao, Z. Electrical Impedance Tomography of Industrial Two-Phase Flow Based on Radial Basis Function Neural Network Optimized by the Artificial Bee Colony Algorithm. *Sensors* **2023**, *23*, 7645. [CrossRef] [PubMed]
24. Prasser, H.M.; Böttger, A.; Zschau, J. A new electrode-mesh tomograph for gas–liquid flows. *Flow Meas. Instrum.* **1998**, *9*, 111–119. [CrossRef]
25. Hampel, U.; Babout, L.; Banasiak, R.; Schleicher, E.; Soleimani, M.; Wondrak, T.; Vauhkonen, M.; Lähivaara, T.; Tan, C.; Hoyle, B.; et al. A Review on Fast Tomographic Imaging Techniques and Their Potential Application in Industrial Process Control. *Sensors* **2022**, *22*, 2309. [CrossRef] [PubMed]
26. Durán, A.L.; Franco, E.E.; Reyna, C.A.B.; Pérez, N.; Tsuzuki, M.S.G.; Buiochi, F. Water Content Monitoring in Water-in-Crude-Oil Emulsions Using an Ultrasonic Multiple-Backscattering Sensor. *Sensors* **2021**, *21*, 5088. [CrossRef] [PubMed]
27. Allegra, J.R.; Hawley, S.A. Attenuation of Sound in Suspensions and Emulsions: Theory and Experiments. *J. Acoust. Soc. Am.* **1972**, *51*, 1545–1564. [CrossRef]
28. Wu, X.; Chahine, G.L. Development of an acoustic instrument for bubble size distribution measurement. *J. Hydrodyn. Ser. B* **2010**, *22*, 330–336. [CrossRef]
29. Pinfield, V.J. Advances in ultrasonic monitoring of oil-in-water emulsions. *Food Hydrocoll.* **2014**, *42*, 48–55. [CrossRef]
30. de Jong, N.; Emmer, M.; van Wamel, A.; Versluis, M. Ultrasonic characterization of ultrasound contrast agents. *Med. Biol. Eng. Comput.* **2009**, *47*, 861–873. [CrossRef] [PubMed]
31. Kremkau, F.W.; Gramiak, R.; Carstensen, E.L.; Shah, P.M.; Kramer, D.H. Ultrasonic detection of cavitation at catheter tips. *Am. J. Roentgenol.* **1970**, *110*, 177–183. [CrossRef]
32. Nishi, R. Ultrasonic detection of bubbles with doppler flow transducers. *Ultrasonics* **1972**, *10*, 173–179. [CrossRef]

33. Baroni, D.B.; Filho, J.S.C.; Lamy, C.A.; Bittencourt, M.S.Q.; Pereira, C.M.N.A.; Motta, M.S. Determination of size distribution of bubbles in a ubbly column two-phase flows by ultrasound and neural networks. In Proceedings of the 2011 International Nuclear Atlantic Conference—INAC 2011, Brazzilian Asiciation of Nuclear Engineering—ABEN, Belo Horizonte, MG, Brazil, 24–28 October 2011.
34. Cents, A.H.G. Mass Transfer and Hydrodynamics in Stirred Gas-Liquid-Liquid Contactors. Ph.D. Thesis, Universiteit Twente, Enschede, The Netherlands, 2003.
35. Djekoune, A.O.; Messaoudi, K.; Amara, K. Incremental circle hough transform: An improved method for circle detection. *Optik* **2017**, *133*, 17–31. [CrossRef]
36. Lubbers, J.; Graaff, R. A simple and accurate formula for the sound velocity in water. *Ultrasound Med. Biol.* **1998**, *24*, 1065–1068. [CrossRef] [PubMed]
37. Del Grosso, V.A.; Mader, C.W. Speed of Sound in Pure Water. *J. Acoust. Soc. Am.* **1972**, *52*, 1442–1446. [CrossRef]
38. Reyna, C.A.; Franco, E.E.; Tsuzuki, M.S.; Buiochi, F. Water content monitoring in water-in-oil emulsions using a delay line cell. *Ultrasonics* **2023**, *134*, 107081. [CrossRef]
39. Franco, E.E.; Reyna, C.A.B.; Durán, A.L.; Buiochi, F. Ultrasonic Monitoring of the Water Content in Concentrated Water–Petroleum Emulsions Using the Slope of the Phase Spectrum. *Sensors* **2022**, *22*, 7236. [CrossRef]

Disclaimer/Publisher's Note: The statements, opinions and data contained in all publications are solely those of the individual author(s) and contributor(s) and not of MDPI and/or the editor(s). MDPI and/or the editor(s) disclaim responsibility for any injury to people or property resulting from any ideas, methods, instructions or products referred to in the content.

Review

A Review of Preconditioning and Artificial Compressibility Dual-Time Navier–Stokes Solvers for Multiphase Flows

Van-Tu Nguyen * and Warn-Gyu Park *

School of Mechanical Engineering, Pusan National University, Busan 46241, Republic of Korea
* Correspondence: vantunguyen@pusan.ac.kr (V.-T.N.); wgpark@pusan.ac.kr (W.-G.P.)

Abstract: This review paper aims to summarize recent advancements in time-marching schemes for solving Navier–Stokes (NS) equations in multiphase flow simulations. The focus is on dual-time stepping, local preconditioning, and artificial compressibility methods. These methods have proven to be effective in achieving high time accuracy in simulations, as well as converting the incompressible NS equations into a hyperbolic form that can be solved using compact schemes, thereby accelerating the solution convergence and allowing for the simulation of compressible flows at all Mach numbers. The literature on these methods continues to grow, providing a deeper understanding of the underlying physical processes and supporting technological advancements. This paper also highlights the imposition of dual-time stepping on both incompressible and compressible NS equations. This paper provides an updated overview of advanced methods for the CFD community to continue developing methods and select the most suitable two-phase flow solver for their respective applications.

Keywords: Navier–Stokes; multiphase flows; preconditioning; artificial compressibility; dual-time methods; curvilinear coordinates

1. Introduction

Two-phase flows such as gas–gas, gas–liquid, liquid–liquid, and three-phase (liquid, gas, and vapor) flows are of great interest in many natural phenomena, engineering, and industrial applications. Numerical simulations and analyses of two-phase flows to obtain an understanding of the mechanism and physical characteristics of the flows in applications have become routine activities. In design and development, computational fluid dynamic (CFD) programs based on the Navier–Stokes (NS) equations are now considered to be standard numerical tools to predict the nonlinear motions of the interface between two phases (two fluids), its deformations and breaks, phase change, heat transfer, turbulence, shockwaves, and violent interaction with devices/systems [1–5].

Both incompressible and compressible NS systems are commonly used for predicting multiphase flows. The governing equations for multiphase incompressible flows generally consist of the mixture continuity equation, mixture momentum equations, and phasic volume fraction equations, which are given as:

$$\nabla \cdot \mathbf{u} = 0, \tag{1}$$

$$\frac{\partial}{\partial t}(\rho \mathbf{u}) + \nabla \cdot (\rho \mathbf{u}\mathbf{u} + p\mathbf{I}) = \nabla \cdot \left(\mu \left(\nabla \mathbf{u} + \nabla \mathbf{u}^T\right)\right) + \rho \mathbf{g}, \tag{2}$$

$$\frac{\partial \alpha_i}{\partial t} + \nabla \cdot (\mathbf{u} \alpha_i) = 0, \tag{3}$$

where $\mathbf{u}$ is the flow velocity vector, p is the pressure, t is the physical time, α_i is the volume fraction of the ith phasic component and g is the gravity acceleration.

Mixture rules are given as

$$\rho = \sum_i^N \alpha_i \rho_i;\ \mu = \sum_i^N \alpha_i \mu_i;\ \sum_i^N \alpha_i = 1. \qquad (4)$$

To predict the mechanism and physical characteristics of flows with transient effects accurately, the governing equations require advanced temporal discretization and time-marching schemes. For the incompressible system (1–3), the time derivative term does not appear in the continuity Equation (1). The numerical solution of these equations presents a major difficulty, and to overcome this, a special equation was derived for solving pressure, e.g., employing the nonlinear Poisson equation for pressure, and then obtaining other state variables by using prediction and correction procedures [6–8]. However, to apply the well-developed compressible flow algorithms to the incompressible problem, the incompressible NS equation system must be hyperbolic as compressible NS equation systems.

The dual-time preconditioning approach originally is a numerical technique used to simulate unsteady incompressible flows. It was first proposed by Merkle [9] as a modification of the artificial compressibility method developed by Chorin [10]. The approach involves adding pseudo-time derivative terms to the incompressible NS equations and treating the time variation of the incompressible flow as a compressible flow, allowing for the coupling of the velocity and pressure fields in each time iteration [11–15]. One of the key advantages of the dual-time preconditioning approach is the direct coupling of the continuity and momentum equations in the incompressible flow equations, which eliminates the factorization error in factored implicit schemes [2]. Additionally, this approach eliminates errors due to approximations made in the implicit operator, improves numerical efficiency, and eliminates errors due to lagged boundary conditions at both solid and internal fluid boundaries [2,15–17]. By using preconditioned iterative methods, the dual-time preconditioning approach can also achieve a more efficient convergence of the sub-iterations.

The concept of artificial compressibility involves transforming elliptic equations describing incompressible flows into a hyperbolic compressible system, making it amenable to a solution using standard time-marching methods, such as explicit or implicit methods [2,18–20]. This allows the use of established numerical techniques for solving compressible flows to be applied to the simulation of incompressible flows. The specific procedure for adding artificial time derivatives to the incompressible NS equations is discussed in detail in Section 2.1.

Multiphase flows are assumed either incompressible or compressible flows depending on the range of Mach numbers of the flows. A flow can be assumed an incompressible one as the Mach number is less than 0.1, in which the compressibility is small and can be ignored. Examples of small Mach numbers are free-surface flows [21,22], dynamics of rising bubbles [23–25], and boiling flows [26,27]. A flow, at a Mach number greater than 0.3, is usually assumed a compressible one. At supersonic speeds, the Mach number is greater than 1.0, supercavitating flow around projectiles characterizes by shock waves, thermodynamic behavior, and compressibility dominates [28]. The physical aspects also were observed in supersonic flows in nozzles and/or separators [29–31]. Traditional time-marching algorithms that are based on physical time derivatives have been widely used and have proven to be effective in simulating transonic and supersonic flows. These algorithms have been successful in capturing the temporal evolution of the flow and have been widely adopted in many engineering and scientific applications. Despite their success, these algorithms may face challenges in the simulation of unsteady incompressible flows, which require the solution of elliptic equations and may suffer from convergence issues. Weakly compressible flow models were introduced for multiphase flows where the compressibility is not very significant [32–34]. Moreover, in realistic problems, the flows are often mixed flows involving both high and low local Mach numbers, e.g., cavitating flows [35], water entry of objects [36], cavitation bubble collapse [37,38] where the sound speed varies largely, or unsteady flows around accelerating and decelerating objects at all Mach number speeds. A major difficulty encountered in most compressible flow solvers is their inability to efficiently solve the problems of very low Mach numbers in the flows. At the low Mach numbers, most of these numerical solvers encounter degraded convergence speeds due to the wide disparity between the fluids and acoustic wave speeds. To overcome these challenges, alternative numerical techniques, such as the artificial compressibility method and the dual-time preconditioning approach, have been developed to improve the accuracy and efficiency of simulations of incompressible flows.

It is worth noting that there are two types of preconditioning techniques. The first type is the linear system level, which is purely mathematical. This approach solves a linear system to accelerate convergence in the initial iterations by preconditioning the matrix. The second type is the partial differential equation level, in which preconditioning terms are introduced in the partial differential equations to overcome difficulties in solving the equation. In this paper, our focus is mainly on the second type, specifically dual-time preconditioning derivatives introduced to the NS equation system to modify the way the solution evolves in pseudo-time towards convergence. The paper will review dual-time preconditioning methods for both steady and unsteady cases.

The subsequent sections provide a detailed mathematical background on the implementation of artificial time derivatives in both incompressible and compressible NS equations for multiphase flows. These sections present summaries of efficient simulation techniques for various multiphase flows, and serve as an informative resource for interested readers.

Subsequently, we conduct a comprehensive review and analysis of current trends, advanced simulation results, and complex numerical methods reported in the literature. This analysis provides researchers with a comprehensive overview of the successful applications of dual-time preconditioning methods in the field.

We conclude by recommending further research in the improvement of numerical methods for the simulation of complex multiphase flow problems. Additionally, we discuss several ways to enhance the accuracy and conductivity of these methods and provide potential areas for future research. Overall, this study contributes to the growing body of literature on the effective use of dual-time preconditioning methods in the simulation of multiphase flows.

2. Incompressible Multiphase Flows

This section focuses on the review of the mathematical foundations of various forms of preconditioning and artificial compressibility methods for multi-phase flows. The review begins with an overview of the incompressible flow models, including the dual-time homogeneous mixture model with a preconditioning parameter and the artificial compressibility model with a pseudo density and pressure function. The latest research and simulations on multi-phase flow modeling are then discussed and presented.

2.1. Preconditioning Dual-Time Stepping Method

As aforementioned, the ideas of artificial compressibility for incompressible flow are to transform the elliptic incompressible equations into a hyperbolic compressible system, which can be solved by standard, explicit or implicit, time-marching methods. The continuity equation can be rewritten as

$$\widetilde{\delta}\frac{\partial p}{\partial \tau} + \nabla \cdot \mathbf{u} = 0 \tag{5}$$

where the artificial equation of state is $p = \widetilde{\rho}/\widetilde{\delta}$, $\widetilde{\rho}$ is artificial density, and $\widetilde{\delta}$ is artificial compressibility. In order to ensure a consistently high convergence rate, the condition number of the Jacobian matrix of the governing equations system should be as close as possible to one for all flow conditions.

At each physical time step of the numerical solver, a pseudo-time iterative procedure is applied such that the term $\widetilde{\delta}\frac{\partial p}{\partial \tau}$ approaches zero upon convergence. Conversely, when

$\tilde{\delta}\frac{\partial p}{\partial \tau}$ approaches zero, the solution of the artificial equations converts back to the solution of the original equations.

The artificial compressibility method is effective in solving single-phase flows. However, when the computational domain involves more than one fluid, a challenging difficulty arises due to the appearance of an interface that separates fluids and behaves as an additional type of discontinuity. The application of single-phase schemes to multi-phase flows can result in problematic issues related to the mixture of two phases, mixture density, and sound speed, which can ultimately affect the entire flow field. To achieve rapid convergence rates and high computational accuracies near material interfaces in multiphase flow systems, it is necessary to modify the preconditioning formulation for single-phase flow systems due to the significantly different densities of the fluids involved.

In order to solve multiphase flows, various formulations of the preconditioning formulation for the continuity equation have been proposed. One such formulation, suggested by Kunz et al. [39,40], is given by $\frac{1}{\beta \rho^1}\frac{\partial p}{\partial \tau}$. Another formulation, proposed by Owis and Neyfeh [41], is given by $\frac{1}{\beta \rho^0}\frac{\partial p}{\partial \tau}$. The different forms of preconditioning result in different convergence rates and accuracies of the methods. To evaluate these rates and accuracies, Nguyen et al. [2] introduced a general formulation: $\frac{1}{\beta \rho^{\gamma_{tt}}}\frac{\partial p}{\partial \tau}$.

The convergence rates and accuracy of the methods were accessed using a variety of $\rho^{\gamma_{tt}}$. The detailed effects of each formulation can be found in the study.

The dual-time preconditioning formulations can be applied to the mixture continuity, mixture momentum, and phasic volume fraction equations. The mass transfer was considered to model cavitation around the projectiles. In our previous study [2], a dual time-stepping algorithm was developed for the unsteady computation of multiphase flow. The algorithm is based on the NS equations, which are expressed as follows:

$$\frac{1}{\beta \rho^{\gamma_{tt}}}\frac{\partial p}{\partial \tau} + \nabla \cdot \mathbf{u} = 0, \tag{6}$$

$$\frac{\partial}{\partial \tau}(\rho \mathbf{u}) + \frac{\partial}{\partial t}(\rho \mathbf{u}) + \nabla \cdot (\rho \mathbf{u}\mathbf{u} + p\mathbf{I}) = \nabla \cdot \left(\mu\left(\nabla \mathbf{u} + \nabla \mathbf{u}^T\right)\right) + \rho \mathbf{g}, \tag{7}$$

$$\frac{\partial \alpha_i}{\partial \tau} + \left(\frac{\alpha_i}{\beta \rho^{\gamma_{tt}}}\right)\frac{\partial p}{\partial \tau} + \frac{\partial \alpha_i}{\partial t} + \nabla \cdot (\mathbf{u}\alpha_i) = 0, \tag{8}$$

where β is the preconditioning compressibility parameter, p is the pressure, t is the physical time, τ is the pseudo time, $\mathbf{u}$ is the flow velocity vector, α_i is the volume fraction of the ith phasic component and $\mathbf{g}$ denotes the gravity vector.

As described by the governing equations, the dual time-stepping algorithm introduces novel pseudo-time terms into the mixture continuity and phasic volume fraction equations. These terms are presented in the general form $\left(\frac{1}{\beta \rho^{\gamma_{tt}}}\right)$, where γ_{tt} represents an exponential factor of the mixture density. The general forms reduce to the form $\left(\frac{1}{\beta \rho}\right)$ suggested by Kunz et al. [39,40] when $\gamma_{tt} = 1$, and to the form $\left(\frac{1}{\beta}\right)$ modified by Owis and Neyfeh [41] when $\gamma_{tt} = 0$. This exponential factor provides a convenient mechanism for controlling the magnitude of the pseudo-time terms, thereby enabling the adjustment of the convergence rate for a given multiphase flow simulation. Accordingly, the pseudo terms are generalized forms in order to effectively evaluate the numerical stability and computational efficiency of the model.

Thanks to the ideas of artificial compressibility, it can transform the elliptic incompressible equations into a hyperbolic compressible system, allowing solving the system by standard, explicit or implicit, time-marching methods [2,18–20]. The governing Equations (6)–(8) can be rewritten in a compact vector form for two-phase flows as follows:

$$\Gamma \frac{\partial \mathbf{W}}{\partial \tau} + \frac{\partial \mathbf{Q}}{\partial t} + \nabla \cdot \mathbf{F}(\mathbf{W}) = \nabla \cdot \mathbf{G}(\mathbf{W}) + \mathbf{S}(\mathbf{W}), \tag{9}$$

where the preconditioning matrix Γ is given as

$$\Gamma = \begin{bmatrix} 1/\beta\rho^{\gamma_{tt}} & 0 & 0 & 0 & 0 \\ \rho & \rho & 0 & 0 & u\Delta_1 \\ \rho & 0 & \rho & 0 & v\Delta_1 \\ \rho & 0 & 0 & \rho & w\Delta_1 \\ \alpha_1/\beta\rho^{\gamma_{tt}} & 0 & 0 & 0 & 1 \end{bmatrix}, \tag{10}$$

$\mathbf{W} = [p, u, v, w, \alpha_1]^T$, $\mathbf{Q} = [0, \rho u, \rho v, \rho w, \alpha_1]^T$ $\mathbf{F}(\mathbf{W}) = (\mathbf{F}_1, \mathbf{F}_2, \mathbf{F}_3)$ is the flux tensor, $\mathbf{G}(\mathbf{W}) = (\mathbf{G}_1, \mathbf{G}_2, \mathbf{G}_3)$ is viscous terms and $\mathbf{S}(\mathbf{W})$ is the source term.

Here,

$$\mathbf{F}_1 = \begin{pmatrix} u \\ \rho u^2 + p \\ \rho uv \\ \rho uw \\ \alpha_1 u \end{pmatrix}; \mathbf{F}_2 = \begin{pmatrix} v \\ \rho uv \\ \rho v^2 + p \\ \rho vw \\ \alpha_1 v \end{pmatrix}; \mathbf{F}_3 = \begin{pmatrix} w \\ \rho uw \\ \rho vw \\ \rho w^2 + p \\ \alpha_1 w \end{pmatrix}. \tag{11}$$

2.2. Numerical Method and Body-Fitted Curvilinear Coordinate System

The dual-time preconditioning approach treats the time variation of the incompressible flow as a compressible flow, which enables the coupling of the velocity and pressure fields in each time iteration [2,23]. This approach uses sub-iterations in pseudo-time and offers several advantages over traditional time-marching algorithms. For example, it provides a direct coupling of the continuity and momentum equations in the incompressible flow equations, eliminates the factorization error in factored implicit schemes, reduces errors due to approximations made in the implicit operator for improved numerical efficiency, eliminates errors due to lagged boundary conditions at both solid and internal fluid boundaries, and allows for the use of nonphysical, preconditioned iterative methods for more efficient convergence of the sub-iterations. These benefits of the dual-time preconditioning approach make it a promising technique for the numerical simulation of unsteady incompressible flows.

By adding a pseudo-time derivative term into the NS systems, the incompressible NS equation system can be formulated in a hyperbolic form, which can be solved using advanced compact schemes where the eigensystem and eigenvectors can be derived, and the wave propagation can be determined by applying Godunov-type methods and upwind solvers. Solving the NS equation in a system that can be implicitly solved and obtain highly accurate simulations [2,15,16].

The preconditioning dual-time multiphase flow model (9) can be solved on a curvilinear body-fitted grid for complex geometries, as illustrated in Figure 1. Accordingly, the governing equations can be transformed from the physical space (x, y, z, t) to the computational space (ξ, η, ζ, τ) in the general curvilinear coordinate system using the following relations [34]:

$$\tau = t;\ \xi = \xi(t, x, y, z);\ \eta = \eta(t, x, y, z);\ \text{and}\ \zeta = \zeta(t, x, y, z). \tag{12}$$

The Cartesian derivatives can be expressed using the chain rule of differential derivatives:

$$\frac{\partial}{\partial \tau} = \frac{\partial}{\partial t} + \xi_t \frac{\partial}{\partial \xi} + \eta_t \frac{\partial}{\partial \eta} + \zeta_t \frac{\partial}{\partial \zeta};\ \text{and}\ \frac{\partial}{\partial x} = \xi_x \frac{\partial}{\partial \xi} + \eta_x \frac{\partial}{\partial \eta} + \zeta_x \frac{\partial}{\partial \zeta}. \tag{13}$$

For convenience, $\xi_t = \frac{\partial \xi}{\partial t}$; $\eta_t = \frac{\partial \eta}{\partial t}$; $\zeta_t = \frac{\partial \zeta}{\partial t}$; $\xi_x = \frac{\partial \xi}{\partial x}$; $\eta_x = \frac{\partial \eta}{\partial x}$; and $\zeta_x = \frac{\partial \zeta}{\partial x}$; the differential derivatives of y and z are defined similarly. The derivatives of the metrics are given as:

$$\begin{aligned}
&\xi_x = J(y_\eta z_\zeta - y_\zeta z_\eta),\ \xi_y = J(x_\zeta z_\eta - x_\eta z_\zeta),\ \xi_z = J(x_\eta y_\zeta - x_\zeta y_\eta),\ \eta_x = J(y_\zeta z_\zeta - y_\zeta z_\zeta),\\
&\eta_y = J(x_\xi z_\zeta - x_\zeta z_{\zeta\zeta}),\ \eta_z = J(x_\zeta y_\xi - x_\xi y_\zeta),\ \zeta_x = J(y_\xi z_\eta - y_\eta z_\xi),\ \zeta_y = J(x_\eta z_\xi - x_\xi z_\eta),\\
&\zeta_z = J(x_\xi y_\eta - x_\eta y_\xi),\ \xi_t = -(x_\tau \xi_x + y_\tau \xi_y + z_\tau \xi_z),\ \eta_t = -(x_\tau \eta_x + y_\tau \eta_y + z_\tau \eta_z),\ \text{and}\\
&\zeta_t = -(x_\tau \zeta_x + y_\tau \zeta_y + z_\tau \zeta_z),
\end{aligned} \qquad (14)$$

where the Jacobian of the transformation is defined as:

$$J = \det\left[\frac{\partial(\xi,\eta,\zeta)}{\partial(x,y,z)}\right] = \frac{1}{x_\xi(y_\eta z_\zeta - y_\zeta z_\eta) - x_\eta(y_\xi z_\zeta - y_\zeta z_\xi) + x_\zeta(y_\xi z_\eta - y_\eta z_\xi)} \qquad (15)$$

By applying Equations (12)–(15), Equation (1) can be transformed into a general curvilinear coordinate system (ξ,η,ζ). Note that the time $\tau = t$; therefore, the system can be rewritten in vector form as:

$$\Gamma\frac{\partial \hat{\mathbf{W}}}{\partial \tau} + \frac{\partial \hat{\mathbf{Q}}}{\partial t} + \frac{\partial \hat{\mathbf{F}}_1}{\partial \xi} + \frac{\partial \hat{\mathbf{F}}_2}{\partial \xi} + \frac{\partial \hat{\mathbf{F}}_3}{\partial \xi} = \frac{\partial \hat{\mathbf{G}}_1}{\partial \xi} + \frac{\partial \hat{\mathbf{G}}_2}{\partial \xi} + \frac{\partial \hat{\mathbf{G}}_3}{\partial \xi} + \hat{\mathbf{S}} \qquad (16)$$

where $\hat{\mathbf{W}} = \frac{1}{J}\mathbf{W}, \hat{\mathbf{Q}} = \frac{1}{J}\mathbf{Q}$ is the state vector, $\hat{\mathbf{S}} = \frac{1}{J}\mathbf{S}$, and convective flux vector $\hat{\mathbf{F}}_1$ and viscous flux vector $\hat{\mathbf{G}}_1$ given as;

$$\hat{\mathbf{F}}_1 = \frac{1}{J}\begin{pmatrix} U \\ \rho u U + p\xi_x \\ \rho v U + p\xi_y \\ \rho w U + p\xi_z \\ \alpha_1 U \end{pmatrix}; \ \hat{\mathbf{G}}_1 = \frac{1}{J}\begin{pmatrix} 0 \\ \xi_x \tau_{xx} + \xi_y \tau_{xy} + \xi_z \tau_{xz} \\ \xi_x \tau_{xy} + \xi_y \tau_{yy} + \xi_z \tau_{yz} \\ \xi_x \tau_{xz} + \xi_y \tau_{yz} + \xi_z \tau_{zz} \\ 0 \end{pmatrix} \qquad (17)$$

To build a moving grid algorithm, the grid velocities ξ_t, η_t, and ζ_t are introduced, and the contravariant velocities are defined as:

$$U = \xi_t + \xi_x u + \xi_y v + \xi_z w;\ V = \eta_t + \eta_x u + \eta_y v + \eta_z w;\ W = \zeta_t + \zeta_x u + \zeta_y v + \zeta_z w. \qquad (18)$$

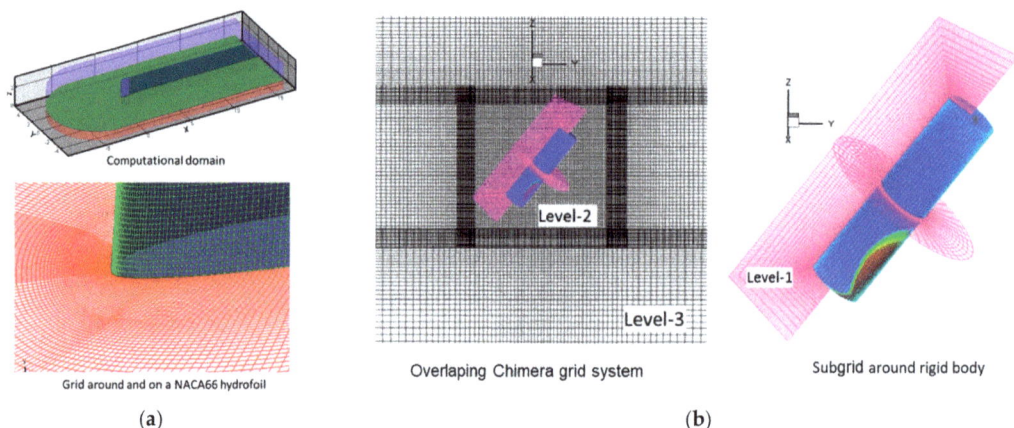

Figure 1. Body-Fitted grid system. (**a**) Grid around a NACA66 hydrofoil (From [1]) and (**b**) Overlapping Chimera grids (From [2]).

2.3. Coupling Artificial Density-Based Dual-Time Method and Sharp Interface Methods

The pseudo-compressibility method used in this study is based on the fundamental principles of classical methods, but it introduces pseudo-time derivative terms by replacing the true density with a pseudo-density, denoted by $\tilde{\rho}$, and calculating the pressure as a function of the pseudo-density, referred to as the pseudo-law of state [1,21,42]. This pseudo-law of state provides a mechanism for controlling the magnitude of the pseudo-time terms and adjusting the convergence behavior for a given simulation, as described in the governing equations. The introduction of pseudo-density and the use of a pseudo-law of state provide a means of effectively and efficiently simulating unsteady incompressible flows in a manner that couples the velocity and pressure fields in each time iteration.

$$p = \rho U_0^2 \ln\left(\frac{\tilde{\rho}}{\rho_\infty}\right) + p_\infty \tag{19}$$

where the parameters are set in accordance to $U_0 = \sqrt{U_\infty^2}$ or $U_0 = \sqrt{u^2 + v^2 + w^2}$, in which u, v, and w, are equal to the local values of the respective velocities obtained at a previous iteration.

Using the pseudo-law of state (19), the incompressible NS Equations (1) and (2) can be solved in a hyperbolic form using density-based solvers, where the pseudo-density and velocity field can be obtained at each time step and, subsequently the pressure can be obtained from Equation (19). The governing equations can be expressed as follows:

$$\frac{\partial \tilde{\rho}}{\partial \tau} + \nabla \cdot (\rho \mathbf{u}) = 0, \tag{20}$$

$$\frac{\partial}{\partial \tau}(\tilde{\rho}\mathbf{u}) + \frac{\partial}{\partial t}(\rho\mathbf{u}) + \nabla \cdot (\rho \mathbf{u}\mathbf{u} + p\mathbf{I}) = \nabla \cdot \left(\mu\left(\nabla \mathbf{u} + \nabla \mathbf{u}^T\right)\right) + \rho \mathbf{g}, \tag{21}$$

For the simulation of immiscible fluid flows, such as free surface flows and bubble dynamics, which involve a sharp interface separating the two fluids, it is crucial to maintain the accuracy and sharpness of the interface. In order to achieve this, a volume of fluid (VOF) equation is used to track the interface position. The advection equation is solved using a known velocity field to update the density and viscosity fields in the next time step. The VOF equation takes the form of:

$$\frac{\partial \alpha}{\partial t} + \mathbf{u}\nabla \cdot \alpha = 0. \tag{22}$$

This equation ensures that the interface is captured accurately, even in the presence of complex flow physics, and allows for the simulation of multiphase flows with sharp interface transitions. The use of the VOF equation in conjunction with the governing equations of the flow helps to accurately model and simulate the behavior of immiscible fluid systems. The geometric VOF/PLIC methods [21] or algebraic VOF methods [22] have been successfully coupled to incompressible flow solutions for free surface flows.

2.4. Simulations of Incompressible Multiphase Flows

2.4.1. Modeling Two-Phase Flows with Sharp Interface

The dual-time preconditioning and artificial compressibility methods have been utilized in the simulation of multiphase flows in the incompressible NS equations. These techniques have led to successful reports of two-phase flow simulations. Extensive numerical analyses have been conducted to study the free surface and fluid-structure interactions with moving bodies [1,2,20–22,42–45]. A novel numerical model that integrates a fully 3D, dual-time, pseudo-compressibility model with a VOF interface tracking algorithm was introduced for multiphase flows [21]. The model is designed for the analysis of free surface flows and water impact problems. The numerical solver is validated through simulations of various water impact problems, including the water entries of free-falling hemisphere and cone, the initial stages of the dam-break problem with dynamic pressure loads, and a long-term simulation of the wave impact on a container and a tall structure. A typical simulation of a three-dimensional falling water column on a container and impact pressures at four positions in the container is illustrated in Figure 2. The time sequence of a falling water column on a container in 3D shows complex and violent fluid dynamics, including wave impact, breaking, jets, mixing, and entrapment of fluids. The simulation results of pressure peak, initial slope, water level, and arrival times of the primary wave show good agreement with experimental data. These findings demonstrate the model's potential for accurate analysis and prediction of fluid dynamics in free surface flows and water impact problems.

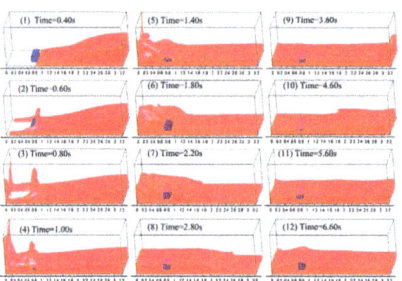
(a) Time evolution of a 3D dam-break

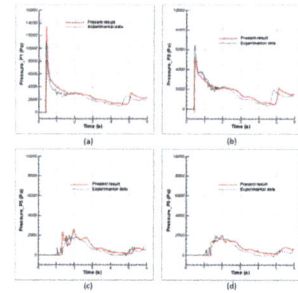
(b) Impact pressures at four positions in the container

Figure 2. Simulation of a 3D falling water column in a container (This figure has been adapted from [21]).

A research article was published describing an advanced technique that uses a volume-of-fluid (VOF)-based method and a dual-time, pseudo-compressibility method on overset grids for 3D free-surface flow problems in engineering [21]. The technique employs overset grids to enable simulations of complex geometries and object motions, enhancing computational efficiency by utilizing the flexibility of these grids. The nonlinear six degrees of freedom (6DOF) motion equations are strongly coupled with the flow solver to enable the simultaneous solution of rigid body motion. A range of complex problems, featuring a broad variety of Froude numbers and large density ratios, was selected to showcase the method's capabilities on a complex, moving overset grid system. Figure 3 illustrates simulations of free-surface wave profiles surrounding a NACA0024 foil using the free-surface flow solver. The numerical results are well compared with experimental photographs from [46]. The numerical simulations of a bubble-bursting phenomenon in two tandem bubbles at the free surface are conducted to explore the influence of another bubble behind it [47]. The problems of water entry and exit of rigid bodies have been numerically simulated to study the slamming effect on structures near the free surface [1]. Figure 4 shows a 3D simulation of an oblique cylinder entering the water. The dual-time preconditioning method was used in ANSYS Fluent software to model flow-front advancement during the impregnation of woven fabrics of a 3D curved mold for a fillet L-shaped structure [48]. The studies show the capability of the methods for simulation and analysis of incompressible two-phase flows with sharp interfaces.

2.4.2. Modeling Cavitating Flows

To model phase change in cavitating flows, vaporization and condensation are taken into the account [2,20]. Accordingly, the mass transfer rates are added in the source terms of the continuity and phasic volume fraction equations as follows.

$$\frac{1}{\beta \rho^{\gamma_{tt}}} \frac{\partial p}{\partial \tau} + \nabla \cdot \mathbf{u} = \dot{m} \left(\frac{1}{\rho_l} - \frac{1}{\rho_v} \right), \tag{23}$$

$$\frac{\partial \alpha_l}{\partial \tau} + \left(\frac{\alpha_l}{\beta \rho^{\gamma_u}}\right)\frac{\partial p}{\partial \tau} + \frac{\partial \alpha_l}{\partial t} + \nabla \cdot (\mathbf{u}\alpha_l) = \frac{\dot{m}}{\rho_l}, \quad (24)$$

The source term for the mass transfer rate in multiphase flows is represented as the sum of two terms: $\dot{m} = \dot{m}^+ + \dot{m}^-$. The term $\dot{m}^+$ represents the mass rate of vapor generation, while $\dot{m}^-$ represents the mass rate of condensation. The source term of the mass transfer rate takes into account the generation and condensation of vapor, which are essential processes in the simulation of multiphase flows. The accurate representation of the mass transfer rate is crucial in ensuring the accuracy of the simulation results and the proper prediction of the behavior of the flow. The source term of the mass transfer rate has been widely studied in the literature and various models have been proposed to account for the generation and condensation of vapor in multiphase flows [26,34].

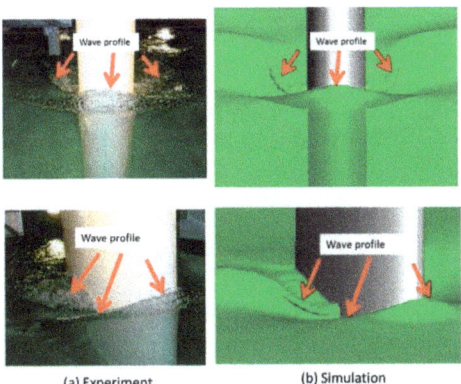

Figure 3. Simulated and experimental free-surface wave profiles around a surface-piercing NACA0024 foil (This figure has been adapted from [1]).

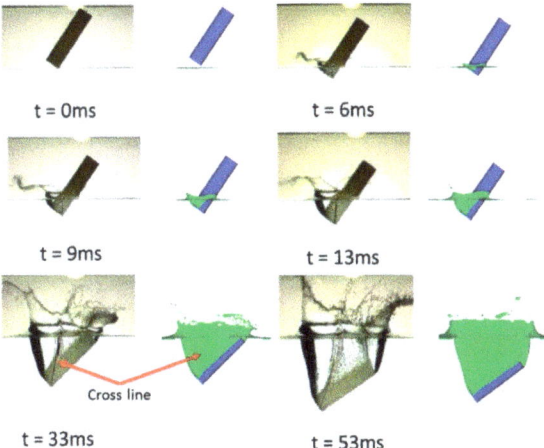

Figure 4. Water entry of an oblique cylinder (This figure has been adapted from [1]).

The mass transfer rates can be characterized and modeled based on the vapor pressure p_v, and the scaling theory of static critical phenomena [2,35,49] as follows:

$$\dot{m}^- = -k_1 \frac{\alpha_l \rho_l}{t_\infty} \min\left\{1, \max\left(0, \frac{p_v - p}{k_p p_v}\right)\right\}, \quad (25)$$

$$\dot{m}^+ = k_2 \frac{\alpha_v \rho_v}{t_\infty} \min\left\{1, \max\left(0, \frac{p - p_v}{k_p p_v}\right)\right\}, \tag{26}$$

where the constant coefficients, k_1 and k_2, represent the transfer rates from fluid 2 to fluid 1 and from fluid 1 to fluid 2, respectively. The damping parameter, k_p, accounts for the pressure difference between the vapor and liquid pressures.

Alternative models of cavitation based on the vapor pressure p_v, and the scaling theory of static critical phenomena can be also employed in this governing equation system such as the Merkle cavitation model [2,50], Kunz cavitation model [39,40], Schnerr and Sauer cavitation model [51,52] Singhal cavitation model [53], and ZGB cavitation model [54,55].

The cavitation number is defined as follows:

$$\sigma = \frac{p_\infty - p_v}{0.5 \rho_\infty u_\infty^2} \tag{27}$$

where the p_∞ is the initial pressure in air, p_v is the vapor pressure, ρ_∞ is the liquid density, and u_∞ is the initial velocity.

This technique was also employed in many studies to analyze cavitating flows. The super- and partial-cavitating flow over submerged projectiles were modeled and reported [2,41,56–58]. Using the general form of the dual-time preconditioning terms [2], the cavitating flows around the projectile with different angles of attack have been simulated as shown in Figure 5. Furthermore, in the study, the simulations of water entry for high-speed projectiles involve modeling the fluid flow as a mixture of liquid, condensable vapor, and non-condensable gases. Separate volume fraction equations are used for each species and solved to capture their behavior. The interface-capturing method can distinguish between the species and is crucial for accurately modeling supercavitation, which occurs when a cavity of vapor is created around the projectile. This enables a more accurate prediction of supercavitating water entry and provides valuable design insights.

The dual-time preconditioning NS solver was employed to study the cavitation behavior of projectiles moving below a free surface and exiting it [20]. The results of the simulation can be used to improve the design of underwater systems and make them more resilient to high-pressure peaks caused by collapsing bubbles. Figure 6 demonstrates the asymmetric cavitation shedding and collapse during the exit of an axisymmetric projectile at an angle of attack of 4 degrees. The simulation accurately captured pressure distribution, cavitation growth, shedding, collapse, velocity magnitude, streamlines, and cavity evolution. These findings can aid in the design and control of submerged vehicles or projectiles operating in similar fluid flows, as they provide insight into the interaction between these objects and their environment.

The dual-time preconditioning method has also been employed with a cut-cell method utilizing 2D cartesian meshes with embedded boundaries to simulate steady-state, turbulent, and cavitating flows over isolated hydrofoils. The numerical solver can characterize two hydrofoils featuring mid-chord and leading-edge cavitation and the simulation results show satisfactory agreement with numerical and experimental data [59]. The cavitation of many different hydrofoils is widely analyzed because of the important application of this shape in ship engineering. A 3D NACA66 hydrofoil fixed at an $8°$ angle of attack and sheet/cloud cavitating conditions has been simulated and compared with experimental data as shown in Figure 7. The unsteady behavior of transient cavitating flow, the results at eight typical instants are presented and the formation, growth, and breakdown of the cavity well agree with the experimental photograph. The dual-time preconditioning method has been used to predict instabilities due to cavitation in turbopump inducers [60]. The method was successful to analyze steady cavitating flows around the NACA 0012 and NACA 66 (MOD) hydrofoils and also an axisymmetric hemispherical fore-body under different conditions, and the results are compared with the available numerical and experimental data [61]. A high-order nodal discontinuous Galerkin method was applied to solve the dual-time preconditioned multiphase NS equations for cavitating flows on unstructured grids, showing accurate results of the cavitation around hydrofoil and axisymmetric bodies [62]. The artificially ventilated supercavities covering an underwater vehicle are shown in Figure 8. The numerical method shows its capability and is a powerful tool for computation of the complex multiphase flows.

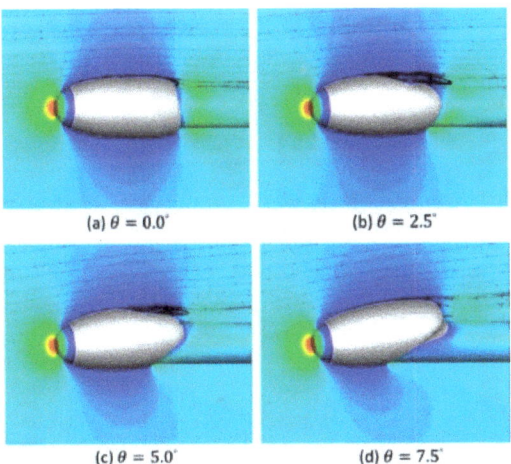

Figure 5. Simulation of cavitating flows over a hemispherical body with different angles of attack (This figure has been adapted from [2]).

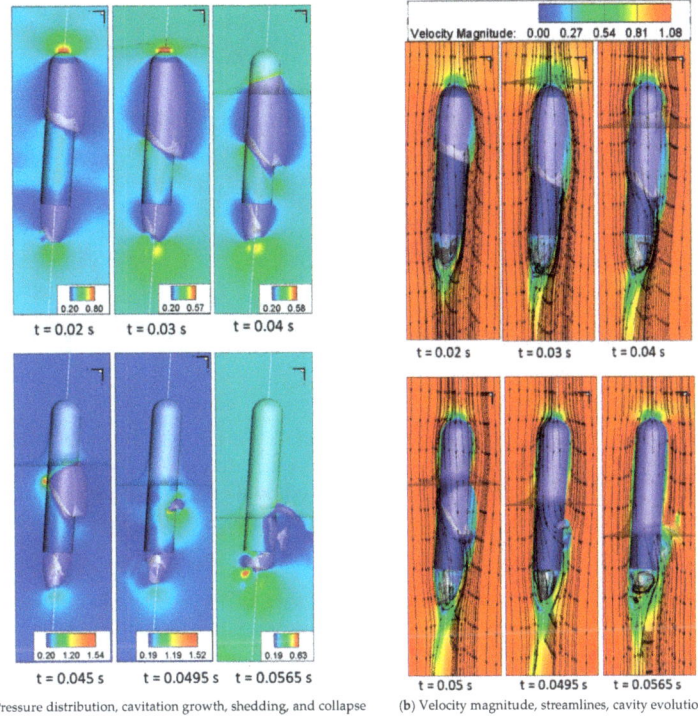

Figure 6. Cavitation evolution during oblique water exits of a projectile (This figure has been adapted from [20]).

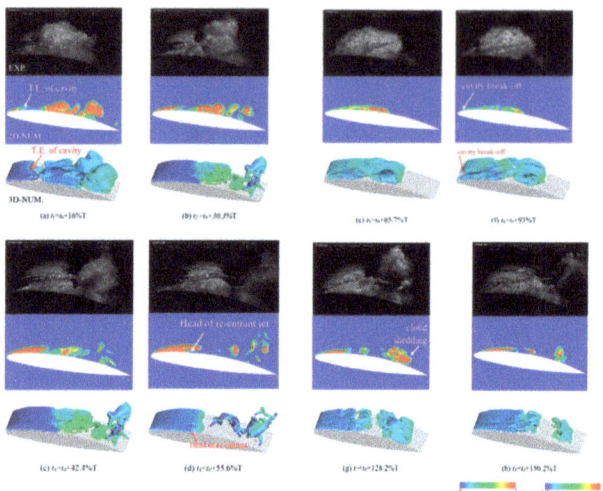

Figure 7. Cavitation about a 3D NACA66 hydrofoil fixed at an 8° angle of attack (This figure has been adapted from [63]).

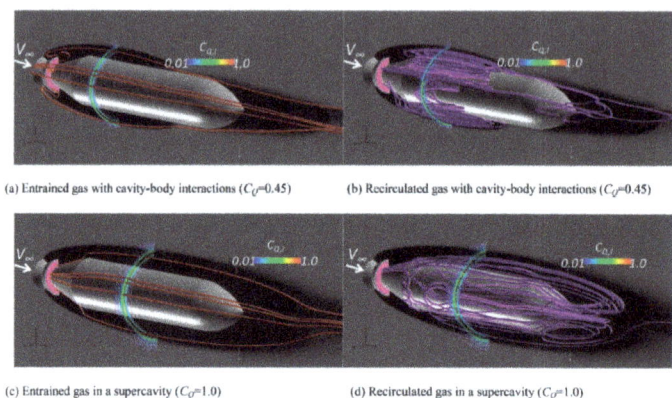

Figure 8. Ventilated cavitating flow over a body at $Fr_N = 26.7$, $Re_N = 56,000$ (This figure has been adapted from [64]).

3. Compressible Multiphase Flows

This section provides a review of the mathematical foundation of dual-time preconditioning techniques for compressible multiphase flows. The focus will be on the description of the dual-time stepping methods and preconditioning approach for the compressible NS equation system, as well as the computation of multiphase flows for a range of Mach numbers.

3.1. Dual-Time Preconditioning Method for Compressible NS Equation System

Almost numerical solvers for compressible flows can well simulate transonic and supersonic flows, however, for problems with mixed flows of very different Mach numbers, the solver can overcome the stiffness of the compressible flows at low Mach numbers and the large discrepancy of wave propagation in the hyperbolic system of the compressible NS equations. Local preconditioning techniques with dual-time stepping methods were introduced to the compressible NS equation to control the wave propagation velocities of the various modes. The preconditioning techniques eliminate the inconsistent scaling behavior of numerical flux functions of Godunov-like schemes. The techniques should be improved and applied to full two-fluid models and DNS-like flow simulations for realistic flow problems [65].

For the compressible homogeneous mixture flow model, the preconditioning matrix is introduced into the compressible Euler/NS equations with pseudo-time derivative terms [66]. The dual-time preconditioning methods not only moderate the stiffness of the system of equations but also improve the accuracy of simulations of flows at low Mach numbers. The preconditioning technique involves the alteration of the time derivatives used in time-marching CFD methods with the primary objective of enhancing their convergence. The preconditioning is introduced by multiplying the time derivative in Equation (17) by a preconditioning matrix Γ as follows [36]:

$$\Gamma \frac{\partial \mathbf{W}}{\partial \tau} + \frac{\partial \mathbf{F}}{\partial x} = 0 \qquad (28)$$

In formulation (17), the alteration of the time derivative, as the time t is replaced by a pseudo time τ, can negatively impact the time accuracy, particularly for transient problems. However, the time accuracy can be restored by using a dual time-stepping method.

To improve the efficiency of numerical simulations, the equation is transformed from the conservative variables Q to a more suitable set of variables, to reduce the system Jacobian matrix to a sparse and easier-to-manipulate form. The derivation of the preconditioning matrix starts with this transformation.

In the case of a mixture of multi-species real gases, the primitive variables are considered the most suitable set of variables. After the preconditioning steps, a transformation back to the conservative set of variables is necessary for numerical implementation. The preconditioning matrix Γ will be modified in different forms according to the numerical models to which is applied [67–72]. Here, the formulation of the dual-time preconditioning NS equation system of a homogeneous mixture flow model based on mass fraction variables is presented. The system is given in a vector form in curvilinear coordinates as follows [36]:

$$\Gamma \frac{\partial \hat{\mathbf{W}}}{\partial \tau} + \frac{\partial \hat{\mathbf{Q}}}{\partial t} + \frac{\partial \hat{\mathbf{F}}_j}{\partial \tilde{\xi}_j} = \frac{\partial \hat{\mathbf{G}}_j}{\partial \tilde{\xi}_j} + \hat{\mathbf{S}}, \qquad (29)$$

where $\hat{\mathbf{W}} = \frac{1}{J}[p, u, v, w, T, Y_2]^T$, $\hat{\mathbf{Q}} = \frac{1}{J}[Y_1\rho, \rho u, \rho v, \rho w, \rho H - p, Y_2\rho]^T$ is the state vector, $\hat{\mathbf{S}} = \frac{1}{J}\mathbf{S}$, and convective flux vector $\hat{\mathbf{F}}_1$ and viscous flux vector $\hat{\mathbf{G}}_1$ given as;

$$\hat{\mathbf{F}}_1 = \frac{1}{J}\begin{pmatrix} Y_1\rho U \\ \rho u U + p\tilde{\xi}_x \\ \rho v U + p\tilde{\xi}_y \\ \rho w U + p\tilde{\xi}_z \\ (\rho H)U \\ Y_2\rho U \end{pmatrix}; \hat{\mathbf{F}}_2 = \frac{1}{J}\begin{pmatrix} Y_1\rho V \\ \rho u V + p\eta_x \\ \rho v V + p\eta_y \\ \rho w V + p\eta_z \\ (\rho H)V \\ Y_2\rho V \end{pmatrix}; \hat{\mathbf{F}}_3 = \frac{1}{J}\begin{pmatrix} Y_1\rho W \\ \rho u W + p\zeta_x \\ \rho v W + p\zeta_y \\ \rho w W + p\zeta_z \\ (\rho H)W \\ Y_2\rho W \end{pmatrix};$$

$$\hat{\mathbf{G}}_1 = \frac{1}{J}\begin{pmatrix} 0 \\ \xi_x\tau_{xx} + \xi_y\tau_{xy} + \xi_z\tau_{xz} \\ \xi_x\tau_{xy} + \xi_y\tau_{yy} + \xi_z\tau_{yz} \\ \xi_x\tau_{xz} + \xi_y\tau_{yz} + \xi_z\tau_{zz} \\ \Psi_\xi \\ 0 \end{pmatrix}; \hat{\mathbf{G}}_2 = \frac{1}{J}\begin{pmatrix} 0 \\ \eta_x\tau_{xx} + \eta_y\tau_{xy} + \eta_z\tau_{xz} \\ \eta_x\tau_{xy} + \eta_y\tau_{yy} + \eta_z\tau_{yz} \\ \eta_x\tau_{xz} + \eta_y\tau_{yz} + \eta_z\tau_{zz} \\ \Psi_\eta \\ 0 \end{pmatrix}; \hat{\mathbf{G}}_3 = \qquad (30)$$

$$\frac{1}{J}\begin{pmatrix} 0 \\ \zeta_x\tau_{xx} + \zeta_y\tau_{xy} + \zeta_z\tau_{xz} \\ \zeta_x\tau_{xy} + \zeta_y\tau_{yy} + \zeta_z\tau_{yz} \\ \zeta_x\tau_{xz} + \zeta_y\tau_{yz} + \zeta_z\tau_{zz} \\ \Psi_\zeta \\ 0 \end{pmatrix},$$

where $\Psi_\zeta = \zeta_x(u\tau_{xx} + v\tau_{xy} + w\tau_{xz} - q_x) + \zeta_y(u\tau_{xy} + v\tau_{yy} + w\tau_{yz} - q_y) + \zeta_z(u\tau_{xz} + v\tau_{yz} + w\tau_{zz} - q_z)$.

The preconditioning matrix Γ is given as

$$\Gamma = \begin{bmatrix} Y_1 \frac{\partial \rho}{\partial p} & 0 & 0 & 0 & Y_1 \frac{\partial \rho}{\partial T} & \Phi_1 \\ u\frac{\partial \rho}{\partial p} & \rho & 0 & 0 & u\frac{\partial \rho}{\partial T} & u\frac{\partial \rho}{\partial Y_2} \\ v\frac{\partial \rho}{\partial p} & 0 & \rho & 0 & v\frac{\partial \rho}{\partial T} & v\frac{\partial \rho}{\partial Y_2} \\ w\frac{\partial \rho}{\partial p} & 0 & 0 & \rho & w\frac{\partial \rho}{\partial T} & w\frac{\partial \rho}{\partial Y_2} \\ \Phi_p & u\rho & v\rho & w\rho & \Phi_T & \Phi_H \\ Y_2\frac{\partial \rho}{\partial p} & 0 & 0 & 0 & Y_2\frac{\partial \rho}{\partial T} & \Phi_2 \end{bmatrix}, \quad (31)$$

where, $\Phi_p = H\frac{\partial \rho}{\partial p} + \rho\frac{\partial H}{\partial p} - 1$; $\Phi_T = H\frac{\partial \rho}{\partial T} + \rho\frac{\partial H}{\partial T}$; $\Phi_H = H\frac{\partial \rho}{\partial Y_2} + \rho\frac{\partial H}{\partial Y_2}$; $\Phi_1 = -\rho + Y_1\frac{\partial \rho}{\partial Y_2}$; $\Phi_2 = \rho + Y_2\frac{\partial \rho}{\partial Y_2}$. The characteristics of the preconditioned system result in a well-conditioned dissipation formulation and ensure reliable accuracy, in which the pseudo-time derivative, $\frac{\partial \tilde{\rho}}{\partial p} = \frac{\partial \rho}{\partial p} + \frac{1}{V_p^2} - \frac{1}{c^2}$; V_p is defined as the pseudo-speed of sound and this control the wave propagation velocities of the various modes of the system.

The fluids' properties can be modeled based on equations of state (EOS) from ideal to real fluids. Examples of EOS equations are adiabatic/Isentropic EOS, Tait EOS, Stiffened Gas EOS, Noble-Abel Stiffened-Gas EOS, and Industrial Association for the Properties of Water and Steam (IAPWS).

3.2. Numerical Solution Procedures

The Dual-Time Preconditioned System (18) can be discretized on a structured grid utilizing a subclass of the lower-upper symmetric Gauss-Seidel (LU-SGS) method [73]. An upwind non-MUSCL total variation diminishing (TVD) algorithm is employed in conjunction with a suitable limiter function to eliminate the generation of unphysical solutions in the vicinity of strong gradients [36]. The discretization of the convective flux terms is accomplished through this approach. Meanwhile, the viscous flux terms are treated using a second-order accurate central difference scheme. The physical-time derivative is approximated using a second-order accurate backward difference method, while the pseudo-time derivative is determined through the application of the Euler finite-difference formula. The pseudo-time step is established based on the largest eigenvalue of the system, and the local pseudo-time step is defined accordingly. For steady-state solutions, the physical-time step is set to infinity, while in unsteady computations, the physical-time step is set to the global pseudo-time step at each pseudo-time level. A hybrid formulation, combining a conservative preconditioned Roe method and a nonconservative preconditioned characteristic-based method, is presented to extend the method to transonic and supersonic flows with the presence of shocks. This hybrid approach allows for a more comprehensive and accurate treatment of flow physics in these challenging regimes [74].

Alternatively, the application of the linearization procedure to System (18) results in the derivation of an implicit unfactored numerical scheme. The solution to this implicit unfactored scheme can be obtained through the utilization of the alternating direction implicit (ADI) method [75]. Furthermore, the dual-time preconditioned system can alternatively be discretized by employing the advection upstream splitting method (AUSM), and the multi-dimensional limiting process (MLP) limiter was utilized to ensure efficient and accurate computation of convective fluxes [76,77].

3.3. Simulations of Compressible Multiphase Flows

The initial preconditioning algorithm designed for the computation of compressible single-phase flow [66] was expanded to accommodate multiphase flow simulations [36]. This study presents a numerical solver for multiphase flows capable of simulating various fluid-structure interactions in underwater environments, such as cavitation flows over underwater projectiles, transonic flow past an underwater projectile, water impact of a circular cylinder, and water entry of a hemisphere with one degree of freedom. The solver has been shown to provide results that are in good agreement with available experimental data or previously published results for various quantities of interest, such as surface pressure coefficients, water impact forces, vertical accelerations, and impact velocities. Figure 9 presents typical results of the contours of pressure, density, and temperature for a transonic flow over a projectile, where the free stream velocity is 1540 m/s. Additionally, the study also examines key aspects of supercavitating flows over axisymmetric projectiles during entry and exit from the water, including the shape of the cavity, phase topography, and drag coefficients.

Given that a significant portion of cavitating flows involve both an incompressible liquid region and a compressible cavity region, it is imperative to employ a system pre-conditioning technique that appropriately scales the numerical speed of sound in order to improve the convergence rate in low Mach number flow simulations [35,67,75,78–80].

Furthermore, the Dual-Time Preconditioned method has been demonstrated to be highly effective in the simulation of cavitation bubble dynamics using real fluid properties of IAPWS [37,81]. The cavitation bubble dynamics, which refers to the formation of vapor-filled bubbles in a liquid, is a potent phenomenon with significant ramifications in various fields such as nature, science, and industrial engineering and technologies. In applications such as hydrofoils, propellers, pipes, control valves and nozzles, and ultrasonic cavitation, repeated instances of cavitation bubble collapse over time can result in high local energy and cause detrimental effects on the mechanical components. Conversely, the repeated occurrences of cavitation bubble collapse can also generate high local energy that can be harnessed for practical applications, such as hydrodynamic cavitation processes, surface cleaning, extraction of natural products, industrial production of food and beverages, microbial inactivation, and others. During the collapse phase of a cavitation bubble, high local energy can be generated through shock waves and high-speed microjets, resulting in hot spots of high temperature and pressure.

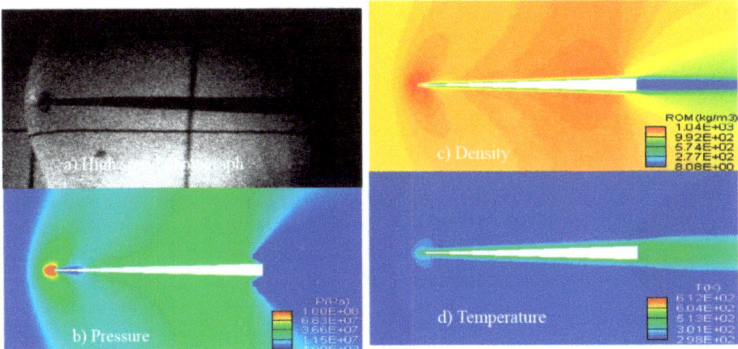

Figure 9. The contours of pressure, density, and temperature for a transonic flow over a projectile, where the free stream velocity is 1540 m/s (This figure has been adapted from [36]).

The numerical computations can analyze the thermodynamic effects on single cavitation bubble dynamics under various ambient temperature conditions [37,38]. The results of these studies can be used to improve our understanding of how different temperatures affect the behavior and intensity of bubbles in a variety of applications as simulations in Figure 10, such as medical imaging or industrial processes that involve fluid flow with high-pressure gradients. The maximum bubble radius, first minimum bubble radius, and collapsing time increase with an increase in ambient temperature. However, the peak values of internal pressure and internal temperature decrease as the ambient temperature increases. Generally speaking, bubbles collapse less violently at higher temperatures than the lower ones.

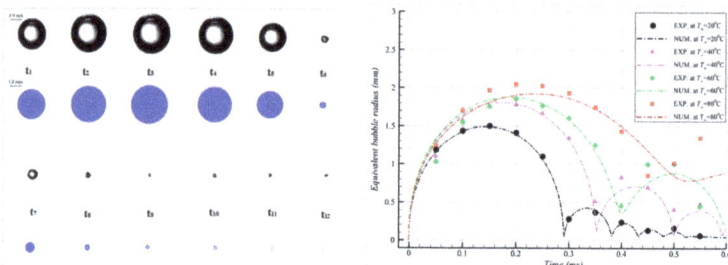

(**a**) Results for the bubble shape evolution at 20 °C. (**b**) The bubble radius under different ambient water temperatures.

Figure 10. Thermodynamic effects on single cavitation bubble dynamics (This figure has been adapted from [37]).

For capturing a sharper interface between two fluids, novel level-set approaches [82], and tangent of hyperbola for interface-capturing methods (THINC) [83,84] were introduced for the simulation of all-Mach multiphase flows. The proposed all-Mach level-set method utilizes the concept of signed distance functions within the framework of a species mass conservation equation to provide accurate evolution of the interface. The method also encompasses multiple reinitialization techniques to address subgrid scale interfacial fragmentation, ensuring the maintenance of accuracy. From a practical perspective, this study has the potential to enhance simulations related to high-speed marine vehicles, particularly in the context of supercavitation, which refers to the formation of gaseous cavities around moving objects such as ships or submarines at very high velocities. The other works have evaluated several approaches for sharp interface capturing in computations of multi-phase mixture flows using the preconditioning method [83]. The practical implications are that these strategies can be used to accurately capture the volume fraction discontinuities, which is important when simulating complex flow phenomena such as Rayleigh Taylor instability and axisymmetric jet instabilities. These methods also provide an accurate representation of fluid properties at interfaces between different phases such as the evolution of vapor volume fraction in an aerated-liquid injector shown in Figure 11. The inception of bubbles occurs in the fluid upstream of the discharge tube, which is characterized by a relatively low flow velocity. As these bubbles progress through the discharge tube, they experience rapid deformation and fragmentation due to the presence of shearing stresses. This allows engineers to better understand how fluids interact with each other under various conditions.

Using the preconditioning method, a comprehensive examination of the behavior of sheet cavitation in 3D Venturi geometries has been studied [85]. The research provides a deeper understanding of the dynamics of sheet cavitation, which is crucial for improving design decisions in hydraulic applications. The correlation between the numerical results and the experimental data for the capture of the re-entrant jet is found to be substantial in Figure 12 This is evidenced by the accurate determination of negative velocity values as depicted in the velocity profiles. Through a comparison of numerical results obtained under both sidewall and periodic conditions, the researchers were able to identify 3D effects that are not directly tied to the presence or absence of walls.

The homogeneous mixture model simulation of compressible multi-phase flows at all Mach numbers was also applied for a numerical simulation of compressible multi-phase flows utilizing the Noble–Abel stiffened gas (NASG) equation of state for fluid properties [86]. The simulation takes into account the effects of gravity and surface tension forces, which allows for the examination of oscillations in elliptical drops and unsteady water surfaces in dam break scenarios. The preconditioning technique was utilized to achieve improved convergence at low speeds without introducing numerical dissipation. The results obtained from this research were compared against analytic solutions and experimental data, such as Schlieren images, and were found to be in good agreement. The method provided a highly resolved result of the vortex formed around the bubble in Figure 13, showing the bubble transformed into an elliptical shape. Additionally, the simulation was able to accurately predict the deformation and evolution of bubbles during air/helium shock interactions, as well as the mixing and heat transfer between liquid and gas phases in underwater explosion scenarios, under various conditions.

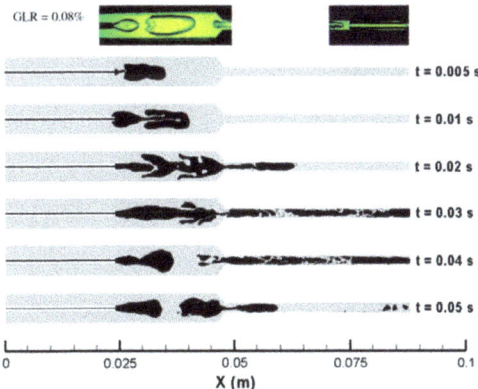

Figure 11. Evolution of vapor volume fraction in an aerated-liquid injector. Numerical computation using the preconditioning method and interface sharpening technique (This figure has been adapted from [83]).

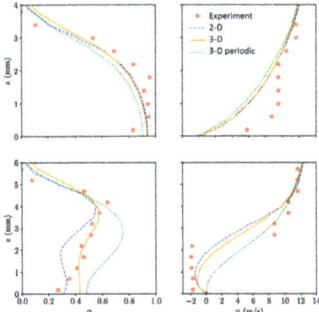

Figure 12. Dynamics of partial cavitation developing in a 3D Venturi geometry and the interaction with sidewalls. Time-averaged comparison at midspan at stations S1 (top) and S2 (bottom) (This figure has been adapted from [85]).

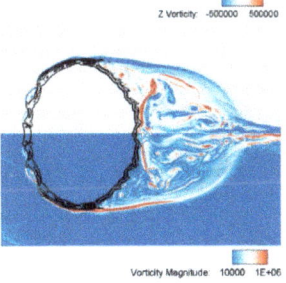

Figure 13. Shock/water-column interaction problem. The z-vorticity and vorticity magnitude contour results at 194 µs (This figure has been adapted from [86]).

Furthermore, the preconditioning method has been applied to various multiphase flows and physics. A study presents the results of a numerical simulation investigating the interaction between an ultrasound wave and a bubble [87]. It provides three different numerical methods which are assessed with one-dimensional spherical benchmarks, showing that the compressible projection method is most suitable when considering spatial accuracy and time step stability. Finally, moving deformable bubbles interacting with plane waves have been simulated successfully demonstrating the ability of this new technique in more complex situations. The preconditioning technique was also applied to the development of a two-phase compressible flow model with stiff mechanical relaxation [70,88]. These studies present numerical results for two-dimensional liquid gas channel flows, shocks, and real cavitating flows in Venturi nozzles, which show that the proposed preconditioning techniques are effective in improving accuracy at low Mach number regimes. The order of pressure fluctuations generated by these methods is consistent with theoretical predictions from an asymptotic analysis on continuous relaxed two-phase flow models. A comprehensive and unified approach to general fluid thermodynamics was developed to account for fluid flows across the entire thermodynamic state [89]. The proposed method was validated through several test cases that examined the vaporization of supercritical droplets in both quiescent and convective environments. These test cases demonstrate the efficacy of the current algorithm in accurately modeling the specified physical phenomena. This technique was also applied to a modified flamelet-progress-variable model for combustion under supercritical conditions. The validity of the model is established through comparisons with experimental data from both laminar flames and turbulent combustion. Additionally, the model is used to investigate the influence of pressure on the coaxial injection and combustion of LOx/methane, as well as on the swirling injection and combustion of LOx/kerosene [90]. An algorithm based on the integration of time-derivative preconditioning techniques with low-diffusion upwinding methods was presented and applied to the simulation of multiphase, compressible flows commonly found in the motion of underwater projectiles [91]. The efficacy of the algorithm is demonstrated through the presentation of results from several multiphase shock tube calculations. Furthermore, calculations are presented for a high-speed axisymmetric supercavitating projectile during the crucial water entry phase of flight. an extension of the preconditioned advection upstream splitting method was introduced for the simulation of 3D two-phase flows in circulating fluidized beds [92]. The results of the calculations performed on a straight tube geometry demonstrate that the behavior of the flow, as reported in previous studies, is accurately modeled. The analysis specifically focuses on the effects of inelastic particle-particle collisions, which result in a fluctuating flow field. Another study focuses on the development of simultaneous solution algorithms for Eulerian–Eulerian gas–solid flow models, analyzing the stability and convergence behavior of both a point solver and a plane solver [93].

4. Conclusions

In this review, we analyzed the advancements in dual-time preconditioning and artificial compressibility methods for solving NS equations in both incompressible and compressible multiphase flows. We discussed the latest progress in different forms of dual-time preconditioning and artificial compressibility terms integrated into the NS equations on structured and unstructured grids, overlapping grid systems, and curvilinear coordinates.

These methods have been proven to be robust and widely used for the analysis of multiphase flows. We also highlighted outstanding issues in the simulation of free-surface, fluid-structure interaction, slamming, water entry, and exit of rigid bodies, and cavitation flows and shock waves. The relative advantages of these techniques for simulating mixed flows of different speeds were pointed out, and it was concluded that they are a powerful tool for the computation of multiphase flows at all Mach numbers. Overall, the dual-time preconditioning and artificial compressibility methods are valuable tools for CFD simulations in the field of multiphase flows. However, the dual-time, pseudo-compressibility method can be still improved in several ways:

(i) Incorporating more accurate and efficient numerical schemes for solving dual-time, pseudo-compressibility NS equations.

(ii) Considering the effects of turbulence by implementing various accurate turbulence models and other physical phenomena that may affect the multiphase flow behaviors.
(iii) Additionally, the method can be extended to handle more complex and realistic boundary conditions and geometries. Accordingly, efforts should be focused on reducing the computational cost of the method while preserving its accuracy and robustness.

Funding: This research was funded by the Basic Science Research Program through the National Research Foundation of Korea (NRF) funded by the Ministry of Education, grant number 2020R1I1A1A01072475.

Conflicts of Interest: The authors declare no conflict of interest.

Nomenclature

α	Phasic volume fraction
Y	Phasic mass fraction
β	Preconditioning parameter
ρ	Density
γ_{tt}	Exponential factor of mixture density
Γ	Preconditioning matrix
Γ_e	Jacobian of physical time derivatives
μ	Molecular viscosity
τ	Pseudo time
$\Delta\tau$	Pseudo time step
Δt	Physical time step
σ	Cavitation number
g	Gravity
p	Pressure
t	Physical time
T	Transformation matrix
u	x-direction velocity of fluid
U	Contravariant velocity in x-direction
U_0	Parameter reference to velocity
U_∞	Free-stream velocity
v	y-direction velocity of fluid
V	Contravariant velocity in y-direction
w	z-direction velocity of fluid
W	Contravariant velocity in z-direction
$\xi, \eta, and\ \zeta$	computational space in the general curvilinear coordinate system
$\dot{m}$	Mass transfer rate
$\tilde{\rho}$	Artificial density
∞	Reference to free stream

References

1. Nguyen, V.-T.; Park, W.-G. Enhancement of Navier–Stokes solver based on an improved volume-of-fluid method for complex interfacial-flow simulations. *Appl. Ocean. Res.* **2018**, *72*, 92–109. [CrossRef]
2. Nguyen, V.-T.; Vu, D.-T.; Park, W.-G.; Jung, C.-M. Navier–Stokes solver for water entry bodies with moving Chimera grid method in 6DOF motions. *Comput. Fluids* **2016**, *140*, 19–38. [CrossRef]
3. Palomino Solis, D.A.; Piscaglia, F. Toward the Simulation of Flashing Cryogenic Liquids by a Fully Compressible Volume of Fluid Solver. *Fluids* **2022**, *7*, 289. [CrossRef]
4. Yao, J.; Yao, Y. Transient CFD Modelling of Air–Water Two-Phase Annular Flow Characteristics in a Small Horizontal Circular Pipe. *Fluids* **2022**, *7*, 191. [CrossRef]
5. Fayed, H.; Bukhari, M.; Ragab, S. Large-Eddy Simulation of a Hydrocyclone with an Air Core Using Two-Fluid and Volume-of-Fluid Models. *Fluids* **2021**, *6*, 364. [CrossRef]
6. Harlow, F.H.; Welch, J.E. Numerical Calculation of Time-Dependent Viscous Incompressible Flow of Fluid with Free Surface. *Phys. Fluids* **1965**, *8*, 2182–2189. [CrossRef]
7. Puckett, E.G.; Almgren, A.S.; Bell, J.B.; Marcus, D.L.; Rider, W.J. A High-Order Projection Method for Tracking Fluid Interfaces in Variable Density Incompressible Flows. *J. Comput. Phys.* **1997**, *130*, 269–282. [CrossRef]

8. Yu, J.D.; Sakai, S.; Sethian, J.A. A coupled level set projection method applied to ink jet simulation. *Interface Free. Bound.* **2003**, *5*, 459–482. [CrossRef]
9. Merkle, C. Time-accurate unsteady incompressible flow algorithms based on artificial compressibility. In Proceedings of the 8th Computational Fluid Dynamics Conference, Honolulu, HI, USA, 9–11 June 1987.
10. Chorin, A.J. A Numerical Method for Solving Incompressible Viscous Flow Problems. *J. Comput. Phys.* **1997**, *135*, 118–125. [CrossRef]
11. Štrubelj, L.; Tiselj, I. Modeling of Rayleigh-Taylor instability with conservative level set method. In Proceedings of the ICMF 2007, 6th International Conference on Multiphase Flow, Leipzig, Germany, 9–13 July 2007.
12. Elmahi, I.; Gloth, O.; Hänel, D.; Vilsmeier, R. A preconditioned dual time-stepping method for combustion problems. *Int. J. Comput. Fluid Dyn.* **2008**, *22*, 169–181. [CrossRef]
13. Lin, H.; Jiang, C.; Hu, S.; Gao, Z.; Lee, C.-H. Disturbance region update method with preconditioning for steady compressible and incompressible flows. *Comput. Phys. Commun.* **2023**, *285*, 108635. [CrossRef]
14. Li, X.-s.; Gu, C.-w.; Xu, J.-z. Development of Roe-type scheme for all-speed flows based on preconditioning method. *Comput. Fluids* **2009**, *38*, 810–817. [CrossRef]
15. Muradoglu, M.; Gokaltun, S. Implicit Multigrid Computations of Buoyant Drops Through Sinusoidal Constrictions. *J. Appl. Mech.* **2005**, *71*, 857–865. [CrossRef]
16. Kinzel, M.P. *Computational Techniques and Analysis of Cavitating-Fluid Flows*; A Dissertation in Aerospace Engineering; The Pennsylvania State University: State College, PA, USA, 2008.
17. Maia, A.A.G.; Kapat, J.S.; Tomita, J.T.; Silva, J.F.; Bringhenti, C.; Cavalca, D.F. Preconditioning methods for compressible flow CFD codes: Revisited. *Int. J. Mech. Sci.* **2020**, *186*, 105898. [CrossRef]
18. Toro, E.F. *Riemann Solvers and Numerical Methods for Fluid Dynamics. A Practical Introduction*; Springer: Berlin/Heidelberg, Germany, 2009.
19. Nguyen, V.-T.; Phan, T.-H.; Duy, T.-N.; Kim, D.-H.; Park, W.-G. Fully compressible multiphase model for computation of compressible fluid flows with large density ratio and the presence of shock waves. *Comput. Fluids* **2022**, *237*, 105325. [CrossRef]
20. Nguyen, V.-T.; Phan, T.-H.; Duy, T.-N.; Park, W.-G. Unsteady cavitation around submerged and water-exit projectiles under the effect of the free surface: A numerical study. *Ocean. Eng.* **2022**, *263*, 112368. [CrossRef]
21. Nguyen, V.-T.; Park, W.-G. A free surface flow solver for complex three-dimensional water impact problems based on the VOF method. *Int. J. Numer. Methods Fluids* **2016**, *82*, 3–34. [CrossRef]
22. Nguyen, V.-T.; Park, W.-G. A volume-of-fluid (VOF) interface-sharpening method for two-phase incompressible flows. *Comput. Fluids* **2017**, *152*, 104–119. [CrossRef]
23. Nguyen, V.-T.; Thang, V.-D.; Park, W.-G. A novel sharp interface capturing method for two- and three-phase incompressible flows. *Comput. Fluids* **2018**, *172*, 147–161. [CrossRef]
24. Ansari, M.R.; Nimvari, M.E. Bubble viscosity effect on internal circulation within the bubble rising due to buoyancy using the level set method. *Ann. Nucl. Energy* **2011**, *38*, 2770–2778. [CrossRef]
25. Komrakova, A.E.; Eskin, D.; Derksen, J.J. Lattice Boltzmann simulations of a single n-butanol drop rising in water. *Phys. Fluids* **2013**, *25*, 042102. [CrossRef]
26. Magnini, M.; Pulvirenti, B.; Thome, J.R. Numerical investigation of the influence of leading and sequential bubbles on slug flow boiling within a microchannel. *Int. J. Therm. Sci.* **2013**, *71*, 36–52. [CrossRef]
27. Tomar, G.; Biswas, G.; Sharma, A.; Agrawal, A. Numerical simulation of bubble growth in film boiling using a coupled level-set and volume-of-fluid method. *Phys. Fluids* **2005**, *17*, 112103. [CrossRef]
28. Nguyen, V.-T.; Park, W.-G. Numerical study of the thermodynamics and supercavitating flow around an underwater high-speed projectile using a fully compressible multiphase flow model. *Ocean. Eng.* **2022**, *257*, 111686. [CrossRef]
29. Ding, H.; Zhang, Y.; Yang, Y.; Wen, C. A modified Euler-Lagrange-Euler approach for modelling homogeneous and heterogeneous condensing droplets and films in supersonic flows. *Int. J. Heat Mass. Tran.* **2023**, *200*, 123537. [CrossRef]
30. Chen, J.; Huang, Z.; Li, A.; Gao, R.; Jiang, W.; Xi, G. Numerical simulation of carbon separation with shock waves and phase change in supersonic separators. *Process. Saf. Environ.* **2023**, *170*, 277–285. [CrossRef]
31. Heydar Rajaee Shooshtari, S.; Honoré Walther, J.; Wen, C. Combination of genetic algorithm and CFD modelling to develop a new model for reliable prediction of normal shock wave in supersonic flows contributing to carbon capture. *Sep. Purif. Technol.* **2022**, *309*, 122878. [CrossRef]
32. Munz, C.D.; Roller, S.; Klein, R.; Geratz, K.J. The extension of incompressible flow solvers to the weakly compressible regime. *Comput. Fluids* **2003**, *32*, 173–196. [CrossRef]
33. Kajzer, A.; Pozorski, J. A weakly Compressible, Diffuse-Interface Model for Two-Phase Flows. *Flow Turbul. Combust.* **2020**, *105*, 299–333. [CrossRef]
34. Nguyen, V.-T.; Nguyen, N.T.; Phan, T.-H.; Park, W.-G. Efficient three-equation two-phase model for free surface and water impact flows on a general curvilinear body-fitted grid. *Comput. Fluids* **2019**, *196*, 104324. [CrossRef]
35. Phan, T.-H.; Shin, J.-G.; Nguyen, V.-T.; Duy, T.-N.; Park, W.-G. Numerical analysis of an unsteady natural cavitating flow around an axisymmetric projectile under various free-stream temperature conditions. *Int. J. Heat Mass. Tran.* **2021**, *164*, 120484. [CrossRef]
36. Nguyen, V.-T.; Ha, C.-T.; Park, W.-G. Multiphase Flow Simulation of Water-entry and -exit of axisymmetric bodies. In Proceedings of the ASME International Mechanical Engineering Congress and Exposition, Diego, CA, USA, 15–21 November 2013.

37. Phan, T.-H.; Kadivar, E.; Nguyen, V.-T.; el Moctar, O.; Park, W.-G. Thermodynamic effects on single cavitation bubble dynamics under various ambient temperature conditions. *Phys. Fluids* **2022**, *34*, 023318. [CrossRef]
38. Phan, T.-H.; Nguyen, V.-T.; Duy, T.-N.; Kim, D.-H.; Park, W.-G. Numerical study on simultaneous thermodynamic and hydrodynamic mechanisms of underwater explosion. *Int. J. Heat Mass. Tran.* **2021**, *178*, 121581. [CrossRef]
39. Kunz, R.F.; Boger, D.A.; Stinebring, D.R.; Chyczewski, T.S.; Lindau, J.W.; Gibeling, H.J.; Venkateswaran, S.; Govindan, T.R. A preconditioned Navier-Stokes method for two-phase flows with application to cavitation prediction. *Comput. Fluids* **2000**, *29*, 849–875. [CrossRef]
40. Kunz, R.F.; Boger, D.A.; Chyczewski, T.S.; Stinebring, D.; Gibeling, H.; Govindan, T. Multi-phase CFD analysis of natural and ventilated cavitation about submerged bodies. In Proceedings of the 3rd ASME-JSME Joint Fluids Engineering Conference, San Francisco, CA, USA, 18–23 July 1999; Volume 99.
41. Owis, F.M.; Nayfeh, A.H. Numerical simulation of 3-D incompressible, multi-phase flows over cavitating projectiles. *Eur. J. Mech. B-Fluid.* **2004**, *23*, 339–351. [CrossRef]
42. de Jouëtte, C.; Laget, O.; Le Gouez, J.M.; Viviand, H. A dual time stepping method for fluid–structure interaction problems. *Comput. Fluids* **2002**, *31*, 509–537. [CrossRef]
43. Helluy, P.; Golay, F.; Caltagirone, J.-P.; Lubin, P.; Vincent, S.; Drevard, D.; Marcer, R.; Fraunié, P.; Seguin, N.; Grilli, S.; et al. Numerical simulations of wave breaking. *ESAIM Math. Model. Numer. Anal. Model. Math. Anal. Numer.* **2005**, *39*, 591–607. [CrossRef]
44. Nourgaliev, R.R.; Dinh, T.N.; Theofanous, T.G. A pseudocompressibility method for the numerical simulation of incompressible multifluid flows. *Int. J. Multiphas Flow.* **2004**, *30*, 901–937. [CrossRef]
45. Ha, C.-T.; Lee, J.H. A modified monotonicity-preserving high-order scheme with application to computation of multi-phase flows. *Comput. Fluids* **2020**, *197*, 104345. [CrossRef]
46. Metcalf, B.; Longo, J.; Ghosh, S.; Stern, F. Unsteady free-surface wave-induced boundary-layer separation for a surface-piercing NACA 0024 foil: Towing tank experiments. *J. Fluid. Struct.* **2006**, *22*, 77–98. [CrossRef]
47. GeolLee, C.; YoubLee, S.; Ha, C.-T.; HwaLee, J. Bursting jet in two tandem bubbles at the free surface. *Phys. Fluids* **2022**, *34*, 083309.
48. Alotaibi, H.; Abeykoon, C.; Soutis, C.; Jabbari, M. Numerical Simulation of Two-Phase Flow in Liquid Composite Moulding Using VOF-Based Implicit Time-Stepping Scheme. *J. Compos. Sci.* **2022**, *6*, 330. [CrossRef]
49. Nguyen, V.-T.; Phan, T.-H.; Duy, T.-N.; Park, W.-G. Numerical simulation of supercavitating flow around a submerged projectile near a free surface. In Proceedings of the 11th International Symposium on Cavitation, Daejon, Korea, 10–13 May 2021.
50. Merkle, C.L.; Feng, J.; Buelow, P. Computational modeling of the dynamics of sheet cavitation. In Proceedings of the Third International Symposium on Cavitation, Grenoble, France, 7–10 April 1998.
51. Schnerr, G.H.; Sauer, J. Physical and numerical modeling of unsteady cavitation dynamics. In Proceedings of the Fourth International Conference on Multiphase Flow, New Orleans, LA, USA, 27 May–1 June 2001.
52. Ullas, P.K.; Chatterjee, D.; Vengadesan, S. Prediction of unsteady, internal turbulent cavitating flow using dynamic cavitation model. *Int. J. Numer. Methods Heat* **2022**, *32*, 3210–3232. [CrossRef]
53. Singhal, A.K.; Athavale, M.M.; Li, H.; Jiang, Y. Mathematical Basis and Validation of the Full Cavitation Model. *J. Fluids Eng.* **2002**, *124*, 617–624. [CrossRef]
54. Zwart, P.J.; Gerber, A.G.; Belamri, T. A two-phase flow model for predicting cavitation dynamics. In Proceedings of the Fifth International Conference on Multiphase Flow, Yokohama, Japan, 30 May–4 June 2004.
55. Tauviqirrahman, M.; Jamari, J.; Susilowati, S.; Pujiastuti, C.; Setiyana, B.; Pasaribu, A.H.; Ammarullah, M.I. Performance Comparison of Newtonian and Non-Newtonian Fluid on a Heterogeneous Slip/No-Slip Journal Bearing System Based on CFD-FSI Method. *Fluids* **2022**, *7*, 225. [CrossRef]
56. Hejranfar, K.; Ezzatneshan, E.; Fattah-Hesari, K. A comparative study of two cavitation modeling strategies for simulation of inviscid cavitating flows. *Ocean. Eng.* **2015**, *108*, 257–275. [CrossRef]
57. Lindau, J.W.; Kunz, R.F.; Boger, D.A.; Stinebring, D.R.; Gibeling, H.J. High Reynolds number, unsteady, multiphase CFD modeling of cavitating flows. *J. Fluid. Eng.* **2002**, *124*, 607–616. [CrossRef]
58. Lindau, J.W.; Kunz, R.F.; Mulherin, J.M.; Dreyer, J.J.; Stinebring, D.R. Fully coupled, 6-DOF to URANS, modeling of cavitating flows around a supercavitating vehicle. In Proceedings of the Fifth International Symposium on Cavitation (CAV2003), Osaka, Japan, 1–4 November 2003.
59. Vrionis, Y.-P.G.; Samouchos, K.D.; Giannakoglou, K.C. Implementation of a conservative cut-cell method for the simulation of two-phase cavitating flows. In Proceedings of the 10th International Conference on Computational Methods (ICCM2019), Singapore, 9–13 July 2019; pp. 440–452.
60. Coutier-Delgosha, O.; Fortes-Patella, R.; Reboud, J.L.; Hakimi, N.; Hirsch, C. Stability of preconditioned Navier–Stokes equations associated with a cavitation model. *Comput. Fluids* **2005**, *34*, 319–349. [CrossRef]
61. Hejranfar, K.; Fattah-Hesary, K. Assessment of a central difference finite volume scheme for modeling of cavitating flows using preconditioned multiphase Euler equations. *J. Hydrodyn. Ser. B.* **2011**, *23*, 302–313. [CrossRef]
62. Hajihassanpour, M.; Hejranfar, K. A high-order nodal discontinuous Galerkin method to solve preconditioned multiphase Euler/Navier-Stokes equations for inviscid/viscous cavitating flows. *Int. J. Numer. Methods Fluids* **2020**, *92*, 478–508. [CrossRef]

63. Tian, B.; Chen, J.; Zhao, X.; Zhang, M.; Huang, B. Numerical analysis of interaction between turbulent structures and transient sheet/cloud cavitation. *Phys. Fluids* **2022**, *34*, 047116. [CrossRef]
64. Kinzel, M.P.; Lindau, J.W.; Kunz, R.F. Gas entrainment from gaseous supercavities: Insight based on numerical simulation. *Ocean. Eng.* **2021**, *221*, 108544. [CrossRef]
65. Saurel, R.; Pantano, C. Diffuse-Interface Capturing Methods for Compressible Two-Phase Flows. *Annu. Rev. Fluid. Mech.* **2018**, *50*, 105–130. [CrossRef]
66. Weiss, J.M.; Smith, W.A. Preconditioning applied to variable and constant density flows. *AIAA J.* **1995**, *33*, 2050–2057. [CrossRef]
67. Venkateswaran, S.; Lindau, J.W.; Kunz, R.F.; Merkle, C.L. Computation of multiphase mixture flows with compressibility effects. *J. Comput. Phys.* **2002**, *180*, 54–77. [CrossRef]
68. Lindau, J.; Kunz, R.; Venkateswaran, S.; Merkle, C. Development of a fully-compressible multi-phase Reynolds-averaged Navier-Stokes model. In Proceedings of the 15th AIAA Computational Fluid, Dynamics Conference, Anaheim, CA, USA, 11–14 June 2001.
69. Braconnier, B.; Nkonga, B. An all-speed relaxation scheme for interface flows with surface tension. *J. Comput. Phys.* **2009**, *228*, 5722–5739. [CrossRef]
70. Pelanti, M. Low Mach number preconditioning techniques for Roe-type and HLLC-type methods for a two-phase compressible flow model. *Appl. Math. Comput.* **2017**, *310*, 112–133. [CrossRef]
71. Murrone, A.; Guillard, H. Behavior of upwind scheme in the low Mach number limit: III. Preconditioned dissipation for a five equation two phase model. *Comput. Fluids* **2008**, *37*, 1209–1224. [CrossRef]
72. Gupta, A. Preconditioning Methods for Ideal and Multiphase Fluid Flows. Ph.D. Thesis, The University of Tennessee at Chattanooga, Chattanooga, TN, USA, 2013.
73. Yoon, S.; Jameson, A. Lower-upper Symmetric-Gauss-Seidel method for the Euler and Navier-Stokes equations. *AIAA J.* **1988**, *26*, 1025–1026. [CrossRef]
74. Housman, J.A.; Kiris, C.C.; Hafez, M.M. Time-Derivative Preconditioning Methods for Multicomponent Flows—Part I: Riemann Problems. *J. Appl. Mech.* **2009**, *76*, 021210. [CrossRef]
75. Ha, C.-T.; Park, W.-G. Evaluation of a new scaling term in preconditioning schemes for computations of compressible cavitating and ventilated flows. *Ocean. Eng.* **2016**, *126*, 432–466. [CrossRef]
76. Kim, H.; Kim, C. A physics-based cavitation model ranging from inertial to thermal regimes. *Int. J. Heat Mass. Tran.* **2021**, *181*, 121991. [CrossRef]
77. Yoo, Y.-L.; Sung, H.-G. A hybrid AUSM scheme (HAUS) for multi-phase flows with all Mach numbers. *Comput. Fluids* **2021**, *227*, 105050. [CrossRef]
78. Kadioglu, S.Y.; Sussman, M.; Osher, S.; Wright, J.P.; Kang, M. A second order primitive preconditioner for solving all speed multi-phase flows. *J. Comput. Phys.* **2005**, *209*, 477–503. [CrossRef]
79. Shin, B.R.; Yamamoto, S.; Yuan, X. Application of Preconditioning Method to Gas-Liquid Two-Phase Flow Computations. *J. Fluids Eng.* **2004**, *126*, 605–612. [CrossRef]
80. Goncalves, E.; Patella, R.F. Numerical simulation of cavitating flows with homogeneous models. *Comput. Fluids* **2009**, *38*, 1682–1696. [CrossRef]
81. Phan, T.-H.; Nguyen, V.-T.; Duy, T.-N.; Kim, D.-H.; Park, W.-G. Influence of phase-change on the collapse and rebound stages of a single spark-generated cavitation bubble. *Int. J. Heat Mass. Tran.* **2022**, *184*, 122270. [CrossRef]
82. Kinzel, M.P.; Lindau, J.W.; Kunz, R.F. A multiphase level-set approach for all-Mach numbers. *Comput. Fluids* **2018**, *167*, 1–16. [CrossRef]
83. Cassidy, D.A.; Edwards, J.R.; Tian, M. An investigation of interface-sharpening schemes for multi-phase mixture flows. *J. Comput. Phys.* **2009**, *228*, 5628–5649. [CrossRef]
84. Kakumanu, N.; Edwards, J.R.; Choi, J.-I. Numerical Simulation of Underwater Burst Events Using Sharp Interface Capturing Methods. In Proceedings of the AIAA Propulsion and Energy 2019 Forum, Indianapolis, IN, USA, 19–22 August 2019; American Institute of Aeronautics and Astronautics: Reston, VA, USA, 2019.
85. Gouin, C.; Junqueira-Junior, C.; Goncalves Da Silva, E.; Robinet, J.-C. Numerical investigation of three-dimensional partial cavitation in a Venturi geometry. *Phys. Fluids* **2021**, *33*, 063312. [CrossRef]
86. Yoo, Y.-L.; Kim, J.-C.; Sung, H.-G. Homogeneous mixture model simulation of compressible multi-phase flows at all Mach number. *Int. J. Multiphas Flow.* **2021**, *143*, 103745. [CrossRef]
87. Huber, G.; Tanguy, S.; Béra, J.-C.; Gilles, B. A time splitting projection scheme for compressible two-phase flows. Application to the interaction of bubbles with ultrasound waves. *J. Comput. Phys.* **2015**, *302*, 439–468. [CrossRef]
88. LeMartelot, S.; Nkonga, B.; Saurel, R. Liquid and liquid–gas flows at all speeds. *J. Comput. Phys.* **2013**, *255*, 53–82. [CrossRef]
89. Meng, H.; Yang, V. A unified treatment of general fluid thermodynamics and its application to a preconditioning scheme. *J. Comput. Phys.* **2003**, *189*, 277–304. [CrossRef]
90. Huang, D.; Wang, Q.; Meng, H. Modeling of supercritical-pressure turbulent combustion of hydrocarbon fuels using a modified flamelet-progress-variable approach. *Appl. Therm. Eng.* **2017**, *119*, 472–480. [CrossRef]
91. Neaves, M.D.; Edwards, J.R. All-Speed Time-Accurate Underwater Projectile Calculations Using a Preconditioning Algorithm. *J. Fluids Eng.* **2005**, *128*, 284–296. [CrossRef]

92. De Wilde, J.; Heynderickx, G.J.; Vierendeels, J.; Dick, E.; Marin, G.B. An extension of the preconditioned advection upstream splitting method for 3D two-phase flow calculations in circulating fluidized beds. *Comput. Chem. Eng.* **2002**, *26*, 1677–1702. [CrossRef]
93. De Wilde, J.; Vierendeels, J.; Heynderickx, G.J.; Marin, G.B. Simultaneous solution algorithms for Eulerian–Eulerian gas–solid flow models: Stability analysis and convergence behaviour of a point and a plane solver. *J. Comput. Phys.* **2005**, *207*, 309–353. [CrossRef]

Disclaimer/Publisher's Note: The statements, opinions and data contained in all publications are solely those of the individual author(s) and contributor(s) and not of MDPI and/or the editor(s). MDPI and/or the editor(s) disclaim responsibility for any injury to people or property resulting from any ideas, methods, instructions or products referred to in the content.

Article

Application of Deep Learning in Predicting Particle Concentration of Gas–Solid Two-Phase Flow

Zhiyong Wang [1,2], Bing Yan [1,2,3,*] and Haoquan Wang [1]

[1] School of Information and Communication Engineering, North University of China, Taiyuan 030051, China; sz202205016@st.nuc.edu.cn (Z.W.); wanghaoquan12@163.com (H.W.)
[2] China-UK Joint Laboratory of Particle & Two-Phase Flow Measurement Technology, North University of China, Taiyuan 030051, China
[3] Shanxi Key Laboratory of Signal Capturing & Processing, North University of China, Taiyuan 030051, China
* Correspondence: yanbing122530@126.com

Abstract: Particle concentration is an important parameter for describing the state of gas–solid two-phase flow. This study compares the performance of three methods, namely, Back-Propagation Neural Networks (BPNNs), Recurrent Neural Networks (RNNs), and Long Short-Term Memory (LSTM), in handling gas–solid two-phase flow data. The experiment utilized seven parameters, including temperature, humidity, upstream and downstream sensor signals, delay, pressure difference, and particle concentration, as the dataset. The evaluation metrics, such as prediction accuracy, were used for comparative analysis by the experimenters. The experiment results indicate that the prediction accuracies of the RNN, LSTM, and BPNN experiments were 92.4%, 92.7%, and 92.5%, respectively. Future research can focus on further optimizing the performance of the BPNN, RNN, and LSTM to enhance the accuracy and efficiency of gas–solid two-phase flow data processing.

Keywords: gas–solid two-phase flow data processing; BPNN; RNN; LSTM

Citation: Wang, Z.; Yan, B.; Wang, H. Application of Deep Learning in Predicting Particle Concentration of Gas–Solid Two-Phase Flow. *Fluids* **2024**, *9*, 59. https://doi.org/10.3390/fluids9030059

Academic Editors: Ricardo Ruiz Baier, Hemant J. Sagar and Nguyen Van-Tu

Received: 13 December 2023
Revised: 29 January 2024
Accepted: 1 February 2024
Published: 27 February 2024

Copyright: © 2024 by the authors. Licensee MDPI, Basel, Switzerland. This article is an open access article distributed under the terms and conditions of the Creative Commons Attribution (CC BY) license (https://creativecommons.org/licenses/by/4.0/).

1. Introduction

Gas–solid two-phase flow is a common and complex fluid state in industries such as energy and chemical engineering [1]. Particle concentration is a key parameter that determines the flow characteristics of gas–solid two-phase flow and plays a crucial role in investigating these characteristics and optimizing industrial production processes. Various techniques, including microwave [2], capacitance [3], acoustic [4], and optical wave fluctuation [5] techniques, have been proposed for measuring the parameters of gas–solid two-phase flow. Particularly, the electrostatic principle has received widespread attention in recent years due to its reliability and high sensitivity. Nonintrusive sensors are widely used for detecting charge in various industrial applications [6,7].

Traditional methods extract useful signals from electrostatic signals, using different algorithms to accurately study the parameters of gas–solid two-phase flow. For instance, Wang et al. decomposed the signal from an electrostatic sensor using harmonic wavelet transform (HWT) [8] and discrete wavelet transform (DWT) [9]. Zhang et al. utilized the Hilbert–Huang transform to obtain the average flow characteristic parameters of particles within the sensor, such as average flow velocity and average mass flow rate [10]. However, a challenge of electrostatic sensing technology is establishing a model between particle concentration, flow rate, and electrostatic current signal. This is due to the complexity of the electrostatic behavior of powder particles, as well as the amount and polarity of charges being related not only to the properties of the particles themselves (shape, size, distribution, roughness, relative humidity, chemical composition, etc.) but also to the material and arrangement of the pipeline, as well as conveying the conditions of particles within the pipeline (pipe size, conveying velocity) [11,12]. Researchers have improved the characteristics of the instrument to mitigate the influence of sensors on flow patterns and enhance the consistency of spatial sensitivity, thereby improving measurement accuracy [9,10]. This compensation for spatial sensitivity is an important step in enhancing measurement accuracy and improving the instrument's suitability for flow patterns.

From a theoretical perspective, it is still challenging to explain the complexity and randomness of gas–solid and other multiphase flow systems. Therefore, it is crucial to acquire a large amount of data through experiments and in actual production processes, study the phenomena using statistical methods, and establish models. Modeling unknown functions of unrelated variables using machine learning techniques is an effective application of electrostatic sensor modeling. Deep learning algorithms can effectively model variables such as signals, concentration, and particle velocity. Furthermore, deep learning [13–15] offers valuable characteristics that allow for efficient learning and processing of complex relationships and non-linear features in data, providing efficient and accurate modeling and prediction capabilities.

In the field of measurement research on gas–solid two-phase flow, Yan et al. have made effective explorations in applying machine learning algorithms to optimize models and improve measurement accuracy [16]. Despite this progress, there has been relatively little research on using deep learning methods to determine parameters in gas–solid two-phase flow.

Nevertheless, deep learning models demonstrate sufficient flexibility to promptly respond to and update based on changes in data in order to adapt to dynamic system variations. This characteristic provides robust support for industrial production optimization, environmental protection, and process safety. Despite the limited research on deep learning for determining parameters in gas–solid two-phase flow, it is foreseeable that deep learning will emerge as a pivotal method in future studies, offering new perspectives and opportunities for addressing related issues.

2. Materials and Methods

Deep learning is a sub-field of artificial intelligence that focuses on developing algorithms and models capable of learning and making predictions or decisions without explicit programming. It involves studying statistical models and algorithms that enable computers to automatically analyze and interpret complex patterns and relationships in data.

At the core of machine learning is the construction of computational models that can learn from data and make predictions or decisions based on that knowledge. These models are trained using large datasets, which consist of input data and corresponding desired outputs or outcomes. During the training process, the models learn to recognize patterns, extract meaningful features, and generalize from the data to make predictions or decisions on new, unseen data.

In the context of complex gas–solid two-phase flow data, this study utilized several models for predicting particle concentration, including RNNs, LSTM, and BPNNs.

2.1. BPNN

The main steps of training a BPNN include parameter initialization, forward propagation, loss computation, back propagation, and parameter updating [17–19]. During forward propagation, input samples are processed through the network to obtain output results. The loss function is then calculated to measure the discrepancy between the output and the target. Subsequently, back propagation is performed to compute the gradients layer by layer and update the weights and biases. This iterative process continuously adjusts the parameters to make the network output approach the target values.

2.1.1. Forward Propagation

During forward propagation, input data are transmitted from the input layer to the output layer. At each layer, the input is multiplied by weights and passed through an activation function to produce the output. The process can be described using the following equations:

$$Z_i = \sum_{i=1}^{n}(W_i A_{i-1} + b_i) \tag{1}$$

$$y_i = f(Z_i) \tag{2}$$

In these equations, Z_i represents the weighted sum of inputs at layer i, W_i represents the weighted matrix connecting layer $i-1$ to layer i, A_{i-1} represents the output of layer $i-1$, b_i represents the bias vector at layer i, $f()$ represents the activation function, and A_i represents the output of layer i.

2.1.2. Backward Propagation

Backward propagation is used to calculate the gradients of the parameters (weights and biases) with respect to the loss function. This allows the neural network to update its parameters and improve its performance. The gradients are calculated using the chain rule of differentiation.

For example, let us consider the output layer. Assuming the activation function is $f()$, the loss function is L, the input to the output layer is Z, and the output is A. The gradient of the loss function with respect to the output can be calculated as:

$$\frac{\partial L}{\partial Z} = \frac{\partial L}{\partial a} f'(Z) \tag{3}$$

Here, f' represents the derivative of the activation function. Using these gradients, the weights and biases can be updated according to the following formulas:

$$w_l = w_l - a\frac{\partial L}{\partial w_l} \tag{4}$$

$$b_l = b_l - a\frac{\partial L}{\partial b_l} \tag{5}$$

In these formulas, a represents the learning rate of the neural network, $\partial L/\partial w_l$ represents the gradient of the loss function with respect to the weights, and $\partial L/\partial b_l$ represents the gradient of the loss function with respect to the biases.

The common activation functions and their expressions are as follows:

$$f_{Sigmoid} = \frac{1}{1+e^{-x}} \tag{6}$$

The Sigmoid function can map real numbers to between 0 and 1 in the input and is usually used for binary classification problems.

$$f_{ReLU} = Max(0, x) \tag{7}$$

The ReLU function returns the input itself for non-negative inputs and returns 0 for negative inputs. ReLU is widely used in deep learning because it can accelerate training and reduce the risk of overfitting.

$$f_{Tanh} = \frac{e^x - e^{-x}}{e^x + e^{-x}} \tag{8}$$

The Tanh function is similar to the Sigmoid function but maps the input to between -1 and 1 and is usually used for multi-classification problems.

These activation functions introduce non-linearity into the neural network, allowing it to learn complex patterns and make predictions. The choice of activation function depends on the nature of the problem and the characteristics of the data.

2.1.3. Advantages and Disadvantages of BPNNs

BPNNs have advantages in handling non-linear relationships, flexibility, and adaptability. By optimizing model parameters, it effectively reduces the loss function and improves performance. However, BPNNs have the drawbacks of long training time, susceptibility to local minimums, and sensitivity to outliers, requiring data preprocessing and parameter tuning to avoid overfitting. When dealing with gas–solid two-phase flow data, additional parameter tuning and preprocessing may be necessary to enhance stability and accuracy.

2.2. RNNs and LSTMs

RNNs and LSTMs are neural network architectures used for processing sequential data [20]. RNNs have recurrent connections that allow information to be passed and shared within a sequence, capturing temporal dependencies and contextual information. However, traditional RNNs suffer from the issues of vanishing and exploding gradients. To address this problem, LSTM was introduced, which incorporates gate mechanisms to selectively update, retain, or discard information, mitigating the gradient problem and better capturing long-term dependencies. Hence, LSTM can be regarded as an enhanced version of RNN designed to improve the handling of long sequential data.

2.2.1. RNNs

An RNN (Recurrent Neural Network) is a type of recursive neural network used for processing sequential data. Its principle can be represented by the following equation:

$$h_t = f(W_{hh} + W_{xh}x_t + b_h) \tag{9}$$

Here, h_t represents the hidden state at time step t, x_t represents an element of the input sequence, W_{hh} is the weight matrix from hidden state to hidden state, W_{xh} is the weight matrix from input to hidden state, b_h is the bias vector, and f is the activation function.

2.2.2. LSTM

LSTM (Long Short-Term Memory) is a variant of Recurrent Neural Networks (RNNs) that effectively handles long-term dependencies. It achieves this by incorporating gate mechanisms to control the flow of information, which primarily consists of input gate, forget gate, and output gate [21,22], as follows:

1. Forget gate.

The forget gate determines which information from the previous time step's memory state should be forgotten. It is calculated using the following formula:

$$g_f = \delta(w_f[h_{t-1}, x_t] + b_f) \tag{10}$$

Here, w_f and b_f are the parameters of the forget gate, and δ represents the Sigmoid activation function; h_{t-1} refers to the previous time step's hidden state, and x_t represents the current input.

2. Input gate.

The input gate determines which information from the current time step should be updated into the memory state. It is calculated using the following formula:

$$i_t = \delta(w_i[h_{t-1}, x_t] + b_i) \tag{11}$$

$$g_t = \tanh(w_g[h_{t-1}, x_t] + b_g) \tag{12}$$

Here, i_t denotes the output of the input gate, g_t represents the candidate memory value at the current time step, and w_i, w_g, b_i, and b_g are the weights and biases of the input gate.

3. Updating memory state (cell state).

The previous memory state, c_{t-1}, is updated based on the outputs of the forget gate and input gate. The computation formula is as follows:

$$c_t = \delta(w_f[h_{t-1}, x_t] + b_f) \tag{13}$$

Here, c_t represents the current memory state.

4. Output gate.

The output gate determines which information from the current hidden state should be output to the next time step or externally. It is calculated using the following formula:

$$o_t = \delta(w_o[h_{t-1}, x_t] + b_o) \tag{14}$$

$$h_t = o_t \tanh(c_t) \tag{15}$$

Here, O_t denotes the output of the output gate, and h_t represents the current hidden state.

At each time step, LSTM utilizes the input, previous hidden state, and memory state to update the memory state and hidden state through the computation of the forget gate, input gate, and output gate. This enables LSTM to model and retain information from sequential data.

2.2.3. Advantages and Disadvantages of RNNs and LSTM

RNNs can effectively capture temporal dependencies in the time-series data of gas–solid two-phase flow, enabling improved prediction and analysis. Its strength lies in its ability to capture sequential time-dependent information, making it suitable for handling time-series data in gas–solid two-phase flow. However, it suffers from the challenge of vanishing or exploding gradients when dealing with long sequences. Consequently, its performance may be limited when handling extremely long sequences.

On the other hand, LSTM's memory units allow for selective retention and forgetting of information, making it a suitable choice for addressing long-term dependencies in gas–solid two-phase flow data. It overcomes the issue of long-term dependencies in RNNs and handles temporal data more effectively in gas–solid two-phase flow. Compared to traditional RNNs, it performs better in handling long sequences and long-term dependencies. Nevertheless, it may require increased computational resources and training time.

3. Data Collection and Processing

3.1. Data Collection

The experiment required the collection of data such as temperature, humidity, pressure difference, and velocity during the running process of the particles. The experimental equipment utilized a gas–solid two-phase flow detection device provided by the laboratory to complete the research work. The experimental platform is shown in Figure 1.

The experimental platform equipment included a separator, a receiving bin, a feeding bin, a blower, a power supply unit, and a digital multimeter, as well as a ring-shaped electrostatic sensor, temperature and humidity sensors, and a pressure difference sensor located near the annular electrostatic sensor. The experiment used fly ash particles with particle sizes ranging from 0.1 mm to 0.9 mm.

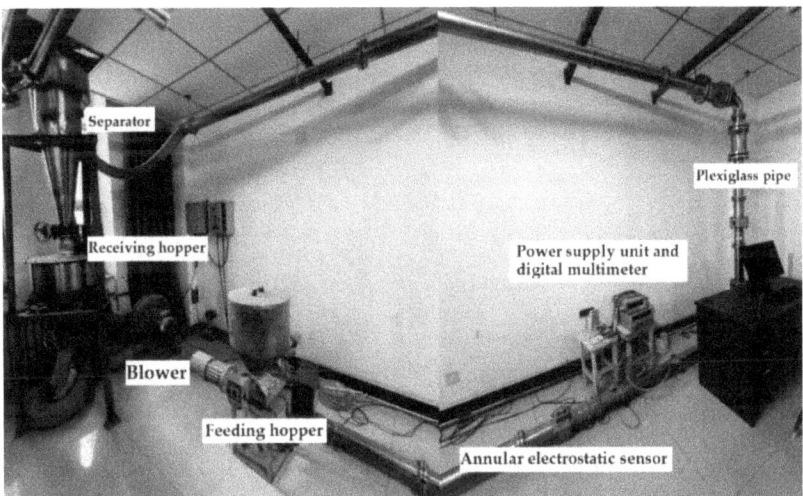

Figure 1. Experimental platform for gas–solid two-phase flow.

The following equipment was used for measurement and monitoring in this experimental platform:

1. KZWSRS485 temperature and humidity transmitter.

The Kunlun Zhongda Company in Beijing, China manufactures this transmitter, which comprises a sensor, signal-processing circuit, and communication interface. It converts the measured temperature and humidity data into standard electrical signal output. The temperature measurement range is from 0 °C to 50 °C, and the humidity measurement range is from 0%RH to 100%RH. The KZWSRS485 temperature and humidity transmitter have high accuracy, reliability, and stability and can operate within a wide range of temperatures and humidity.

2. KZY-808BGA pressure difference sensor.

The sensor, produced by Kunlun Zhongda Company in Beijing, China, is designed for measuring air pressure differences and converting them into corresponding electrical signal outputs. It offers a measurement range from 0 to 3 KPa.

3. Electrostatic sensor.

Two inductive ring-shaped electrostatic sensors and a metal shield were used in the experiment. The sensor electrodes were made of highly sensitive stainless steel material in a ring-shaped structure, providing good wear resistance. To ensure the stability of the signal output, the electrostatic sensor was equipped with a metal shield to reduce the influence of external electromagnetic interference.

4. APS3005S-3D power supply unit.

This device is manufactured by Shenzhen AntaiXin Technology Co., Ltd., located in Shenzhen, China. It provides stable voltage and current outputs for sensors, ensuring the normal operation of the sensors.This device provided stable voltage and current output to the sensors, ensuring their normal operation.

5. GDM-842 digital multimeter.

The manufacturer of the GDM-842 Digital Multimeter is GW Instek (Good Will Instrument Co., Ltd.), located in Taiwan, China. This multimeter is used to measure voltage signals outputted by sensors, ensuring accurate measurement results.

The use of these devices in the experimental platform aimed at achieving precise measurements and data acquisition of environmental conditions, providing reliable data support for the experiment.

Due to the randomness and complexity of fluid motion, the charge signals sensed by electrostatic sensors often exhibit instability. Processing the charge signal itself is relatively challenging. Therefore, the experiments were designed to consider the influence of the sensor's impedance and signal bandwidth, and corresponding conditioning circuits were developed. Through the conditioning circuit, the charge signal was able to be converted into a stable and measurable voltage signal.

The data acquisition system uses an FPGA (Intel's EP4CE40F) as the main control chip, which is produced by Shenzhen Gongshen Electronic Technology Co., Ltd. located in Shenzhen, China. It employs the AD7606 as the ADC with a precision of 16 bits and a sampling frequency set at 10^4 Hz. In addition, to achieve data storage, a NAND flash memory was added, utilizing the H27U1G8F2B chip from Micron Technology Inc., located in Boise, ID, USA. The CY60813A chip from Cypress Semiconductor Corporation in San Jose, CA, USA, was utilized as the core chip for USB transmission. This chip played a crucial role in enabling seamless communication with the PC. The data collection process is illustrated in Figure 2.

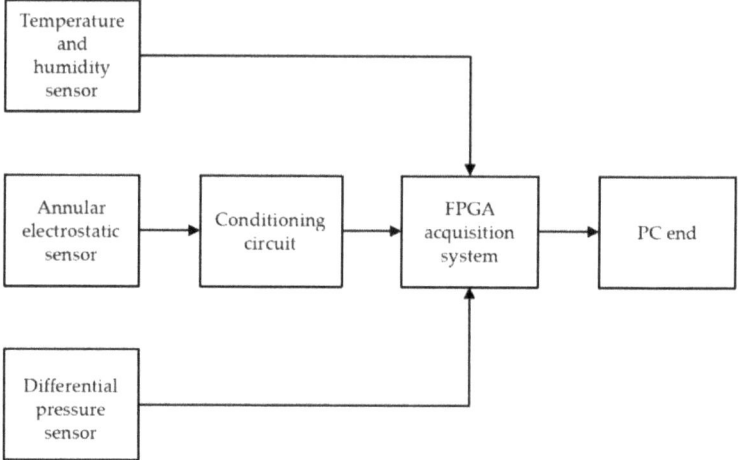

Figure 2. Data collection process diagram.

The experiment utilized a feeder and a blower to control the mass flow rate and velocity of particles, obtaining data on gas–solid two-phase flow at different mass flow rates and velocities within a range of 30–50 m/s for velocity and 33.5–54.5 g/s for mass flow rate.

3.2. Calculation of Particle Velocity and Particle Concentration

The experiment involved the processing of voltage output from two electrostatic sensors. After running the system for a period of time, it was observed that the signal outputs between these two sensors exhibited notable similarities. This trend can be clearly observed in Figure 3.

The experiment measured the difference in the index numbers corresponding to the peak values of signals from the upstream and downstream sensors as the particle delay. Considering the known constant values of the sampling frequency and the distance between the two electrostatic sensors, this delay value can be used as an approximate particle velocity.

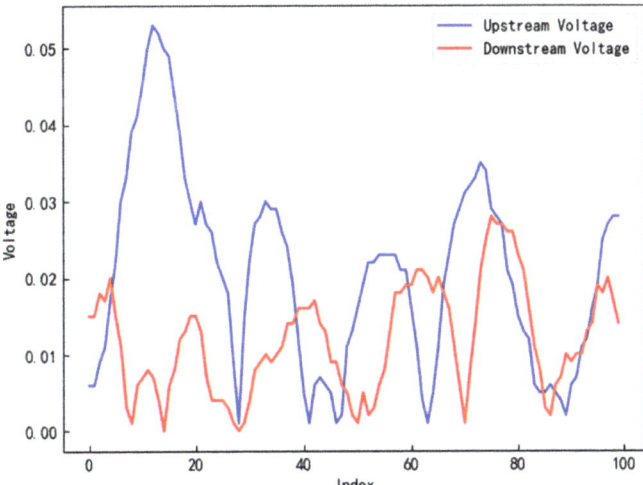

Figure 3. Comparison of signal outputs between upstream and downstream sensors.

After obtaining the velocity, the particle concentration can be calculated using the following formula:

$$C = \frac{q}{AV} \qquad (16)$$

where C is the particle concentration, q is the mass flow rate, V is the particle velocity, and A is the cross-sectional area of the particle transport pipeline.

By following the aforementioned process, the experiment obtained complete data for temperature, pressure difference, humidity, signals from upstream and downstream sensors, velocity, and particle concentration.

3.3. Normalization of Model Data

Data normalization is typically performed in deep learning models to ensure that the input data have similar value ranges across different features. This helps in accelerating model convergence and improving stability. Normalization prevents issues such as feature bias and excessive gradient changes, allowing the model to treat all features fairly and enhance training effectiveness and reliability. It is calculated using the following formula:

$$x' = \frac{x - x_{min}}{x_{max} - x_{min}} \qquad (17)$$

where, x_{min} refers to the minimum value of the original data, and x_{max} refers to the maximum value of the original data. x refers to the value before normalization, and x' refers to the value after normalization.

4. Results

4.1. Experimental Preparation and Environment Configuration

The experiment utilized a two-phase gas–solid flow dataset consisting of seven columns of data, including temperature, humidity, upstream sensor signal, downstream sensor signal, delay, pressure differential, and particle concentration. The training set consisted of 25,500 samples with varying velocities and particle concentrations. The test set was extracted from the remaining samples and contained 5100 samples. The training set was used for model training and parameter optimization, while the test set was used to evaluate the model's predictive performance.

To establish the experimental environment, Jupyter Lab (4.0.3) was chosen as the computational environment, which is a scientific computing tool implemented in an interactive manner. For the selection of the main programming language, Python (3.9) was adopted, which is a high-level programming language widely used in the field of machine learning.

To enhance the reliability of project management and the independence of experiments, the Anaconda platform was utilized. With Anaconda, it is possible to create isolated virtual environments to ensure the isolation of dependencies between different projects. Within the virtual environment, TensorFlow was installed as the machine learning library, which is an open-source framework developed by Google. Along with TensorFlow, other essential machine learning libraries and tools such as Keras, PyTorch, and scikit-learn experiments were installed to enhance the functionality and flexibility of the experiments.

4.2. Model Evaluation Metrics

This experiment set up three evaluation metrics: prediction accuracy (A_F), mean squared error (E_{MSE}), root-mean-square error (E_{RMSE}), and mean absolute error (E_{MAE}). The specific expressions are:

$$A_F = \frac{1}{n}\sum 1 - \frac{|y-y'|}{y} \tag{18}$$

$$E_{RMSE} = \sqrt{\frac{1}{n}\sum (y-y')^2} \tag{19}$$

$$E_{MSE} = \frac{1}{n}\sum (y-y')^2 \tag{20}$$

$$E_{MAE} = \frac{1}{n}\sum |y-y'| \tag{21}$$

where n represents the total number of samples, and y and y' represent the actual value and predicted value, respectively.

4.3. Model Construction and Parameter Determination

In order to construct a model for predicting particle concentration, the present experiment used temperature, humidity, upstream and downstream sensor signals, velocity, and pressure difference as input parameters and the predicted particle concentration as the output of the model. The experiment employed BPNN, RNN, and LSTM models for modeling. Before beginning the modeling process, it was necessary to determine the number of hidden layers and the number of nodes in the hidden layers.

The decision to utilize three hidden layers in the experimental setup was driven by the findings of preliminary experiments, which suggested that a model with three hidden layers exhibits superior capability in capturing the intricate features within the dataset. Conversely, the potential drawbacks associated with introducing a fourth hidden layer, such as an increased risk of overfitting, heightened model complexity without substantial performance enhancements, and escalated computational expenses, were taken into consideration. The primary focus of the experiment was to explore the impact of varying node quantities within the three hidden layers. Starting with 40 nodes, the quantity was incrementally increased to 80 in increments of 10 nodes. Evaluation of each parameter combination was conducted to pinpoint the combination of node quantities that yielded the most favorable results. Throughout the experiment, E_{MSE} was employed as the designated loss function.

Following the completion of training, the model's performance was evaluated using the test set. The prediction accuracy of models with different parameter settings was quantified by calculating the loss function. The experimental results are shown in Figures 4–6.

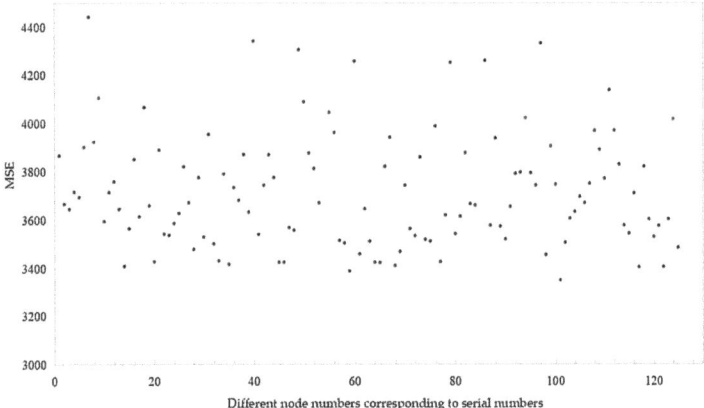

Figure 4. E_{MSE} comparison of different LSTM network structures.

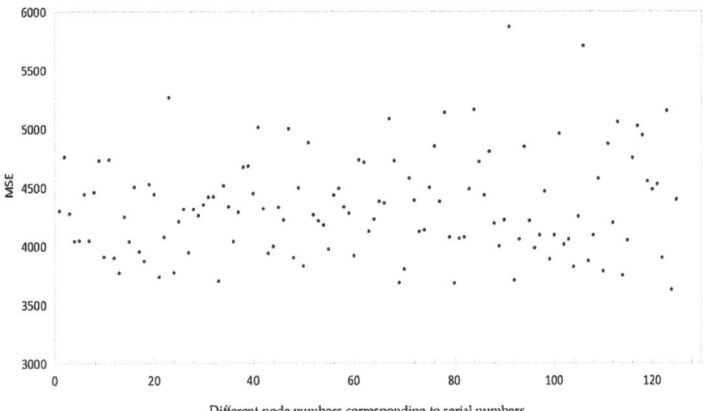

Figure 5. E_{MSE} comparison of different RNN structures.

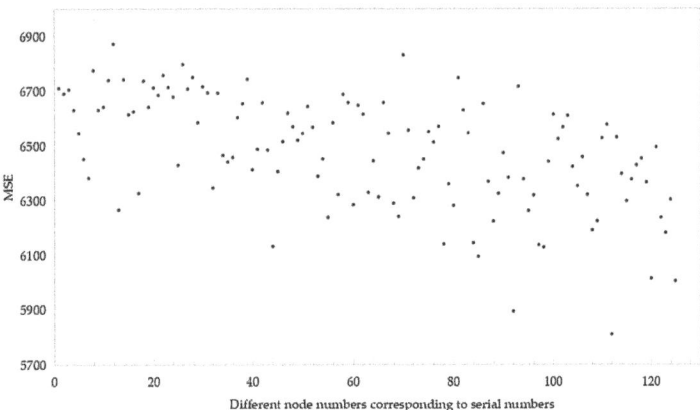

Figure 6. E_{MSE} comparison of different BPNN structures.

Based on the results presented in Figures 4–6, it is evident that the BPNN demonstrated superior predictive performance when configured with three hidden layers and 80, 60, and 50 memory units in each respective layer. Similarly, the LSTM model delivered optimal results with three hidden layers and 80, 80, and 70 memory units, while the RNN model exhibited the best predictive performance when utilizing three hidden layers with 80, 40, and 40 memory units. The decision to set the maximum iteration count to 500 was informed by experimental data indicating that errors across different models converged to a stable level after 500 iterations. The parameters determined for the BPNN, LSTM, and RNN models are presented in Table 1.

Table 1. The key parameters of the model.

Parameter	BPNN	RNN	LSTM
Number of Hidden Layers	3	3	3
Number of Units per Layer	80, 60, 50	80, 40, 40	80, 80, 70
Maximum Number of Iterations	500	500	500
Training Batch Size	32	32	32

4.4. Results

When handling gas–solid two-phase flow data, the present experiment compared the predictive performance of different methods and used three evaluation metrics for comparison, including A_F, E_{RMSE}, and E_{MAE}. Table 2 shows the prediction results of the different models. Figure 7 presents a comparison between the predicted values and actual values for the different models.

Table 2. Comparison of prediction results for different models.

Prediction Model	A_F	R_{MSE}	R_{MAE}
BPNN	92.5	52.22	37.45
RNN	92.4	53.19	38.88
LSTM	92.7	53.43	38.70

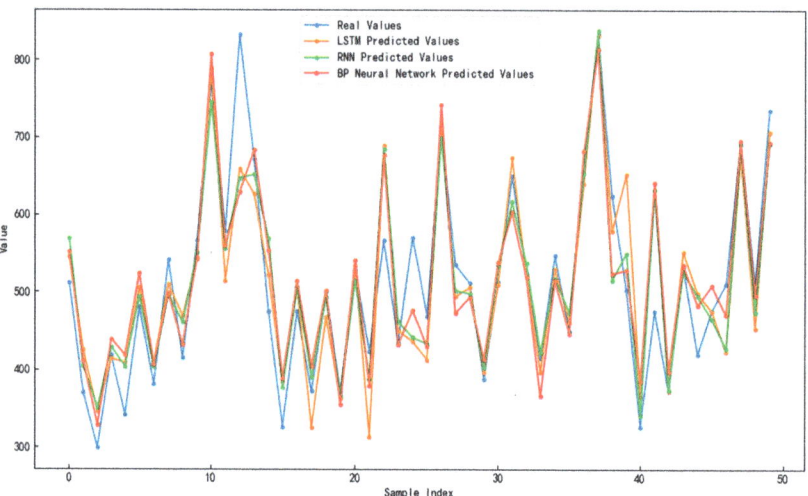

Figure 7. Comparison of predicted values and actual values for different models.

After obtaining the predicted values and actual values of the different models' test sets, the results were divided into different intervals based on the relative error. By evaluating the proportion of low-error-interval samples, the performance of the models was able to be quantitatively assessed. A high proportion of low-error-interval samples indicated that the model could accurately predict the majority of samples, demonstrating a high level of precision. The proportions of samples within different relative error intervals for each model can be observed in Figure 8.

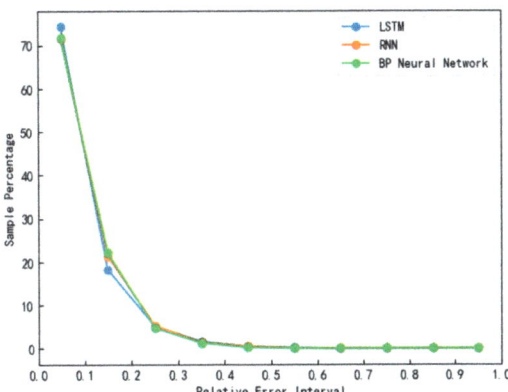

Figure 8. Relative error distribution plots of different models.

Based on the analysis of experimental results, it was evident that there were differences in the performance of the LSTM, RNN, and BPNN models in predicting the relative error of particulate matter concentration in gas–solid two-phase flow. In the relative error range of [0.0, 0.1], the LSTM model demonstrated strong performance, with an accuracy of 74.27%, which was slightly higher than the RNN model (71.57%) and the BPNN model (71.61%). Regarding the overall prediction performance indicators, the three models showed similar values in prediction accuracy, root-mean-square error, and mean absolute error, with no significant advantages observed. The LSTM model exhibited superior memory and long-term dependency capabilities compared to the BPNN and RNN, making it suitable for handling time-series data and long-distance dependency relationships, thereby providing the LSTM model with stronger modeling capabilities for gas–solid two-phase flow data processing. Although the LSTM model required a longer training time and computational resources, its training efficiency and convergence speed in handling complex gas–solid two-phase flow sequence data were higher compared to the BPNN and RNN, as observed in Figures 3–5, where, under the same number of iterations (100) and parameter settings, the LSTM's loss function was relatively smaller than those of the BPNN and RNN.

In summary, according to the experimental results, the LSTM model performs well in the relative error range of [0.0, 0.1], which is possibly attributable to its superior memory and long-term dependency capabilities, as well as its modeling capabilities in handling gas–solid two-phase flow sequence data. However, there were no significant differences observed among the three models in terms of overall prediction performance indicators.

5. Discussion

This experimental study compared the performance of LSTM, RNN, and BPNN models in predicting the concentration of particulate matter in gas–solid two-phase flow. The results showed that within the relative error range of [0.0, 0.1], the LSTM model exhibited the best performance, validating its advantages in handling time-series data and capturing long-term dependencies. Although the overall prediction performance indicators of the three models were similar and showed no significant advantages, they all demonstrated certain predictive capabilities and adaptability. This may be due to the similar challenges and limitations they face in handling the concentration of particulate matter in gas–solid two-phase flow. The complexity and dynamics of gas–solid two-phase flow may lead to noise and uncertainty in the data, posing similar challenges for all models. Additionally, the prediction of particulate matter concentration is influenced by various factors, such as gas flow velocity, particle size distribution, and pipeline or equipment structure, which may have similar impacts on the prediction performance of different models.

Furthermore, these models possess similar modeling and expressive capabilities, leading to similar predictive performance when processing gas–solid two-phase flow data. The quantity and quality of the data may have similar effects on the prediction performance of the three models, resulting in relatively close predictive effects if the dataset features are challenging to all of the models to some extent. Therefore, despite potential differences in handling time-series data and capturing long-term dependencies, these models demonstrate similar predictive effects in specific tasks of predicting the concentration of particulate matter in gas–solid two-phase flow.

To enhance the performance of these models, it is recommendable to conduct further experiments with different parameter settings and optimization strategies, feature selections and engineering, and model integration methods. Future research may involve evaluating other machine learning models (such as CNNs and self-attention mechanism models), improving feature selection and engineering methods, and exploring the interpretability of the models. Interdisciplinary cooperation is crucial to integrate fluid mechanics and deep learning for in-depth research to seek more reliable and accurate data processing models for gas–solid two-phase flow.

Author Contributions: Conceptualization, H.W.; Methodology, Z.W.; Software, Z.W.; Resources, B.Y.; Data curation, B.Y.; Writing—original draft, Z.W.; Writing—review & editing, B.Y.; Supervision, H.W. All authors have read and agreed to the published version of the manuscript.

Funding: This work was supported by the Open Research Fund of the Shanxi Key Laboratory of Signal Capturing and Processing.

Data Availability Statement: Data are contained within the article.

Conflicts of Interest: The authors declare no conflict of interest.

References

1. Wang, Y.; Lyu, X.; Li, W.; Yao, G.; Bai, J.; Bao, A. Investigation on Measurement of Size and Concentration of Solid Phase Particles in Gas-Solid Two Phase Flow. *Chin. J. Electron.* **2018**, *27*, 381–385. [CrossRef]
2. Datta, U.; Dyakowski, T.; Mylvaganam, S. Estimation of Particulate Velocity Components in Pneumatic Transport Using Pixel Based Correlation with Dual Plane ECT. *Chem. Eng. J.* **2007**, *130*, 87–99. [CrossRef]
3. Coghill, P.J. Particle Size of Pneumatically Conveyed Powders Measured Using Impact Duration. *Part. Part. Syst. Charact.* **2007**, *24*, 464–469. [CrossRef]
4. Yan, F.; Rinoshika, A. Application of High-Speed PIV and Image Processing to Measuring Particle Velocity and Concentration in a Horizontal Pneumatic Conveying with Dune Model. *Powder Technol.* **2011**, *208*, 158–165. [CrossRef]
5. Baer, C.; Jaeschke, T.; Mertmann, P.; Pohl, N.; Musch, T. A mmWave Measuring Procedure for Mass Flow Monitoring of Pneumatic Conveyed Bulk Materials. *IEEE Sens. J.* **2014**, *14*, 3201–3209. [CrossRef]
6. Qian, X.; Huang, X.; Yonghui, H.; Yan, Y. Pulverized Coal Flow Metering on a Full-Scale Power Plant Using Electrostatic Sensor Arrays. *Flow Meas. Instrum.* **2014**, *40*, 185–191. [CrossRef]
7. Wang, C.; Zhang, J.; Zheng, W.; Gao, W.; Jia, L. Signal Decoupling and Analysis from Inner Flush-Mounted Electrostatic Sensor for Detecting Pneumatic Conveying Particles. *Powder Technol.* **2017**, *305*, 197–205. [CrossRef]
8. Zhang, W.; Wang, C.; Wang, H. Hilbert–Huang Transform-Based Electrostatic Signal Analysis of Ring-Shape Electrodes With Different Widths. *IEEE Trans. Instrum. Meas.* **2012**, *61*, 1209–1217. [CrossRef]
9. Zhang, J.; Cheng, R.; Yan, B.; Abdalla, M. Improvement of Spatial Sensitivity of an Electrostatic Sensor for Particle Flow Measurement. *Flow Meas. Instrum.* **2020**, *72*, 101713. [CrossRef]
10. Zhang, J.; Cheng, R.; Yan, B.; Abdalla, M. Reweighting Signal Spectra to Improve Spatial Sensitivity for an Electrostatic Sensor. *Sensors* **2019**, *19*, 4963. [CrossRef] [PubMed]
11. Shao, J.; Yan, Y.; Lv, Z. On-Line Non-Intrusive Measurements of the Velocity and Particle Size Distribution of Pulverised Fuel on a Full Scale Power Plant. In Proceedings of the 2011 IEEE International Instrumentation and Measurement Technology Conference, Hangzhou, China, 10–12 May 2011; pp. 1–5.
12. Wang, C.; Zhan, N.; Zhang, J. Induced and Transferred Charge Signals Decoupling Based on Discrete Wavelet Transform for Dilute Gas-Solid Two-Phase Flow Measurement. In Proceedings of the 2017 IEEE International Instrumentation and Measurement Technology Conference (I2MTC), Turin, Italy, 22–25 May 2017; pp. 1–6.

13. LeCun, Y.; Bengio, Y.; Hinton, G.E. Deep Learning. *Nature* **2015**, *521*, 436–444. [CrossRef]
14. Yu, K.; Jin, K.; Deng, X. Review of Deep Reinforcement Learning. In Proceedings of the 2022 IEEE 5th Advanced Information Management, Communicates, Electronic and Automation Control Conference (IMCEC), Chongqing, China, 16–18 December 2022; Volume 5, pp. 41–48.
15. Kawaguchi, K.; Kaelbling, L.P.; Bengio, Y. *Generalization in Deep Learning*; Cambridge University Press: Cambridge, UK, 2022; pp. 112–148.
16. Yan, B.; Zhang, J.; Cheng, R.; Liu, C. Modelling of A Flow Meter through Machine Learning. In Proceedings of the 2020 IEEE SENSORS, Rotterdam, The Netherlands, 25–28 October 2020; pp. 1–4. [CrossRef]
17. Basheer, I.A.; Hajmeer, M. Artificial Neural Networks: Fundamentals, Computing, Design, and Application. *J. Microbiol. Methods* **2000**, *43*, 3–31. [CrossRef]
18. Wang, S.; Zhang, N.; Wu, L.; Wang, Y. Wind Speed Forecasting Based on the Hybrid Ensemble Empirical Mode Decomposition and GA-BP Neural Network Method. *Renew. Energy* **2016**, *94*, 629–636. [CrossRef]
19. Sutskever, I.; Vinyals, O.; Le, Q.V. Sequence to Sequence Learning with Neural Networks. In Proceedings of the Advances in Neural Information Processing Systems 27: Annual Conference on Neural Information Processing Systems 2014, Montreal, QC, Canada, 8–13 December 2014; Ghahramani, Z., Welling, M., Cortes, C., Lawrence, N.D., Weinberger, K.Q., Eds.; Curran Associates, Inc.: Red Hook, NY, USA, 2014; pp. 3104–3112.
20. Yu, Y.; Si, X.; Hu, C.; Zhang, J. A Review of Recurrent Neural Networks: LSTM Cells and Network Architectures. *Neural Comput.* **2019**, *31*, 1235–1270. [CrossRef] [PubMed]
21. Gers, F.A.; Schmidhuber, J.; Cummins, F. Learning to Forget: Continual Prediction with LSTM. In Proceedings of the 1999 Ninth International Conference on Artificial Neural Networks ICANN 99. (Conf. Publ. No. 470), Edinburgh, UK, 7–10 September 1999; Volume 2, pp. 850–855.
22. Zhang, Y.; Tang, J.; Liao, R.; Zhang, M.; Zhang, Y.; Wang, X.; Su, Z. Application of an Enhanced BP Neural Network Model with Water Cycle Algorithm on Landslide Prediction. *Stoch. Environ. Res. Risk Assess.* **2021**, *35*, 1273–1291. [CrossRef]

Disclaimer/Publisher's Note: The statements, opinions and data contained in all publications are solely those of the individual author(s) and contributor(s) and not of MDPI and/or the editor(s). MDPI and/or the editor(s) disclaim responsibility for any injury to people or property resulting from any ideas, methods, instructions or products referred to in the content.

Article

Pressure Drop Estimation of Two-Phase Adiabatic Flows in Smooth Tubes: Development of Machine Learning-Based Pipelines

Farshad Bolourchifard [1], Keivan Ardam [1], Farzad Dadras Javan [1], Behzad Najafi [1], Paloma Vega Penichet Domecq [2], Fabio Rinaldi [1] and Luigi Pietro Maria Colombo [1,*]

[1] Dipartimento di Energia, Politecnico di Milano, Via Lambruschini 4, 20156 Milano, Italy; farshad.bolourchifard@mail.polimi.it (F.B.); keivan.ardam@mail.polimi.it (K.A.); farzad.dadras@polimi.it (F.D.J.); behzad.najafi@polimi.it (B.N.); fabio.rinaldi@polimi.it (F.R.)

[2] Escuela Técnica Superior de Ingenieros Industriales, Universidad Politecnica de Mardrid, c/Jose Gutiérrez Abascal 2, 28006 Madrid, Spain; paloma.vega-penichetd@alumnos.upm.es

* Correspondence: luigi.colombo@polimi.it

Abstract: The current study begins with an experimental investigation focused on measuring the pressure drop of a water–air mixture under different flow conditions in a setup consisting of horizontal smooth tubes. Machine learning (ML)-based pipelines are then implemented to provide estimations of the pressure drop values employing obtained dimensionless features. Subsequently, a feature selection methodology is employed to identify the key features, facilitating the interpretation of the underlying physical phenomena and enhancing model accuracy. In the next step, utilizing a genetic algorithm-based optimization approach, the preeminent machine learning algorithm, along with its associated optimal tuning parameters, is determined. Ultimately, the results of the optimal pipeline provide a Mean Absolute Percentage Error (MAPE) of 5.99% on the validation set and 7.03% on the test. As the employed dataset and the obtained optimal models will be opened to public access, the present approach provides superior reproducibility and user-friendliness in contrast to existing physical models reported in the literature, while achieving significantly higher accuracy.

Keywords: pressure drop; two-phase flow; machine learning; feature selection; pipeline optimization

1. Introduction

Gas and liquid multiphase systems are commonly encountered in pipe flows within the petrochemical, energy, and healthcare industries, where accurate estimation of pressure drop is essential. In applications such as oil and gas transportation [1–3], pipelines handle mixtures of liquids and gases, necessitating precise pressure drop calculations to ensure efficient and cost-effective transportation. Similarly, in cryogenic applications like liquefied natural gas (LNG) transport, understanding pressure drops helps in maintaining optimal flow rates and preventing equipment failure due to excessive pressure differentials, especially during loading and unloading operations [4]. These calculations not only influence the design and operation of systems but also impact safety and reliability, making them essential for achieving desired performance levels without compromising operational integrity. However, estimating the frictional pressure drop of two-phase flow is considered to be a complex problem [5], as the corresponding governing physical phenomena depend on several parameters including the ones corresponding to the flow conditions, properties of the fluids involved, and the specific geometry of the system. This field has been the subject of extensive research since the 20th century. Numerous investigations have been undertaken in this field, leading to the development of multiple empirical correlations suitable for various applications [6]. This research area has specifically received notable attention since the accurate estimation of energy conversion plants, which include two-phase processes, requires an accurate prediction of two-phase flow pressure drop.

Citation: Bolourchifard, F.; Ardam, K.; Dadras Javan, F.; Najafi, B.; Vega Penichet Domecq, P.; Rinaldi, F.; Colombo, L.P.M. Pressure Drop Estimation of Two-Phase Adiabatic Flows in Smooth Tubes: Development of Machine Learning-Based Pipelines. *Fluids* **2024**, *9*, 181. https://doi.org/10.3390/fluids9080181

Academic Editors: D. Andrew S. Rees, Hemant J. Sagar and Nguyen Van-Tu

Received: 13 June 2024
Revised: 30 July 2024
Accepted: 3 August 2024
Published: 11 August 2024

Copyright: © 2024 by the authors. Licensee MDPI, Basel, Switzerland. This article is an open access article distributed under the terms and conditions of the Creative Commons Attribution (CC BY) license (https://creativecommons.org/licenses/by/4.0/).

The literature commonly utilizes two approaches to determine the frictional pressure gradient of two-phase flow in pipes: the homogeneous and separated flow models. These models are chosen for their independence from the specific flow pattern. The first methodology considers the two-phase flow as if it were a single-phase flow, with the physical properties determined through an appropriate weighting of the properties associated with the individual phases. The second approach is instead based on the assumption that the two-phase pressure gradient is correlated with the pressure gradient of each individual phase, which is considered separately. In the field of separated flow models (liquid or gas), the groundbreaking work was conducted by Lockhart and Martinelli [7] with experimental data corresponding to air and liquids such as water, along with various types of oils and organic fluids. Chisholm [8] subsequently pursued this research, presenting a straightforward model relying on Lockhart–Martinelli charts. Additional enhancement on Chisholm's model was conducted by Mishima and Hibiki [9] (for viscous gas, viscous liquid flows), Zhang et al. [10] (for adiabatic and diabatic viscous flows), and Sun and Mishima [11] (refrigerants, water, CO_2, and air). The Lockhart and Martinelli approach can also be successfully applied to non-straight pipes; for instance, Colombo et al. (2015) introduced adjustments to account for centrifugal forces in steam-water two-phase flow within helical tubes [12], incorporating pressure drop estimates from a previous study for single-phase laminar flow [13]. Different models were proposed by Chisholm [14] (transformation of the Baroczy plots [15]) and MullerSteinhagen and Heck [16] (a solely empirical model based on a very broad dataset). Other studies in this approach include the works conducted by Souza and Pimenta [17] based on data obtained from the experimental activity on refrigerants, Friedel [18] (a large dataset taking into account the effects of surface tension and gravity), Cavallini et al. [19] (condensing refrigerants), and Tran et al. [20] (refrigerants during the phase transition).

Most of the proposed models in the above-mentioned studies are simple empirical models (for the sake of reproducibility and ease of use), which are obtained from the regression of the experimentally acquired data. Such data fitting provides acceptable accuracy for datasets with limited flow conditions; however, often the models cannot be easily extended to include different conditions. Furthermore, while attempting to fit large datasets with a wide range of flow and geometric conditions, the accuracy of these types of models is notably reduced. An alternative methodology to the latter approach is utilizing complex machine learning-based models, which can provide significantly higher accuracy compared to simple empirical ones while being fed with large datasets.

Considering their higher accuracy, machine learning-based models have been employed in several studies for simulating the multi-phase flow phenomena [21,22]. An Artificial Neural Network (ANN) was employed to predict the pressure drop in both horizontal and vertical circular pipes conveying a mixture of oil, water, and air from the production well [2,3]. Artificial Neural Network (ANN) was additionally utilized to predict pressure drop in piping components for the conveyance of non-Newtonian fluids [23], predicting the performance of a parallel flow condenser utilizing air and R134a as working fluids [24], and the pressure drop production of R407C two-phase flow inside horizontal smooth tubes [25]. The integration of the Group Method Data Handling with an Artificial Neural Network (ANN) was employed to forecast frictional pressure drops in mini-channel multi-port tubes for flows involving five different refrigerant fluids [26]. Various machine learning-based approaches were employed to predict the pressure drop associated with the evaporation of R407C [27] and through the condensation process of several fluids in inclined smooth tubes [28]. An Artificial Neural Network has been utilized to predict the pressure drop in horizontal long pipes for two-phase flow [29]. In a reverse approach, by measuring pressure drop and other physical parameters, flow rates of the multiphase flow could be predicted [30]. Shaban and Tavoularis [31] implemented the latter approach for predicting the flow rate of two-phase water/air flows in vertical pipes. A new universal correlation for predicting frictional pressure drop in adiabatic and diabatic flow is developed employing machine learning methods [32]. In a work conducted by Faraji et al. [33], a comparison between various ANN models including six multilayer perceptron (MLP) and one Radial Basis Function (RBF) was performed to assess their performance in estimating the pressure drop in two-phase flows. It was shown that the multilayer perceptron neural network incorporated with the genetic algorithm obtained the best results with a Root Mean Square Error of 0.525 and an Average Absolute Relative Error percentage of 6.722. Consequently, a sensitivity analysis was performed to indicate features with positive and negative impacts on the flow pressure drop. Similarly, Moradkhani et al. [34] investigated the performance of MLP, RBF, and Gaussian process regression (GPR) in developing dimensionless predictive models based on the separated model suggested by Lockhart and Martinelli, where GPR was shown to be the most effective model with a coefficient of determination of 99.23%. Additionally, the order of importance of features was established using a sensitivity analysis. The investigation into the prediction performance of frictional pressure drop in helically coiled tubes at different conditions and orientations was detailed in a work by Moradkhani et al. [35]. Employing the dataset from 64 published papers, including 25 distinct fluids and various operating conditions, Nie et al. [32] attempted to implement the ML methods to develop a universal correlation for predicting the frictional pressure drop in adiabatic and diabatic flow. ANN and extreme gradient boosting (XGBoost) were employed, obtaining an MARD of 8.59%. It was shown that the newly established correlations can offer more reliable predictive accuracy compared to the current correlations for the employed database with a MARD of 24.84%. Ardam et al. [22] investigated the application of machine learning-based pipelines with a focus on feature selection and pipeline optimization in estimating the pressure drop in R134a flow in micro-fin tube setup and obtained a MARD of 18.08% on the test set. Alternatively, the pressure drop of two-phase adiabatic air–water flow was investigated by Najafi et al. [21] in a set-up of horizontal micro-finned tubes.

Research Gap and Contributions of the Present Work

The present study aims to develop optimized ML-based pipelines to estimate pressure drop in a two-phase adiabatic flow of water/air mixture using experimentally obtained data in horizontal smooth copper tubes. Each row of the dataset used to train the pipelines corresponds to an individual experiment performed at various flow conditions. In the next step, the physical phenomena-based models, which have been proposed in the studies available in the literature, have been implemented and the corresponding obtained accuracies are compared. Taking into account the MAPE (Mean Absolute Percentage Error) as the key accuracy index, the best model has been determined. Next, pipelines utilizing machine learning algorithms are implemented to predict pressure drop, with two-phase flow multipliers considered as targets and non-dimensional parameters (chosen according to the physics of the subject) serving as inputs.

Within the existing body of literature, a gap emerges in the domain of pipeline optimization and feature selection specific to the case study at hand. This gap underscores a need for a more refined methodology aimed at optimizing state-of-the-art machine learning (ML) as well as a systematic approach for selecting the features in the investigated setup with smooth tubes. By employing a systematic framework for feature selection, it is possible to identify and eliminate those features that make negligible contributions, thus enhancing the interpretability of results and providing a deeper understanding of the complex dynamics inherent within the system.

Therefore, in order to obtain optimal pipelines, a comprehensive search grid is first implemented to determine the optimal Random Forest as the benchmark algorithm employing all the features. Next, a feature selection procedure has been conducted in order to identify the most promising set of features, resulting in the highest accuracy. In this approach, the features are first ordered based on their ranking according to their correlation with the target. Next, the features are added incrementally, making predictions at each step, and tracking the resulting accuracy. After choosing the initial most promising combination of features, a feature elimination procedure is carried out to evaluate the potential increase in accuracy by adding further features and removing the existing features that do not enhance the accuracy. Following the above process, the features yielding the maximum accuracy are identified. Finally, employing the chosen features as inputs, an optimization approach based on genetic algorithms is utilized to identify the optimal ML pipeline. The optimized pipeline and the utilized dataset are shared as an open-source tool, aiming to improve the accessibility and reproducibility of the proposed models, while ensuring elevated accuracy.

2. Frictional Pressure Gradient in Two-Phase Flow

Owing to the increased complexity of the motion, predicting frictional pressure drop in two-phase flows is more intricate compared to single-phase flows. The phases flowing together show a variety of arrangements, called flow patterns, strongly affecting their mutual interaction, as well as the interaction between each phase and the pipe's wall. For the sake of simplicity, models designated as "flow pattern independent" have been introduced despite their considerable approximation. In this context, two different approaches are possible:

1. **Homogeneous flow model**: The two-phase mixture is modeled as an equivalent single-phase fluid, using properties that are averaged to reflect the characteristics of both phases.
2. **Separated flow model**: The two-phase mixture is presumed to consist of two independent single-phase streams flowing separately. The resulting pressure gradient is determined through an appropriate combination of the pressure gradients from each individual single-phase stream.

The specifics of the two approaches are outlined in the subsequent section.

2.1. Homogeneous Models

The homogeneous approach determines the frictional pressure gradient of a two-phase flow as

$$-\frac{dp_f}{dz} = \frac{2f_{tp}G_{tp}^2}{\rho_b D} \quad (1)$$

$$\rho_b = \rho_g x_v + \rho_l(1 - x_v) = \left(\frac{x}{\rho_g} + \frac{1-x}{\rho_l}\right)^{-1} \quad (2)$$

where f_{tp} is typically assessed as a function of a two-phase Reynolds number, incorporating the definition of an average two-phase dynamic viscosity denoted as μ_{tp}. Table 1 showcases some of the most promising predictive models proposed in the literature [36–38] that have utilized the homogeneous approach.

Table 1. Selected models, proposed in the literature, which have utilized the homogeneous approach.

Author(s) and the Respective Reference(s)	Equation	Eq. No
McAdams et al. [36]	$\frac{1}{\mu_{tp}} = \frac{x}{\mu_g} + \frac{1-x}{\mu_l}$	(3)
Beattie and Whalley [37]	$\mu_{tp} = \mu_l(1-\beta)(1+2.5\beta) + \mu_g\beta$ $\beta = \dfrac{x}{x + (1-x)\dfrac{\rho_g}{\rho_l}}$	(4)
Awad and Muzychka [38]	$\mu_{tp} = \mu_g \dfrac{2\mu_g + \mu_l - 2(\mu_g - \mu_l)(1-x)}{2\mu_g + \mu_l + (\mu_g - \mu_l)(1-x)}$	(5)

2.2. Separated Flow Models

It is frequently convenient to establish a dimensionless factor by referencing the two-phase frictional pressure gradient to a single-phase counterpart. The following conventions can be employed for this purpose:

1. *l*—**liquid alone:** refers to the liquid moving independently, specifically at the liquid superficial velocity.
2. *lo*—**liquid only:** refers to the liquid moving with the total flow rate, specifically at the mixture velocity.
3. *g*—**gas alone:** refers to the gas moving alone, specifically at the gas superficial velocity.
4. *go*—**gas only:** refers to the gas flowing with the total flow rate, specifically at the mixture velocity.

Accordingly, the four distinct, but equivalent, two-phase multiplier factors are defined as follows:

$$\Phi_l^2 = \frac{\frac{dp_f}{dz}}{\left(\frac{dp_f}{dz}\right)_l}, \quad \Phi_{lo}^2 = \frac{\frac{dp_f}{dz}}{\left(\frac{dp_f}{dz}\right)_{lo}}, \quad \Phi_g^2 = \frac{\frac{dp_f}{dz}}{\left(\frac{dp_f}{dz}\right)_g}, \quad \Phi_{go}^2 = \frac{\frac{dp_f}{dz}}{\left(\frac{dp_f}{dz}\right)_{go}} \quad (6)$$

Likewise, Table 2 lists some of the most promising predictive models from the literature [8–11,14,16–20] that have employed the separated flow approach.

Table 2. Selected models that have employed the separated flow approach.

Author(s) and the Respective Reference(s)	Equation	Eq. No
Chisholm [8]	$\Phi_l^2 = 1 + \frac{C}{X} + \frac{1}{X^2}$ $X^2 = \frac{\left(\frac{\Delta p}{\Delta z}\right)_l}{\left(\frac{\Delta p}{\Delta z}\right)_g}$ $C = 5$ for vv, $C = 10$ for tv $C = 12$ for vt, $C = 20$ for tt	(7)
Mishima and Hibiki [9]	$C = 21[1 - \exp(-0.319 D_{int})]$	(8)
Zhang et al. [10]	$C = 21\left[1 - \exp\left(-\frac{a}{La}\right)\right]$ $La = \frac{\sqrt{\frac{\sigma}{g(\rho_l - \rho_g)}}}{D_{int}}$ For gas and liquid $a = 0.647$ For vapor and liquid $a = 0.142$	(9)
Sun and Mishima [11]	For viscous flow: $C = 26(1 + \frac{Re_l}{1000})\left[1 - \exp\left(\frac{-0.153}{0.8 + 0.27 La}\right)\right]$ For turbulent flow: $\Phi_l^2 = 1 + \frac{C}{X^{1.19}} + \frac{1}{X^2}$ $C = 1.79\left(\frac{Re_g}{Re_l}\right)^{0.4} \sqrt{\frac{1-x}{x}}$ where $Re_g = \frac{G_{tp} x D_{int}}{\mu_g}$, $Re_l = \frac{G_{tp}(1-x) D_{int}}{\mu_l}$	(10)

Table 2. Cont.

Author(s) and the Respective Reference(s)	Equation	Eq. No
Chisholm [14]	$\Phi_{lo}^2 = 1 + (Y^2 - 1)\{B[x(1-x)]^{0.875} + x^{1.75}\}$ $Y^2 = \dfrac{(\frac{\Delta p}{\Delta z})_{go}}{(\frac{\Delta p}{\Delta z})_{lo}}$ if $0 < Y < 9.5$, $B = \begin{cases} 4.8 & G_{tp} \leq 500 \\ 2400 & 500 < G_{tp} < 1900 \\ \dfrac{55}{G_{tp}^{0.5}} & G_{tp} \geq 1900 \end{cases}$ if $9.5 < Y < 28$, $B = \begin{cases} \dfrac{520}{YG_{tp}^{0.5}} & G_{tp} \leq 600 \\ \dfrac{21}{Y} & G_{tp} > 600 \end{cases}$ if $Y > 28$, $B = \dfrac{15000}{Y^2 G_{tp}^{0.5}}$	(11)
Muller-Steinhagen and Heck [16]	$\Phi_{lo}^2 = Y^2 x^3 + (1-x)^{0.333}[1 + 2x(Y^2 - 1)]$	(12)
Souza and Pimenta [17]	$\Phi_{lo}^2 = 1 + (\Gamma^2 - 1)x^{1.75}(1 + 0.9524\,\Gamma\,X_{tt}^{0.4126})$ $\Gamma = (\frac{\rho_l}{\rho_g})^{0.5}(\frac{\mu_g}{\mu_l})^{0.125},\ X_{tt} = \frac{1}{\Gamma}(\frac{1-x}{x})^{0.875}$	(13)
Friedel [18]	$\Phi_{lo}^2 = (1-x)^2 + x^2 \dfrac{\rho_l f_{go}}{\rho_g f_{lo}} + \dfrac{3.24\, x^{0.78}(1-x)^{0.224}\, H}{Fr_{tp}^{0.045}\, We_{tp}^{0.035}}$ $H = (\frac{\rho_l}{\rho_g})^{0.91}(\frac{\mu_g}{\mu_l})^{0.19}(1-\frac{\mu_g}{\mu_l})^{0.7}$ $We_{tp} = \dfrac{G_{tp}^2 D_{int}}{\sigma \rho_{tp}},\ Fr_{tp} = \dfrac{G_{tp}^2}{g D_{int} \rho_{tp}^2},\ \dfrac{1}{\rho_{tp}} = \dfrac{x}{\rho_g} + \dfrac{1-x}{\rho_l}$	(14)
Cavallini et al. [19]	$\Phi_{lo}^2 = (1-x)^2 + x^2 \dfrac{\rho_l f_{go}}{\rho_g f_{lo}} + \dfrac{1.262\, x^{0.6987}\, H}{We_{go}^{0.1458}}$ $H = (\frac{\rho_l}{\rho_g})^{0.3278}(\frac{\mu_g}{\mu_l})^{-1.181}(1-\frac{\mu_g}{\mu_l})^{3.477}$ $We_{go} = \dfrac{G_{tp}^2 D_{int}}{\sigma \rho_g}$	(15)
Tran et al. [20]	$\Phi_{lo}^2 = 1 + (4.3\,Y^2 - 1)\{[x(1-x)]^{0.875}\, La + x^{1.75}\}$	(16)

3. Experimental Procedures and Utilized Dataset

The measurement procedure for frictional pressure drop was performed on the adiabatic stream of the air–water mixture at various flow rates [39], within a horizontal, smooth copper pipe.

3.1. Overview of the Laboratory Setup

The simplified configuration of the laboratory setup is depicted in Figure 1. Water is pumped from the storage tank, passing through the temperature, pressure, and volume flow rate measurement section, then proceeding to the mixing section and subsequently to the test section. Finally, it flows back to the tank. The volumetric flow rate is assessed using a trio of parallel rotameters, with each specifically designed to accommodate a distinct range of flow rates. Compressed air, provided by the building's auxiliary supply system, passes through the temperature, pressure, and volumetric flow rate measurement section, proceeds to the mixing section, and finally to the test section. Next, it flows to the tank where, after separation from the water, it is vented to the environment. Similarly, the volumetric flow rate of the air is gauged by a set of three parallel rotameters.

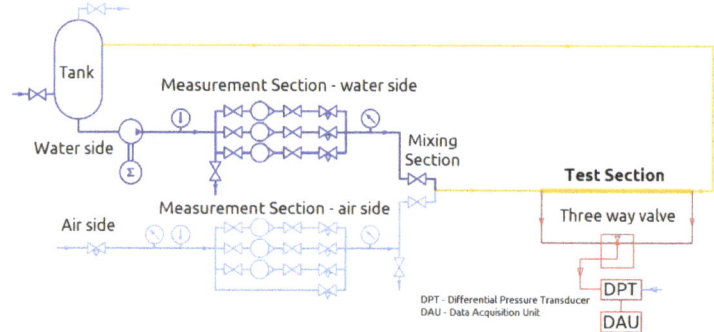

Figure 1. A schematic depiction illustrating the laboratory configuration employed for conducting the experimental procedures.

The test section is a copper tube, the properties of which are provided in Table 3. Pressure taps are installed at both the inlet and outlet of the test section, and these taps are connected to a differential pressure transducer (DPT). The DPT then sends electrical signals to a data-acquisition unit (DAU). Each pressure tap is connected to the pressure transducer through a pair of nylon tubes filled with water, enabling the hydraulic transmission of pressure variations. A bypass valve enables measuring either the pressure drop between the two pressure taps or the pressure difference between each tap and the ambient, which is more convenient for reducing the signal-to-noise ratio under specific operating conditions. The data-acquisition unit operates at a sampling frequency of 1 kHz with an acquisition time of 15 s. The maximum measurable pressure difference is about 70 kPa. Table 4 reports the uncertainties of the employed measurement devices. It is noteworthy to emphasize that, given the adiabatic nature of the studied flow in this investigation, temperature measurements are solely utilized in the data processing phase. Consequently, the inherent uncertainty associated with the thermometer has a negligible influence on the final results.

Table 3. Specification of the test section and experimental setup conditions of the flow.

Parameters	Units	Values
Utility	-	E-C-A
External diameter of the tube	[mm]	9.52
Thickness of the tube	[mm]	0.3
Internal diameter of the tube	[mm]	8.92
Length of the test section	[m]	1.295
Wet perimeter	[mm]	28.02
Cross-section area	[mm^2]	62.49
Number of two-phase data points	-	119
Mass fluxes range of the air	$[\frac{kg}{(m^2 s)}]$	5.30–47.42
Mass fluxes range of the water	$[\frac{kg}{(m^2 s)}]$	44.29–442.91
Mass quality range	[−]	0.01–0.52
Pressure gradient range	$[\frac{Pa}{m}]$	450.05–20,047.16

The investigated range of the water and air volume flow rate were 10 to 100 $[\frac{Nl}{h}]$ with a step of 10 $[\frac{Nl}{h}]$ and 500 to 4000 $[\frac{Nl}{h}]$ with a step of 500 $[\frac{Nl}{h}]$, respectively. The pressure of both water and air in the measurement section was maintained at a constant value of 2.2 [bar] and 3 [bar]. The temperature was measured at the initiation of each test, being insignificantly variable between 22 and 25 °C for water and 19 to 21 °C for air. Starting from the minimum water flow rate, the airflow rate was systematically adjusted to cover the entire range. At each stage, steady-state conditions were achieved through iterative adjustments of the volume flow rates using regulation valves, ensuring constant pressure in the measurement section. Measurements at each step were repeated 5 to 10 times, depending on the observed fluctuations, and averaged as a single data point which resulted in 119 measured data points.

Table 4. Specifications of the employed measurement devices.

Device	Range	Uncertainty
Manometer	0–6 bar (gauge)	0.2 bar
Thermometer	5–120 [°C]	1 [°C]
Differential Pressure Transducer	0–70 kPa	1.5% full scale
Air Flow Meter	4–190 $\frac{Nl}{h}$	3% of the observed value
Air Flow Meter	85–850 $\frac{Nl}{h}$	3% of the observed value
Air Flow Meter	400–4000 $\frac{Nl}{h}$	3% of the observed value
Water Flow Meter	10–100 $\frac{Nl}{h}$	3% of the observed value
Water Flow Meter	40–400 $\frac{Nl}{h}$	3% of the observed value

3.2. Data Processing

All of the fluid properties were evaluated at the inlet temperature and at the arithmetic average between the inlet and outlet pressure. The airflow meter readings are calibrated to normal conditions, defined as $T_0 = 20\ °C$ and $P_0 = 101{,}325$ Pa. According to the manufacturer's specifications, by taking the actual temperature (T_{ms}) and pressure (p_{ms}) at the measurement segment and the average pipe pressure (p_{av}), the actual volume flow rate flowing in the pipe is calculated as

$$Q = Q_0 \sqrt{\frac{p_{ms}\, p_0}{p_{av}^2} \frac{T_{ms}}{T_0}} \quad (17)$$

Consequently, the air superficial velocity in the test section ranges between 5 and 34 $[\frac{m}{s}]$. On the other hand, the water superficial velocity varies between 0.05 and 0.45 $[\frac{m}{s}]$, which causes the observed flow regimes to vary from wavy to annular flow (at low water velocity) and from slug flow to annular/annular-mist flow, approximately in accordance with the Mandhane map [40]. The pressure drop across the test section was determined from the difference between inlet and outlet measured pressures. With the flow being fully developed, the pressure gradient was determined by calculating the ratio of the pressure drop to the distance between the pressure taps. A summary of the operating conditions are provided with the test section specifications (Table 3).

4. Machine Learning

4.1. Overall Framework

Since, in the present study, the machine learning algorithms are utilized for predicting a known target, which is a continuous value, the overall framework is a supervised regression problem. Accordingly, the machine learning models are employed in order to predict the target value, while being provided a set of features (input parameters). The provided dataset is randomly divided into two subsets: a training subset used to optimize the machine learning algorithms, and a test subset which comprises 20% of the dataset and is utilized to determine the corresponding accuracy. In order to evade the dependence of the determined accuracy on the choice of training and validation subsets, the k-fold cross-validation methodology is employed (with the k = 10). In this procedure, the training is divided into k subsets, and each subset is once used as the validation set, while the remaining subsets are used for the training. In an iterative procedure, all of the subsets will then play the role of the validation set thus predictions, are provided for all of the samples of the training set.

In order to compare the accuracy of different models, several metrics including the Mean Percentage Error (MPE) (Equation (18)), and the Mean Absolute Percentage Error

(MAPE) (Equation (19)), were considered. MPE and MAPE were chosen as the primary metrics since the current study aims to compare its achieved performance with the one obtained using existing physical models. As the performance of these physical (empirical) models, in the previous studies, has been reported using MPE and MAPE, for the sake of coherence, these metrics have also been utilized in the present study. Additionally, these metrics are among the most commonly used evaluation metrics in other studies focused on data-driven models in this area, which thus facilitates performing comparisons with previously proposed models. The Mean Absolute Percentage Error (MAPE) is then utilized as the primary accuracy metric for selecting the most promising models.

$$MPE = \frac{1}{N} \sum_{i=1}^{N} \frac{y_{i,pred} - y_{i,exp}}{y_{i,exp}} \quad [\%] \quad (18)$$

$$MAPE = \frac{1}{N} \sum_{i=1}^{N} \frac{|y_{i,pred} - y_{i,exp}|}{y_{i,exp}} \quad [\%] \quad (19)$$

4.2. Machine Learning Algorithms

In the present study, the machine learning pipelines have been first developed employing Random Forest [41] as the benchmark algorithm, which is an ensemble method built upon decision trees.

4.3. Optimisation of Machine Learning Pipelines

Following the implementation of machine learning-based pipelines using the selected benchmark algorithms, a genetic algorithm-based optimization procedure is carried out in order to improve the model accuracy by selecting the most effective combination of preprocessing steps and machine learning models with their corresponding optimal hyperparameters.

The optimization procedure is conducted employing TPOT, a Tree-Based Pipeline Optimization Tool [42], while defining a custom objective function that considers MAPE as the key accuracy metric. This AutoML employs various combinations of preprocessing methods, feature selection techniques, and machine learning models to find the best pipeline.

4.3.1. Optimization Settings

The corresponding key settings include the following:

- Generations (generations = 100): number of iterations for the genetic algorithm.
- Population Size (population size = 100): number of candidate pipelines in each generation.
- Scoring Function (MAPE): Mean Absolute Percentage Error used to evaluate pipeline performance.
- Cross-Validation (cv = 10): 10-fold cross-validation to assess pipeline generalizability.

4.3.2. Pool of Hyperparameters

The hyperparameters search pool includes a variety of preprocessors and machine learning algorithms [43] with their respective hyperparameters such as the depth of decision trees, the number of trees in a Random Forest, regularization parameters in linear models, and learning rates in gradient boosting machines. The major optimized hyperparameters for the considered machine learning algorithms are provided in Table 5.

The pipeline structure optimization enhances the overall pipeline, refining the combination and sequence of preprocessing steps and model training components. Although feature selection methods are included in the optimization process, they are not implemented in this study since a more comprehensive feature selection procedure has been performed in the previous step.

Table 5. Algorithm search pool and their corresponding hyperparameter range.

Model	Hyperparameters
ElasticNetCV	• l1_ratio: $\{0.0, 0.05, 0.1, \ldots, 1.0\}$ • tol: $\{1 \times 10^{-5}, 1 \times 10^{-4}, 1 \times 10^{-3}, 1 \times 10^{-2}, 1 \times 10^{-1}\}$
ExtraTreesRegressor	• n_estimators: 100 • max_features: $\{0.05, 0.1, \ldots, 1.0\}$ • min_samples_split: $\{2, 3, \ldots, 20\}$ • min_samples_leaf: $\{1, 2, \ldots, 20\}$ • bootstrap: True or False
GradientBoostingRegressor	• n_estimators: 100 • loss: {"ls", "lad", "huber", "quantile"} • learning_rate: $\{0.001, 0.01, 0.1, 0.5, 1.0\}$ • max_depth: $\{1, 2, \ldots, 10\}$ • min_samples_split: $\{2, 3, \ldots, 20\}$ • min_samples_leaf: $\{1, 2, \ldots, 20\}$ • subsample: $\{0.05, 0.1, \ldots, 1.0\}$ • max_features: $\{0.05, 0.1, \ldots, 1.0\}$ • alpha: $\{0.75, 0.8, 0.85, 0.9, 0.95, 0.99\}$
AdaBoostRegressor	• n_estimators: 100 • learning_rate: $\{0.001, 0.01, 0.1, 0.5, 1.0\}$ • loss: {"linear", "square", "exponential"}
DecisionTreeRegressor	• max_depth: $\{1, 2, \ldots, 10\}$ • min_samples_split: $\{2, 3, \ldots, 20\}$ • min_samples_leaf: $\{1, 2, \ldots, 20\}$
KNeighborsRegressor	• n_neighbors: $\{1, 2, \ldots, 100\}$ • weights: {"uniform", "distance"} • p: $\{1, 2\}$
LassoLarsCV	• normalize: True or False
LinearSVR	• loss: {"squared_epsilon_insensitive", "epsilon_insensitive"} • dual: True or False • tol: $\{1 \times 10^{-5}, 1 \times 10^{-4}, 1 \times 10^{-3}, 1 \times 10^{-2}, 1 \times 10^{-1}\}$ • C: $\{1 \times 10^{-4}, \ldots, 1 \times 10^{-1}, 0.5, 1.0, 5.0, 10.0, 15.0, 20.0, 25.0\}$ • epsilon: $\{1 \times 10^{-4}, 1 \times 10^{-3}, 1 \times 10^{-2}, 1 \times 10^{-1}, 1.0\}$
RandomForestRegressor	• n_estimators: 100 • max_features: $\{0.05, 0.1, \ldots, 1.0\}$ • min_samples_split: $\{2, 3, \ldots, 20\}$ • min_samples_leaf: $\{1, 2, \ldots, 20\}$ • bootstrap: True or False
XGBRegressor	• n_estimators: 100 • max_depth: $\{1, 2, \ldots, 10\}$ • learning_rate: $\{0.001, 0.01, 0.1, 0.5, 1.0\}$ • subsample: $\{0.05, 0.1, \ldots, 1.0\}$ • min_child_weight: $\{1, 2, \ldots, 20\}$ • nthread: 1 • objective: {"reg:squarederror"}
SGDRegressor	• loss: {"epsilon_insensitive", "squared_loss", "huber"} • penalty: {"elasticnet"} • alpha: $\{0.0, 0.01, 0.001\}$ • learning_rate: {"invscaling", "constant"}

Table 5. *Cont.*

Model	Hyperparameters
SGDRegressor	• `fit_intercept`: True or False • `l1_ratio`: $\{0.25, 0.0, 1.0, 0.75, 0.5\}$ • `eta0`: $\{0.1, 1.0, 0.01\}$ • `power_t`: $\{0.5, 0.0, 1.0, 0.1, 100.0, 10.0, 50.0\}$
RidgeCV	
Preprocessors	
Model	Hyperparameters
Binarizer	• `threshold` set to $\{0.0, 0.05, 0.1, \ldots, 1.0\}$
FastICA	• `tol` set to $\{0.0, 0.05, 0.1, \ldots, 1.0\}$
FeatureAgglomeration	• `linkage` set to {"ward", "complete", "average"} • `affinity` set to {"euclidean", "l1", "l2", "manhattan", "cosine"}
MaxAbsScaler	
MinMaxScaler	
Normalizer	• `norm` set to {"l1", "l2", "max"}
Nystroem	• `kernel` set to {"rbf", "cosine", "chi2", "laplacian", "polynomial", "poly", "linear", "additive_chi2", "sigmoid"} • `gamma` set to $\{0.0, 0.05, 0.1, \ldots, 1.0\}$ • `n_components` set to $\{1, 2, \ldots, 10\}$
PCA	• `svd_solver` set to {"randomized"} • `iterated_power` set to $\{1, 2, \ldots, 10\}$
PolynomialFeatures	• `degree` set to 2 • `include_bias` set to False • `interaction_only` set to False
RBFSampler	• `gamma` set to $\{0.0, 0.05, 0.1, \ldots, 1.0\}$
RobustScaler	
StandardScaler	
ZeroCount	
OneHotEncoder	• `minimum_fraction` set to $\{0.05, 0.1, 0.15, 0.2, 0.25\}$ • `sparse` set to False • `threshold` set to 10

5. Methodology and Implemented Pipelines

Various machine learning-based pipelines are employed to identify the most promising one that yields the most accurate prediction of the pressure drop, given the considered features. In the initial stage, a comprehensive grid search (GridSearchCV [43]) was conducted to determine the optimal Random Forest as the benchmark algorithm by exploring the hyperparameters listed in Table 5. Subsequently, all potential features are fed into the benchmark algorithm, and the corresponding accuracies are recorded. In the next step, the feature selection procedure, explained in the next sub-section, is executed to ascertain the most promising combination of features, employing the benchmark algorithm, resulting in the highest achievable accuracy. Finally, the pipeline optimization procedure, as explained in Section 4.3, considering the selected features as the input is conducted to identify the optimal machine learning algorithm along with their associated tuning parameters.

5.1. Feature Selection Procedure

As was previously pointed out, the objective of the feature selection procedure is to determine the most promising set of features that leads to the highest accuracy. An ideal solution to achieve the mentioned goal is evaluating all possible combinations of features, though the latter leads to an excessive computational cost, which makes it practically infeasible. Accordingly, an alternative procedure is implemented, in which the features are first ordered based on their rank ((determined by their Pearson's correlation coefficient [44] to the target (two-phase pressure drop multipliers)). The features are then added one by one and the resulting set is provided to the machine learning model as input features and the resulting MAPE value at each step is registered. The employed machine learning model in this procedure is Random Forest, which is considered as the key benchmark algorithm for the studied case. The set of features, that results in the lowest MAPE (initial chosen set), is next selected. The remaining features are divided into two categories: the ones that had increased the MAPE, and the ones that had decreased it. The features that had increased the MAPE are discarded, while the ones that had decreased MAPE are sorted in descending order (based on the MAPE achieved while adding them) and added at the beginning of the initial chosen set. The obtained chosen set is again provided to the Random Forest algorithm as the set of features; the features are then dropped one by one, and the resulting MAPE at each step is registered again. The set of features that leads to the lowest possible MAPE value is chosen as the updated set of selected features. Subsequently, the Random Forest is initially fitted with each selected feature individually (as a single input), and the one yielding the lowest Mean Absolute Percentage Error (MAPE) is chosen as the first feature. In the next step, the model is trained utilizing solely two variables (the first one is already chosen), where each of the remained features is given as the second feature. Similarly, the one that leads to the lowest MAPE is chosen as the second feature. This procedure persists until all selected features are incorporated. Evidently, the result obtained in the last step is the same as the one provided in the previous step. The final set of selected features is the one that results in the minimum MAPE value. Finally, it is noteworthy that only the dimensionless numbers that represent the governing phenomena are employed in this work, which results in a rather limited number of features. Thus, executing the proposed gradual selection method results in a limited computational cost. Therefore, taking into account the fact that the feature selection procedure is performed offline (and not in real-time or in-operando), the corresponding execution time does not limit the application of the method in the present work's context.

5.2. Feature Selection Based on Random Forest-Derived Feature Importance

To compare the effectiveness of the utilized feature selection procedure in the current study, another feature selection approach based on the feature importance of Random Forest algorithm [45] has been applied. The Random Forest algorithm quantifies the significance of a feature by assessing the extent to which the model's accuracy decreases when that feature is randomly permuted or shuffled while maintaining all other features unchanged [45]. In this approach, features are initially ranked by their importance employing the optimal benchmark. The process begins by using the most important feature for predictions and recording the accuracy achieved. Subsequently, each additional feature is sequentially incorporated, and predictions are made at each step. This iterative process continues until all features have been included. The elbow method is then applied to determine the optimal set of features. Finally, the performance of the model using this optimal feature set is evaluated on both the training/validation set and the test set, and the results are reported accordingly.

6. Results and Discussions

6.1. Accuracy of Existing Standard Physical Models

The accuracies of pressure gradient prediction from various models for the two-phase dataset are presented in Table 6. The model proposed by Muller-Steinhagen and Heck [16] is identified as the most accurate, resulting in a Mean Absolute Percentage Error (MAPE) of 15.79%.

Table 6. Results derived from the conventional physical phenomena-based models (the deviation between the estimated pressure gradient ($\frac{dp}{dz}$ [$\frac{Pa}{m}$]) and the experimentally acquired values).

Author(s) and the Respective Reference(s)	MPE [%]	MAPE [%]
McAdams et al. [36]	−10.05	18.56
Bettiel and Whalley [37]	−28.27	28.38
Awad and Muzychka [38]	6.67	17.58
Chisholm [8]	−13.25	16.69
Mishima and Hibiki [9]	−4.37	17.13
Zhang et al. [10]	−8.75	18.29
Sun and Mishima [11]	−35.94	36.52
Chisholm [14]	65.91	66.18
Muller-Steinhagen and Heck [16]	7.80	15.79
Souza and Pimenta [17]	32.93	54.93
Friedel [18]	23.97	36.04
Cavallini et al. [19]	−89.03	89.03
Tran et al. [20]	−65.45	65.45

6.2. Implemented Hybrid Data-Driven/Physical-Based Models

As was previously mentioned, two-phase flow multipliers are taken into account as targets, while the non-dimensional parameters are given as inputs. Consequently, four general machine learning-based pipelines were defined, in which the target wasn defined as a liquid, liquid only, gas, or gas-only multiplier. For each of the above-mentioned targets, the Random Forest-based pipelines were first implemented both with all features and selected features. Next, the optimal pipeline, given the selected features, was determined for each scenario, and the resulting accuracies were compared.

Table 7 represents the prediction accuracies that are obtained while employing the abovementioned pipelines. The most accurate optimal pipeline was determined to be the one with the Φ_l^2 as the target and utilizing selected features which leads to an MAPE of 5.99% on the validation set and 7.03% on the test set.

The comprehensive grid search was conducted for each target employing all the input features (Table 8). The optimal benchmark algorithm for Φ_l^2 is defined as described in Table A1. Table 8, in addition to the list of all of the input features, provides their corresponding Pearson correlation with respect to the targets. These correlations are used initially to rank the features in the feature selection step. The feature selection results in Table 9 demonstrate that the model requires only three input parameters (x, Re_g, and X), significantly simplifying it with respect to physical models. Clearly, the reason behind the fact that only three parameters are selected is that only one pair of fluids (water and air) is utilized in this study; accordingly, the other parameters that depend on the material properties do not provide additional information.

Table 7. Implemented machine learning pipelines and their corresponding results.

Pipeline	Two-Phase Flow Multiplier ($\Phi_i^2[-]$)	Pipeline	Validation (CV)		Test	
			MPE [%]	MAPE [%]	MPE [%]	MAPE [%]
A	1	All Features—Random Forest	3.97	9.89	9.97	18.01
B		Selected Features—Random Forest	1.95	9.16	7.99	15.04
C		Selected Features—Optimal Pipeline	0.40	5.99	2.86	7.03
D	lo	All Features—Random Forest	4.01	11.76	11.71	17.28
E		Selected Features—Random Forest	5.25	10.41	8.44	11.80
F		Selected Features—Optimal Pipeline	−0.31	7.29	3.29	9.33

Table 7. Cont.

Pipeline	Two-Phase Flow Multiplier ($\Phi_i^2[-]$)	Pipeline	Validation (CV)		Test	
			MPE [%]	MAPE [%]	MPE [%]	MAPE [%]
G		All Features—Random Forest	2.07	8.34	5.43	9.97
H	g	Selected Features—Random Forest	0.20	7.49	4.90	9.83
I		Selected Features—Optimal Pipeline	0.45	6.05	4.21	8.68
J		All Features—Random Forest	2.45	9.42	3.08	10.09
K	go	Selected Features—Random Forest	0.81	8.16	−0.20	6.56
L		Selected Features—Optimal Pipeline	0.66	6.09	2.63	7.79

Table 8. Employed input features and their corresponding correlation to each target (sorted by $\phi_{exp,l}^2$).

Features	$\phi_{exp,l}^2$	$\phi_{exp,lo}^2$	$\phi_{exp,g}^2$	$\phi_{exp,go}^2$
x [-]	0.96	0.85	0.60	0.96
Re_g [-]	0.73	0.74	0.53	0.78
X [-]	0.64	0.69	0.98	0.70
f_g [-]	0.63	0.67	0.62	0.68
$\frac{1}{Re_l}$ [-]	0.62	0.46	0.47	0.63
x_v [-]	0.61	0.66	0.98	0.68
$\frac{1-x_v}{x_v}$ [-]	0.60	0.65	0.99	0.66
Re_l [-]	0.60	0.60	0.64	0.60
$\frac{G_l^*}{G_g^*}$ [-]	0.59	0.64	0.98	0.65
$\frac{1-x}{x}$ [-]	0.59	0.64	0.98	0.65
f_l [-]	0.57	0.40	0.42	0.60
$\frac{1}{Re_g}$ [-]	0.56	0.61	0.63	0.61
Re_{go} [-]	0.52	0.52	0.59	0.51
Re_{lo} [-]	0.52	0.52	0.59	0.51
f_{go} [-]	0.49	0.42	0.53	0.50
$\frac{1}{Re_{go}}$ [-]	0.42	0.32	0.45	0.45
$\frac{1}{Re_{lo}}$ [-]	0.42	0.32	0.45	0.45
f_{lo} [-]	0.23	0.02	0.24	0.27
Y [-]	0.09	0.18	0.02	0.15

Table 9. Selected input features associated with each target.

Targets	Selected Features								
$\phi_{exp,l}^2$	x	Re_g	X						
$\phi_{exp,lo}^2$	x_v	Re_g	Re_{lo}	$\frac{1}{Re_g}$	f_l	$\frac{1-x_v}{x_v}$	Y	f_{lo}	$\frac{1}{Re_{lo}}$	Re_l
$\phi_{exp,g}^2$	x_v	f_l	x						
$\phi_{exp,go}^2$	x	Re_g							

Next, the performance of the three-step feature selection performed was compared with the methodology provided in Section 5.2. The accuracy obtained by progressively adding features, based on their importance as determined by the Random Forest algorithm, is depicted in Figure 2. It can be observed that by expanding the list of input features, improvements in results are obtained up to a certain point. The plot illustrates that improvements in results are observed as more input features are included, reaching an optimal point. Using the elbow method, the optimal set of features, consisting of nine features, is identified. Finally, the identified set of features is used to make predictions on both the training/validation and test sets, resulting in accuracies of 9.35% and 15.70% in terms of MAPE, respectively. It is evident that the three-step feature selection method employed in this study (pipeline B) outperforms the feature importance-driven method, achieving higher prediction accuracies with only three selected features.

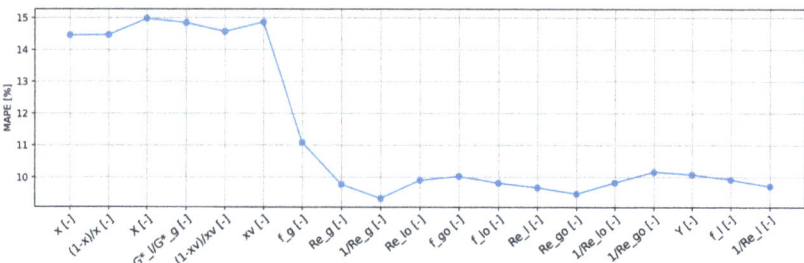

Figure 2. Obtained accuracy by progressively adding features based on Random Forest algorithm-derived importance, focusing on ϕ_l^2 as the target.

Next, the details of the optimal pipeline identified in the current work are provided. Figure 3 provides the ML algorithms of the optimized pipeline and the corresponding additional feature processing. Additionally, Table A2 in Appendix A section provides the attributes and the detailed description of each parameter utilized in the various steps of the optimal pipeline.

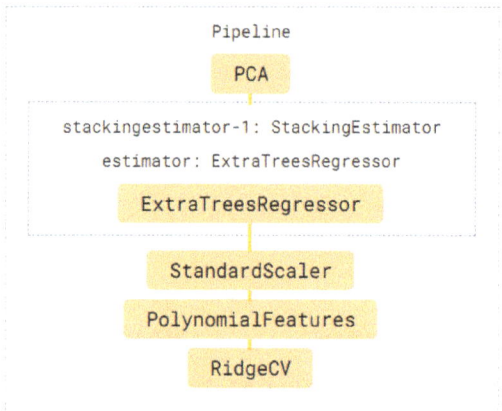

Figure 3. A schematic description of the identified optimal pipeline (Pipeline C).

Figure 4 compares the estimated two-phase multiplier values generated by this pipeline and the corresponding experimental data, with respect to the Lockhart–Martinelli parameter X [-].

Figure 5 illustrates the comparison between the measured pressure gradient values, those estimated by Muller-Steinhagen and Heck [16], and the predictions of the proposed optimal pipeline. Using the obtained Φ_{lo}^2 and employing Equation (6) the pressure gradient values were calculated (as mentioned in Section 6.1) and are compared to experimental data. It can be observed that the model proposed by Muller-Steinhagen and Heck [16] overestimates the low-pressure gradient ranges and underestimates high-pressure gradient values. Additionally, it can be noted that the optimal model demonstrates a significantly higher accuracy in estimating the two-phase pressure drop compared to the Muller-Steinhagen and Heck [16], which is the most precise physical model currently available in the literature.

Furthermore, since both the dataset and the optimal pipeline are made publicly accessible as tools, the latter approach also enhances reproducibility and ease of use.

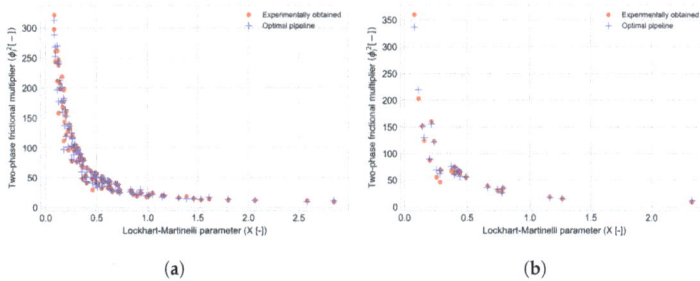

Figure 4. Comparison between the experimentally obtained two-phase flow multiplier and the estimation by the optimal pipeline (Pipeline C) with respect to the Lockhart–Martinelli parameter X. (**a**) Training set (CV). (**b**) Test set.

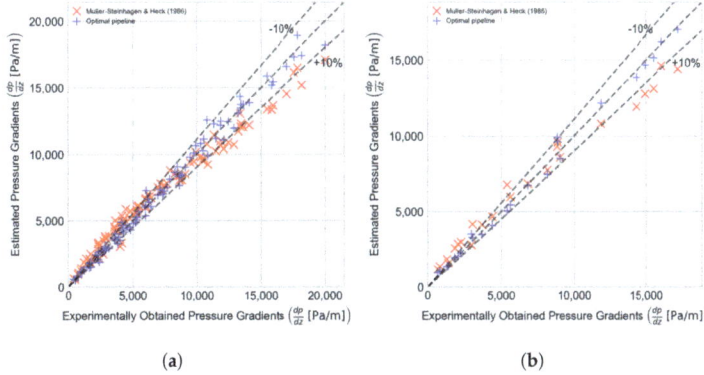

Figure 5. Experimentally acquired pressure gradient vs. the optimal pipeline estimations (Pipeline C) and juxtaposed with the values estimated by Muller-Steinhagen and Heck [16] model. (**a**) Training set (CV). (**b**) Test set.

7. Conclusions

In the present study, an experimental study was first conducted in which the frictional pressure drop of the water–air mixture (two-phase), while passing through a smooth horizontal tube at various flow conditions was measured. In the next step, on the acquired dataset, the state-of-the-art standard physical models, proposed in the literature, were applied and the corresponding accuracy, being implemented to the experimentally acquired dataset, was calculated and the most accurate models were determined. Next, machine learning-based pipelines, while following dimensionless approaches, were implemented. In this approach, the two-phase flow multipliers were employed as the targets while only non-dimensional parameters were employed as features. Feature selection and pipeline optimization procedures were applied to each pipeline in order to determine the most promising set of features along with the most accurate machine learning algorithm (and the associated tuning parameters), respectively. The optimal pipeline was determined to be the one where the liquid two-phase multiplier serves as the target, and three dimensionless parameters (chosen from 19 provided features) are utilized as input features. Employing the latter pipeline, a MAPE of 5.99% on the validation set and 7.03% on the test set can be achieved that is notably more accurate than the most promising physical model (Muller-Steinhagen and Heck [16] leading to a MAPE of 15.79% on the whole dataset). It can be mentioned that other methods such as Bayesian optimization are also available for pipeline optimization, and while AutoML methods that are based on Bayesian optimization methods, such as those implemented in tools like auto-sklearn [46], offer advantages in certain aspects, including efficient handling of continuous parameters and better performance with limited data, they may not always outperform genetic algorithms in terms of scalability and ability to handle diverse types of hyper-parameters. However, there are Bayesian optimization methods designed for scalability and handling mixed variables, thus eliminating the need for the extensive evaluations required by genetic algorithms. Therefore, it is recommended for future work to explore Bayesian optimization methods for optimizing machine learning-based pipelines in predicting pressure drops in smooth tubes. This approach could significantly enhance the efficiency and accuracy of the optimization process.

The feature selection procedure provided in the current work effectively assessed the performance of various combinations of input features in predicting the pressure drop, allowing for the identification of the most promising set. This study demonstrated that a significant reduction in the number of input parameters can be achieved while maintaining accuracy, thereby enhancing the interpretability of the results. Nevertheless, other methods regarding explainable machine learning such as the Sindys [47,48] via Neural Networks can be employed to achieve/enhance the explainability of the model in future works. Finally, with the public accessibility of both the dataset and the optimal pipelines, the proposed model offers superior reproducibility and user-friendliness compared to state-of-the-art physical models.

Author Contributions: F.B.: Software, Formal analysis, Methodology, Validation, Data Curation, Writing—Original Draft; K.A.: Formal analysis, Methodology, Validation, Data Curation, Writing—Original Draft; F.D.J.: Software, Data Curation, Writing—Original Draft; B.N.: Conceptualization, Methodology, Supervision, Writing—Review & Editing; P.V.P.D.: Validation, Data Curation; F.R.: Supervision, Funding acquisition; L.P.M.C.: Supervision, Conceptualization, Methodology. All authors have read and agreed to the published version of the manuscript.

Funding: This research received no external funding.

Data Availability Statement: The processed dataset and the resulting optimal pipelines, including the most promising feature sets and the most suitable algorithms, are available in an online repository: https://github.com/FarzadJavan/Pressure-Drop-Estimation-of-Two-Phase-Adiabatic-Flows-in-Smooth-Tubes (accessed on 12 June 2024).

Conflicts of Interest: The authors declare that they have no known competing financial interests or personal relationships that could have appeared to influence the work reported in this paper.

Nomenclature

Δp	Pressure drop [Pa]
$\frac{\Delta p}{\Delta z}$	Pressure gradient $[\frac{Pa}{m}]$
$\frac{Nl}{h}$	Normal litter per hour
AI	Artificial intelligence
ANN	Artificial Neural Network
CV	Cross-Validation
D_{int}	Pipe internal diameter [m]
f	Friction factor (by Fanning) [-]
Fr	Froude number [-]
G	Mass flux $[\frac{kg}{m^2 s}]$
G^*	Apparent mass flux $[\frac{kg}{m^2 s}]$
J	Superficial velocity $\frac{m}{s}$
La	Laplace constant [-]
$MAPE$	Mean Absolute Percentage Error [%]
ML	Machine learning
MPE	Mean Percentage Error [%]
p	Pressure [Pa]

Q	Volume flow rate $[\frac{m^3}{s}]$	
Re	Reynolds number [-]	
RF	Random Forest algorithm	
S	Internal wetted perimeter [m]	
SVM	Support Vector Machines	
U	Phase velocity $\frac{m}{s}$	
We	Weber number [-]	
X	Lockhart–Martinelli parameter [-]	
x	Average mass quality [-]	
x_v	Average volume quality [-]	
Y	Chisholm parameter [-]	

Greek symbols

α	Void fraction [-]
Γ	Mass flow rate $[\frac{kg}{s}]$, non-dimensional parameter (Equation (13))
μ	Dynamic viscosity [Pa s]
Ω	Cross-section [m^2]
Φ^2	Two-phase flow friction multiplier [-]
ρ	Density $[\frac{kg}{m^3}]$
σ	Surface tension $[\frac{N}{m}]$
τ_w	Shear stress [Pa]

Subscripts

a	Accelerative
av	Average
b	Bulk
exp	Experimental value
f	Frictional
go	Gas only
l	Liquid
lo	Liquid only
m	Micro-finned
ms	Manufacturer's specifications
$pred$	Predicted value
s	Smooth
tp	Two-phase
tt	Turbulent liquid, turbulent gas flow
tv	Turbulent liquid, laminar gas flow
vt	Laminar liquid, turbulent gas flow
vv	Laminar liquid, Laminar gas flow

Appendix A. Optimal Pipeline

Table A1. Attributes of the identified optimal Random Forest (pipelines A and B) along with the description of each parameter (defined by scikit-learn guidelines [43]).

Optimal Benchmark Algorithm	Arguments	Definitions	Values
RandomForestRegressor	bootstrap	Whether bootstrap samples are used when building trees	False
	max-features	The number of features to consider when looking for the best split	0.2
	min-samples-leaf	The minimum number of samples required to be at a leaf n	1
	min-samples-split	The minimum number of samples required to split an internal node	2
	n-estimators	The number of trees in the forest	100

Table A2. Attributes of the identified optimal pipeline (Pipeline C) along with the description of each parameter (defined by scikit-learn guidelines [43]).

Optimal Pipeline Steps	Arguments	Definitions	Values
Step 1: PCA [1]	iterated-power	The number of iterations used by the randomized SVD solver to improve accuracy	3
	svd-solver	Selects the algorithm for computing SVD, balancing speed and accuracy	randomized
Step 2: StackingEstimator: estimator = ExtraTreesRegressor	bootstrap	Whether bootstrap samples are used when building trees	False
	max-features	The number of features to consider when looking for the best split	0.95
	min-samples-leaf	The minimum number of samples required to be at a leaf n	15
	min-samples-split	The minimum number of samples required to split an internal node	8
	n-estimators	The number of trees in the forest	100
Step 3: StandardScaler [2]	-	-	-
Step 4: PolynomialFeatures [3]	degree	The degree of the polynomial features	2
	include-bias	bias column is considered	False
	interaction-only	interaction features are produced	False
Step 5: RidgeCV	-	-	-

[1] A pre-processing step that reduces dimensionality by transforming data into a set of orthogonal principal components that capture the maximum variance. [2] A pre-processing step that standardizes features by removing the mean and scaling them to unit variance. [3] A pre-processing step that generates a new feature matrix consisting of all polynomial combinations of the features with degree less than or equal to the specified degree.

References

1. Szilas, A.P. *Production and Transport of Oil and Gas*; Elsevier: Amsterdam, The Netherlands, 2010.
2. Ebrahimi, A.; Khamehchi, E. A robust model for computing pressure drop in vertical multiphase flow. *J. Nat. Gas Sci. Eng.* **2015**, *26*, 1306–1316. [CrossRef]
3. Osman, E.S.A.; Aggour, M.A. Artificial neural network model for accurate prediction of pressure drop in horizontal and near-horizontal-multiphase flow. *Pet. Sci. Technol.* **2002**, *20*, 1–15. [CrossRef]
4. Taccani, R.; Maggiore, G.; Micheli, D. Development of a process simulation model for the analysis of the loading and unloading system of a CNG carrier equipped with novel lightweight pressure cylinders. *Appl. Sci.* **2020**, *10*, 7555. [CrossRef]
5. Xu, Y.; Fang, X. A new correlation of two-phase frictional pressure drop for evaporating flow in pipes. *Int. J. Refrig.* **2012**, *35*, 2039–2050. [CrossRef]
6. Xu, Y.; Fang, X.; Su, X.; Zhou, Z.; Chen, W. Evaluation of frictional pressure drop correlations for two-phase flow in pipes. *Nucl. Eng. Des.* **2012**, *253*, 86–97. [CrossRef]
7. Lockhart, R.; Martinelli, R. Proposed correlation of data for isothermal two-phase, two-component flow in pipes. *Chem. Eng. Prog.* **1949**, *45*, 39–48.
8. Chisholm, D. A theoretical basis for the Lockhart-Martinelli correlation for two-phase flow. *Int. J. Heat Mass Transf.* **1967**, *10*, 1767–1778. [CrossRef]
9. Mishima, K.; Hibiki, T. Some characteristics of air-water two-phase flow in small diameter vertical tubes. *Int. J. Multiph. Flow* **1996**, *22*, 703–712. [CrossRef]
10. Zhang, W.; Hibiki, T.; Mishima, K. Correlations of two-phase frictional pressure drop and void fraction in mini-channel. *Int. J. Heat Mass Transf.* **2010**, *53*, 453–465. [CrossRef]
11. Sun, L.; Mishima, K. Evaluation analysis of prediction methods for two-phase flow pressure drop in mini-channels. In Proceedings of the International Conference on Nuclear Engineering, Orlando, FL, USA, 11–15 May 2008; Volume 48159, pp. 649–658.
12. Colombo, M.; Colombo, L.P.; Cammi, A.; Ricotti, M.E. A scheme of correlation for frictional pressure drop in steam–water two-phase flow in helicoidal tubes. *Chem. Eng. Sci.* **2015**, *123*, 460–473. [CrossRef]
13. De Amicis, J.; Cammi, A.; Colombo, L.P.; Colombo, M.; Ricotti, M.E. Experimental and numerical study of the laminar flow in helically coiled pipes. *Prog. Nucl. Energy* **2014**, *76*, 206–215. [CrossRef]
14. Chisholm, D. Pressure gradients due to friction during the flow of evaporating two-phase mixtures in smooth tubes and channels. *Int. J. Heat Mass Transf.* **1973**, *16*, 347–358. [CrossRef]
15. Baroczy, C. Systematic Correlation for Two-Phase Pressure Drop. *Chem. Eng. Progr. Symp. Ser.* **1966**, *62*, 232–249.
16. Müller-Steinhagen, H.; Heck, K. A simple friction pressure drop correlation for two-phase flow in pipes. *Chem. Eng. Process. Process. Intensif.* **1986**, *20*, 297–308. [CrossRef]
17. Lobo de Souza, A.; de Mattos Pimenta, M. Prediction of pressure drop during horizontal two-phase flow of pure and mixed refrigerants. *ASME-Publ.-FED* **1995**, *210*, 161–172.

18. Friedel, L. Improved friction pressure drop correlation for horizontal and vertical two-phase pipe flow. In Proceedings of the European Two-Phase Group Meeting, Ispra, Italy, 5–8 June 1979.
19. Cavallini, A.; Censi, G.; Del Col, D.; Doretti, L.; Longo, G.A.; Rossetto, L. Condensation of halogenated refrigerants inside smooth tubes. *Hvac&R Res.* **2002**, *8*, 429–451.
20. Tran, T.; Chyu, M.C.; Wambsganss, M.; France, D. Two-phase pressure drop of refrigerants during flow boiling in small channels: An experimental investigation and correlation development. *Int. J. Multiph. Flow* **2000**, *26*, 1739–1754. [CrossRef]
21. Najafi, B.; Ardam, K.; Hanušovský, A.; Rinaldi, F.; Colombo, L.P.M. Machine learning based models for pressure drop estimation of two-phase adiabatic air-water flow in micro-finned tubes: Determination of the most promising dimensionless feature set. *Chem. Eng. Res. Des.* **2021**, *167*, 252–267. [CrossRef]
22. Ardam, K.; Najafi, B.; Lucchini, A.; Rinaldi, F.; Colombo, L.P.M. Machine learning based pressure drop estimation of evaporating R134a flow in micro-fin tubes: Investigation of the optimal dimensionless feature set. *Int. J. Refrig.* **2021**, *131*, 20–32. [CrossRef]
23. Bar, N.; Bandyopadhyay, T.K.; Biswas, M.N.; Das, S.K. Prediction of pressure drop using artificial neural network for non-Newtonian liquid flow through piping components. *J. Pet. Sci. Eng.* **2010**, *71*, 187–194. [CrossRef]
24. Tian, Z.; Gu, B.; Yang, L.; Liu, F. Performance prediction for a parallel flow condenser based on artificial neural network. *Appl. Therm. Eng.* **2014**, *63*, 459–467. [CrossRef]
25. Garcia, J.J.; Garcia, F.; Bermúdez, J.; Machado, L. Prediction of pressure drop during evaporation of R407C in horizontal tubes using artificial neural networks. *Int. J. Refrig.* **2018**, *85*, 292–302. [CrossRef]
26. López-Belchí, A.; Illan-Gomez, F.; Cano-Izquierdo, J.M.; García-Cascales, J.R. GMDH ANN to optimise model development: Prediction of the pressure drop and the heat transfer coefficient during condensation within mini-channels. *Appl. Therm. Eng.* **2018**, *144*, 321–330. [CrossRef]
27. Khosravi, A.; Pabon, J.; Koury, R.; Machado, L. Using machine learning algorithms to predict the pressure drop during evaporation of R407C. *Appl. Therm. Eng.* **2018**, *133*, 361–370. [CrossRef]
28. Zendehboudi, A.; Li, X. A robust predictive technique for the pressure drop during condensation in inclined smooth tubes. *Int. Commun. Heat Mass Transf.* **2017**, *86*, 166–173. [CrossRef]
29. Shadloo, M.S.; Rahmat, A.; Karimipour, A.; Wongwises, S. Estimation of Pressure Drop of Two-Phase Flow in Horizontal Long Pipes Using Artificial Neural Networks. *J. Energy Resour. Technol.* **2020**, *142*, 112110. [CrossRef]
30. Yan, Y.; Wang, L.; Wang, T.; Wang, X.; Hu, Y.; Duan, Q. Application of soft computing techniques to multiphase flow measurement: A review. *Flow Meas. Instrum.* **2018**, *60*, 30–43. [CrossRef]
31. Shaban, H.; Tavoularis, S. Measurement of gas and liquid flow rates in two-phase pipe flows by the application of machine learning techniques to differential pressure signals. *Int. J. Multiph. Flow* **2014**, *67*, 106–117. [CrossRef]
32. Nie, F.; Yan, S.; Wang, H.; Zhao, C.; Zhao, Y.; Gong, M. A universal correlation for predicting two-phase frictional pressure drop in horizontal tubes based on machine learning. *Int. J. Multiph. Flow* **2023**, *160*, 104377. [CrossRef]
33. Faraji, F.; Santim, C.; Chong, P.L.; Hamad, F. Two-phase flow pressure drop modelling in horizontal pipes with different diameters. *Nucl. Eng. Des.* **2022**, *395*, 111863. [CrossRef]
34. Moradkhani, M.; Hosseini, S.; Song, M.; Abbaszadeh, A. Reliable smart models for estimating frictional pressure drop in two-phase condensation through smooth channels of varying sizes. *Sci. Rep.* **2024**, *14*, 10515. [CrossRef] [PubMed]
35. Moradkhani, M.; Hosseini, S.H.; Mansouri, M.; Ahmadi, G.; Song, M. Robust and universal predictive models for frictional pressure drop during two-phase flow in smooth helically coiled tube heat exchangers. *Sci. Rep.* **2021**, *11*, 20068. [CrossRef] [PubMed]
36. McAdams, W. Vaporization inside horizontal tubes-II, Benzene oil mixtures. *Trans. ASME* **1942**, *64*, 193–200. [CrossRef]
37. Beattie, D.; Whalley, P. A simple two-phase frictional pressure drop calculation method. *Int. J. Multiph. Flow* **1982**, *8*, 83–87. [CrossRef]
38. Awad, M.; Muzychka, Y. Effective property models for homogeneous two-phase flows. *Exp. Therm. Fluid Sci.* **2008**, *33*, 106–113. [CrossRef]
39. Vega-Penichet Domecq, P. Pressure Drop Measurement for Adiabatic Single and Two Phase Flows Inside Horizontal Micro-Fin Tubes. Available online: https://oa.upm.es/52749/ (accessed on 12 June 2024).
40. Mandhane, J.; Gregory, G.; Aziz, K. A flow pattern map for gas–liquid flow in horizontal pipes. *Int. J. Multiph. Flow* **1974**, *1*, 537–553. [CrossRef]
41. Breirnan, L. Arcing classifiers. *Ann. Stat.* **1998**, *26*, 801–849.
42. Olson, R.S.; Bartley, N.; Urbanowicz, R.J.; Moore, J.H. Evaluation of a tree-based pipeline optimization tool for automating data science. In Proceedings of the Genetic and Evolutionary Computation Conference 2016, Denver, CO, USA, 20–24 July 2016; pp. 485–492.
43. Pedregosa, F.; Varoquaux, G.; Gramfort, A.; Michel, V.; Thirion, B.; Grisel, O.; Blondel, M.; Prettenhofer, P.; Weiss, R.; Dubourg, V.; et al. Scikit-learn: Machine learning in Python. *J. Mach. Learn. Res.* **2011**, *12*, 2825–2830.
44. Pearson, K. VII. Note on regression and inheritance in the case of two parents. *Proc. R. Soc. Lond.* **1895**, *58*, 240–242.
45. Breiman, L. Random forests. *Mach. Learn.* **2001**, *45*, 5–32. [CrossRef]
46. Feurer, M.; Klein, A.; Eggensperger, K.; Springenberg, J.; Blum, M.; Hutter, F. Efficient and robust automated machine learning. *Adv. Neural Inf. Process. Syst.* **2015**, *28*. Available online: https://proceedings.neurips.cc/paper/2015/hash/11d0e6287202fced83f79975ec59a3a6-Abstract.html (accessed on 12 June 2024).

47. Brunton, S.L.; Proctor, J.L.; Kutz, J.N. Discovering governing equations from data by sparse identification of nonlinear dynamical systems. *Proc. Natl. Acad. Sci. USA* **2016**, *113*, 3932–3937. [CrossRef] [PubMed]
48. Loiseau, J.C.; Brunton, S.L. Constrained sparse Galerkin regression. *J. Fluid Mech.* **2018**, *838*, 42–67. [CrossRef]

Disclaimer/Publisher's Note: The statements, opinions and data contained in all publications are solely those of the individual author(s) and contributor(s) and not of MDPI and/or the editor(s). MDPI and/or the editor(s) disclaim responsibility for any injury to people or property resulting from any ideas, methods, instructions or products referred to in the content.

MDPI AG
Grosspeteranlage 5
4052 Basel
Switzerland
Tel.: +41 61 683 77 34

Fluids Editorial Office
E-mail: fluids@mdpi.com
www.mdpi.com/journal/fluids

Disclaimer/Publisher's Note: The statements, opinions and data contained in all publications are solely those of the individual author(s) and contributor(s) and not of MDPI and/or the editor(s). MDPI and/or the editor(s) disclaim responsibility for any injury to people or property resulting from any ideas, methods, instructions or products referred to in the content.

www.ingramcontent.com/pod-product-compliance
Lightning Source LLC
LaVergne TN
LVHW070428100526
838202LV00014B/1546